STUDENT'S SOLUTIONS MANUAL

Edgar N. Reyes
Southeastern Louisiana University

College Algebra
—— Third Edition ——

MARK DUGOPOLSKI

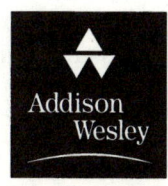

Addison
Wesley

Boston San Francisco New York
London Toronto Sydney Tokyo Singapore Madrid
Mexico City Munich Paris Cape Town Hong Kong Montreal

Reproduced by Addison-Wesley from camera-ready copy supplied by the authors.

Copyright © 2003 Pearson Education, Inc.

Printed in the United States of America.

ISBN   0-201-78573-0

3 4 5 6 7 8 9 10 VG 05 04 03

TABLE OF CONTENTS

For Thought

1. False, 0 is not an irrational number.

2. True   **3.** True

4. False, since 0 has no multiplicative inverse.

5. True

6. False, since the statement fails when $a = w = z$.

7. False, since $a - (b - c) = a - b + c$.

8. False, the distance is $|a - b|$.

9. False, a TI-92 can give an exact answer like π.

10. False, the opposite of $a + b$ is $-a - b$.

P.1 Exercises

1. **e**, true since $\sqrt{2}$ is an irrational number and all irrational numbers are real numbers.

3. **h**, true since 0 is a rational number.

5. **g**, true since each integer is a real number.

7. **c**, true since an irrational number is not an element of the set of rational numbers.

9. All

11. $\{-\sqrt{2}, \sqrt{3}, \pi, 5.090090009...\}$

13. $\{0, 1\}$ **15.** $x + 7$

17. $5x + 15$ **19.** $5(x + 1)$

21. $(-13 + 4) + x$ **23.** $\dfrac{1}{0.125} = 8$

25. $\sqrt{3}$ **27.** $y^2 - x^2$ **29.** 7.2

31. $\sqrt{5}$ **33.** $|13 - 8| = 5$

35. $|17 - (-5)| = 22$

37. $|-18 - (-6)| = |-12| = 12$

39. $\left|\dfrac{1}{4} - \left(-\dfrac{1}{2}\right)\right| = \left|\dfrac{3}{4}\right| = \dfrac{3}{4}$

41. $2^3 = 2 \cdot 2 \cdot 2 = 8$

43. $-7^2 = -(7 \cdot 7) = -49$

45. $(-4)^2 = (-4) \cdot (-4) = 16$

47. $\left(-\dfrac{1}{4}\right)\left(-\dfrac{1}{4}\right)\left(-\dfrac{1}{4}\right) = -\dfrac{1}{64}$

49. $10 - 18 = -8$

51. $|3 - 20| - 5 = |-17| - 5 = 17 - 5 = 12$

53. $12 - 15 = -3$ **55.** $\dfrac{-2 + 6}{-5 + 9} = \dfrac{4}{4} = 1$

57. $4 - 5 \cdot 9 = 4 - 45 = -41$

59. $3 - 4 + 5 - 7 - 4 = -1 + 5 - 7 - 4 = 4 - 7 - 4 = -3 - 4 = -7$

61. $18 + 8 = 26$

63. $\dfrac{26}{5} \cdot \dfrac{2}{1} \cdot 5 = 52$

65. $(12 - 1)(1 + 8) = 11 \cdot 9 = 99$

67. $2 - 3|3 - 24| = 2 - 3 \cdot 21 = -61$

69. $49 - 2(18) = 49 - 36 = 13$

71. $3^2 - 4(-2)(4) = 9 - (-32) = 41$

73. $\dfrac{-2 - 4}{3 - 4} = \dfrac{-6}{-1} = 6$

75. $4 - 9 = -5$

77. $(-5)(1) = -5$

79. $(-5)(4 - 6 + 9) = -35$

81. $(-8) + (64) = 56$

83. $-2x$ **85.** $0.85x$

87. $-6xy$ **89.** $3 - 2x$

91. $\dfrac{6x}{2} - \dfrac{2y}{2} = 3x - y$ **93.** $-3x - 6$

95. $-8 + 2x - 9 + 9x = 11x - 17$

97. **a)** $132 - 0.60a + 0.40r$

b) Substitute 20 into a. Then
$132 - 0.6(20) + 0.4(70) = 148$ beats/min.

c) 123-0.3a-0.6r+r=123-0.3a+0.4r

99. Arranged from smallest to largest, we get

$$-\frac{1}{2}, -\frac{5}{12}, -\frac{1}{3}, 0, \frac{1}{3}, \frac{5}{12}, \frac{1}{2}.$$

To compare the numbers , rewrite each number as a fraction with denominator 12.

For Thought

1. True, since $\frac{1}{2} + \frac{1}{2} = 1$.

2. True, since $2^{100} = (2^2)^{50}$.

3. True, since $9^{16} = (9^2)^8 = 81^8$.

4. True

5. False, since $\frac{5^{10}}{5^{-12}} = 5^{22}$.

6. False, since $2 \cdot 2 \cdot 2 \cdot 2^{-1} = 4$.

7. True

8. True

9. False, since $10^{-4} = 0.0001$

10. False, since $98.6 \times 10^8 = 9.86 \times 10^9$.

P.2 Exercises

1. $\frac{1}{3}$ **3.** $-\frac{1}{16}$ **5.** $2^3 = 8$

7. $\left(\frac{2}{3}\right)^3 = \frac{8}{27}$ **9.** $\left(-\frac{2}{1}\right)^2 = (-2)^2 = 4$

11. $\frac{1}{2} \cdot 16 \cdot \frac{1}{10} = 8 \cdot \frac{1}{10} = \frac{4}{5}$

13. $\frac{6^3}{3^2} = 24$

15. $1 + \frac{1}{2} = \frac{3}{2}$

17. $-\frac{2}{1000} = -\frac{1}{500}$

19. $-6x^{11}y^{11}$ **21.** $x^6 + x^6 = 2x^6$ **23.** $25 \cdot 9 = 225$

25. $\frac{1}{6}x^0y^{-3} = \frac{1}{6y^3}$ **27.** $3x^4$ **29.** $\frac{n^2}{2}$

31. $8a^6 + 9a^6 = 17a^6$ **33.** $-4x^6$

35. $\frac{(-2)^3(x^2)^3}{27} = \frac{-8x^6}{27}$ **37.** $\frac{25}{y^4}$

39. $(x^{3b-3})(x^{8-2b}) = x^{b+5}$

41. $-125a^{6t}b^{-9t} = -\frac{125a^{6t}}{b^{9t}}$ **43.** $\frac{-3y^{6v}}{2x^{5w}}$

45. $\left(a^{-s+5}\right)^4 = a^{-4s+20}$

47. $43,000$ **49.** 0.0000356

51. 5×10^6 **53.** 6.72×10^{-5}

55. 0.000000007 **57.** 2×10^{10}

59. $20 \times 10^{15} = 2 \times 10^{16}$

61. $2 \times 10^{-6+3} = 2 \times 10^{-3}$

63. $40 \times 10^{-30} = 4 \times 10^{-29}$

65. $\frac{(8 \times 10^{18})(5 \times 10^{-6})}{(4 \times 10^{-10})} = \frac{40 \times 10^{12}}{4 \times 10^{-10}} = 1 \times 10^{23}$

67. 9.936×10^{-5}

69. Using $\pi \approx 3.14$, we find
$$\frac{(178.45 \times 10^{-18})(0.00666 \times 10^{-28})}{3.14(79.21 \times 10^{-8})} \approx$$
$$\frac{1.188 \times 10^{-46}}{248.72 \times 10^{-8}} \approx 4.78 \times 10^{-41}$$

71. BMI$= \frac{703(335)}{79^2} \approx 37.7$.

73. Since 1 ton = 2000 pounds, the ratio is
$$D = \frac{2200(2000)175^{-3}10^6}{2240} \approx 367.$$

75. Each person owes $\frac{5.715 \times 10^{12}}{2.846 \times 10^8} = \$20,080.82$

77. The number of seconds in 10 billion years is $60^2 \cdot 24 \cdot 365 \cdot 10^{10} = 3.1536 \times 10^{17}$. So, the amount of mass transformed by the sun in 10 billion years is 5 million $\times 3.1536 \times 10^{17} \approx 1.577 \times 10^{24}$ tons.

79. The earth's radius is $\frac{6.9599 \times 10^5}{109.2} \approx 6379$ km.

81. $\frac{1.989 \times 10^{30}}{5.976 \times 10^{24}} \approx 3.3 \times 10^5$.

83. a) 5365.4

b) 9702.38

c) Let a, b, c, d, e, f be whole numbers. Then

$$a \cdot 10^3 + b \cdot 10^2 + c \cdot 10^1 + d \cdot 10^0 + e \cdot 10^{-1} + f \cdot 10^{-2}$$

is the number $abcdef$ obtained by juxtaposition.

d) $9 \cdot 10^3 + 6 \cdot 10^1 + 3 \cdot 10^0 + 2 \cdot 10^{-1} + 4 \cdot 10^{-2} + 1 \cdot 10^{-3}$

Not unique for $9063.241 = 90 \cdot 10^2 + 63 \cdot 10^1 + 241 \cdot 10^{-3}$.

e) $4 \cdot 10^4 + 3 \cdot 10^3 + 2 \cdot 10^0 + 1 \cdot 10^{-1} + 9 \cdot 10^{-2}$

For Thought

1. False, since $8^{-\frac{1}{3}} = \dfrac{1}{2}$.　**2.** True, since $16^{\frac{1}{4}} = 2 = 4^{\frac{1}{2}}$.　**3.** False, since $\sqrt{\dfrac{4}{6}} = \dfrac{2}{\sqrt{6}}$.　**4.** True

5. False, since $(-1)^{\frac{1}{2}}$ is not a real number.

6. False, since $\sqrt[3]{7^2} = 7^{\frac{2}{3}}$.　**7.** False, since $9^{\frac{1}{2}} = 3$.

8. True, since $\dfrac{1}{\sqrt{3}} \cdot \dfrac{\sqrt{3}}{\sqrt{3}} = \dfrac{\sqrt{3}}{3}$.

9. True, since $\dfrac{2^{\frac{1}{2}}}{2^{\frac{1}{3}}} = 2^{\frac{1}{2}-\frac{1}{3}} = 2^{\frac{1}{6}} = \sqrt[6]{2}$.

10. True, since $\sqrt[3]{7^5} = \sqrt[3]{7^3 \cdot 7^2} = 7\sqrt[3]{49}$.

P.3 Exercises

1. -3　**3.** 8

5. -4　**7.** 81

9. $(8^{1/3})^{-4} = \dfrac{1}{16}$　**11.** $\left(\dfrac{1}{4}\right)^{1/2} = \dfrac{1}{2}$

13. $\left(\dfrac{4}{9}\right)^{3/2} = \left(\left(\dfrac{4}{9}\right)^{1/2}\right)^3 = \left(\dfrac{2}{3}\right)^3 = \dfrac{8}{27}$

15. $|x|$　**17.** a^3　**19.** a^2

21. $x^{3 \cdot (1/3)} y^{6 \cdot (1/3)} = xy^2$

23. $y^{2/3+7/3} = y^{9/3} = y^3$

25. $x^2 y^{1/2}$　**27.** $6a^{1/2+1} = 6a^{3/2}$

29. $3a^{1/2-1/3} = 3a^{1/6}$

31. $a^{2+1/3} b^{1/2-1/2} = a^{7/3}$　**33.** $\dfrac{x^2 y}{z^3}$

35. 30　**37.** -2　**39.** -2

41. $\dfrac{2}{3}$　**43.** 0.1

45. $\dfrac{\sqrt[3]{-8}}{\sqrt[3]{1000}} = \dfrac{-2}{10} = -\dfrac{1}{5}$

47. $\sqrt[4]{(2^4)^3} = \sqrt[4]{2^{12}} = 2^{12 \cdot (1/4)} = 2^3 = 8$

49. $10^{2/3} = \sqrt[3]{10^2}$　**51.** $\dfrac{3}{y^{3/5}} = \dfrac{3}{\sqrt[5]{y^3}}$

53. $x^{-1/2}$　**55.** $(x^3)^{1/5} = x^{3/5}$

57. $4x$　**59.** $2y^3$

61. $\dfrac{\sqrt{xy}}{10}$　**63.** $\dfrac{-2a}{b^{15 \cdot (1/3)}} = -\dfrac{2a}{b^5}$

65. $\sqrt{4(7)} = 2\sqrt{7}$　**67.** $\dfrac{1}{\sqrt{5}} \cdot \dfrac{\sqrt{5}}{\sqrt{5}} = \dfrac{\sqrt{5}}{5}$

69. $\dfrac{\sqrt{x}}{\sqrt{8}} \cdot \dfrac{\sqrt{2}}{\sqrt{2}} = \dfrac{\sqrt{2x}}{\sqrt{16}} = \dfrac{\sqrt{2x}}{4}$

71. $\sqrt[3]{8 \cdot 5} = 2\sqrt[3]{5}$

73. $\sqrt[3]{-250x^4} = \sqrt[3]{(-125)(2)x^3 x} = -5x\sqrt[3]{2x}$

75. $\sqrt[3]{\dfrac{1}{2}} \cdot \sqrt[3]{\dfrac{4}{4}} = \sqrt[3]{\dfrac{4}{8}} = \dfrac{\sqrt[3]{4}}{2}$

77. $2\sqrt{2} + 2\sqrt{5} - 2\sqrt{3}$

79. $-10\sqrt{18} = -30\sqrt{2}$　**81.** $12(5a) = 60a$

83. $(-5)^2(3) = 75$

85. $\sqrt{\dfrac{9}{a^3}} = \dfrac{3}{a\sqrt{a}} \cdot \dfrac{\sqrt{a}}{\sqrt{a}} = \dfrac{3\sqrt{a}}{a^2}$

87. $\dfrac{5}{\sqrt{x}} \cdot \dfrac{\sqrt{x}}{\sqrt{x}} = \dfrac{5\sqrt{x}}{x}$

89. $2x\sqrt{5x} + 3x\sqrt{5x} = 5x\sqrt{5x}$

91. $\sqrt[6]{3^2}\sqrt[6]{2^3} = \sqrt[6]{72}$

93. $\sqrt[12]{3^4}\sqrt[12]{x^3} = \sqrt[12]{81x^3}$

95. $\sqrt[6]{(xy)^3}\sqrt[6]{(2xy)^2} = \sqrt[6]{x^3 y^3}\sqrt[6]{4x^2 y^2} = \sqrt[6]{4x^5 y^5}$

97. $(7^{1/2})^{1/3} = \sqrt[6]{7}$

99. Since $E = \sqrt{\dfrac{2(25)(6000)}{140}} \approx 46.3$, the most economic order quantity is $E = 46$.

101. $S = \dfrac{16(42,700)}{[(2200)(2000)]^{2/3}} \approx 25.4$

103. a) 50%, 30%

   b) If $n = 5$, the depreciation rate is
   $$r = 1 - \left(\frac{200}{5000}\right)^{1/5} \approx 0.47469 \text{ or } 47.5\%.$$

   If $n = 10$, the depreciation rate is
   $$r = 1 - \left(\frac{200}{5000}\right)^{1/10} \approx 0.2752 \text{ or } 27.5\%.$$

105. $D = \sqrt{4^2 + 6^2 + 12^2} = 14$ in.

107. $A = \sqrt{12(6)(5)(1)} = \sqrt{360} \approx 19.0 \text{ ft}^2$

109. Each expression is equivalent to $t^{4/5}$ except for $\sqrt[4]{t^5}$ which is $t^{5/4}$, the expression in (b).

111. No, since $\sqrt{9 + 16} = 5 \neq 7 = \sqrt{9} + \sqrt{16}$.

For Thought

1. False, since polynomials do not have negative exponents.

2. False, since the degree is 4.

3. False, since $(8)^2 \neq 9 + 25$.

4. True, since $50^2 - 1^2 = 2499$.

5. False, since $(x + 3)^2 = x^2 + 6x + 9$.

6. False, since the left-hand side is $-x - 6$.

7. False, since $(a + b)^3 = a^3 + 3a^2b + 3ab^2 + b^3$.

8. False, since divisor times quotient plus the remainder is equal to the dividend.

9. False, since the equation is not defined at $x = -2$.

10. True, since $P(17) + Q(17) = 2310 = S(17)$.

P.4 Exercises

1. Degree 3, leading coefficient 1, trinomial

3. Degree 2, leading coefficient -3, binomial

5. Degree 0, leading coefficient 79, monomial

7. $P(-2) = 4 + 6 + 2 = 12$

9. $M(-3) = 27 + 45 + 3 + 2 = 77$

11. $8x^2 + 3x - 1$

13. $4x^2 - 3x - 9x^2 + 4x - 3 = -5x^2 + x - 3$

15. $4ax^3 - a^2x - 5a^2x^3 + 3a^2x - 3 = (-5a^2 + 4a)x^3 + 2a^2x - 3$

17. $2x - 1$ **19.** $3x^2 - 3x - 6$

21. $-18a^5 + 15a^4 - 6a^3$

23. $(3b^2 - 5b + 2)b - (3b^2 - 5b + 2)3 = 3b^3 - 5b^2 + 2b - 9b^2 + 15b - 6 = 3b^3 - 14b^2 + 17b - 6$

25. $2x(4x^2 + 2x + 1) - (4x^2 + 2x + 1) = 8x^3 + 4x^2 + 2x - 4x^2 - 2x - 1 = 8x^3 - 1$

27. $(x - 4)z + (x - 4)3 = xz - 4z + 3x - 12$

29. $a(a^2 + ab + b^2) - b(a^2 + ab + b^2) = a^3 + a^2b + ab^2 - a^2b - ab^2 - b^3 = a^3 - b^3$

31. $a^2 - 2a + 9a - 18 = a^2 + 7a - 18$

33. $2y^2 + 18y - 3y - 27 = 2y^2 + 15y - 27$

35. $4x^2 + 18x - 18x - 81 = 4x^2 - 81$

37. $4x^2 + 10x + 10x + 25 = 4x^2 + 20x + 25$

39. $6x^4 + 10x^2 + 12x^2 + 20 = 6x^4 + 22x^2 + 20$

41. $(1 + \sqrt{2})(3 + \sqrt{2}) = 3 + \sqrt{2} + 3\sqrt{2} + 2 = 5 + 4\sqrt{2}$

43. $20 - 15\sqrt{2} + 4\sqrt{2} - 3\sqrt{2}\sqrt{2} = 14 - 11\sqrt{2}$

45. $6\sqrt{2}\sqrt{2} - 3\sqrt{6} + 2\sqrt{6} - \sqrt{3}\sqrt{3} = 9 - \sqrt{6}$

47. $\sqrt{5}\sqrt{5} + \sqrt{15} + \sqrt{15} + \sqrt{3}\sqrt{3} = 8 + 2\sqrt{15}$

49. $(3x)^2 + 2(3x)(5) + (5)^2 = 9x^2 + 30x + 25$

51. $(x^n)^2 - 3^2 = x^{2n} - 9$ **53.** $(\sqrt{2})^2 - 5^2 = -23$

55. $(3\sqrt{6})^2 - 2(3\sqrt{6})(1) + 1 = 55 - 6\sqrt{6}$

57. $(3x^3)^2 - 2(3x^3)(4) + 4^2 = 9x^6 - 24x^3 + 16$

59. $\dfrac{5}{1-\sqrt{7}} \cdot \dfrac{1+\sqrt{7}}{1+\sqrt{7}} = \dfrac{5+5\sqrt{7}}{1-7} = \dfrac{-5-5\sqrt{7}}{6}$

61. $\dfrac{\sqrt{10}}{\sqrt{5}-2} \cdot \dfrac{\sqrt{5}+2}{\sqrt{5}+2} = \dfrac{\sqrt{50}+2\sqrt{10}}{5-4} = 5\sqrt{2}+2\sqrt{10}$

63. $\dfrac{\sqrt{6}}{6+\sqrt{3}} \cdot \dfrac{6-\sqrt{3}}{6-\sqrt{3}} = \dfrac{6\sqrt{6}-3\sqrt{2}}{33} = \dfrac{2\sqrt{6}-\sqrt{2}}{11}$

65.

$\dfrac{1+\sqrt{2}}{2-\sqrt{3}} \cdot \dfrac{2+\sqrt{3}}{2+\sqrt{3}} = \dfrac{2+2\sqrt{2}+\sqrt{3}+\sqrt{6}}{4-3} =$

$2+2\sqrt{2}+\sqrt{3}+\sqrt{6}$

67. $\dfrac{36x^6}{-4x^3} = -9x^3$

69. $\dfrac{3x^2}{-3x} + \dfrac{6x}{3x} = -x+2$

71. Factor & simplify: $\dfrac{(x+3)^2}{x+3} = x+3$

73. Factor & simplify: $\dfrac{(a-1)(a^2+a+1)}{a-1} =$

a^2+a+1

75. Quotient $x+5$, remainder 13 since

$$
\begin{array}{r}
x+5 \\
x-2 \overline{\smash{\big)}\,x^2+3x+3} \\
\underline{x^2-2x} \\
5x+3 \\
\underline{5x-10} \\
13
\end{array}
$$

77. Quotient $2x-6$, remainder 13 since

$$
\begin{array}{r}
2x-6 \\
x+3 \overline{\smash{\big)}\,2x^2+0x-5} \\
\underline{2x^2+6x} \\
-6x-5 \\
\underline{-6x-18} \\
13
\end{array}
$$

79. Quotient x^2+x+1, remainder 0 since

$$
\begin{array}{r}
x^2+x+1 \\
x-3 \overline{\smash{\big)}\,x^3-2x^2-2x-3} \\
\underline{x^3-3x^2} \\
x^2-2x \\
\underline{x^2-3x} \\
x-3 \\
\underline{x-3} \\
0
\end{array}
$$

81. $x+1+\dfrac{-1}{x-1}$ since

$$
\begin{array}{r}
x+1 \\
x-1 \overline{\smash{\big)}\,x^2+0x-2} \\
\underline{x^2-x} \\
x-2 \\
\underline{x-1} \\
-1
\end{array}
$$

83. $\dfrac{2x^2}{x} - \dfrac{3x}{x} + \dfrac{1}{x} = 2x-3+\dfrac{1}{x}$

85. $\dfrac{x^2}{x} + \dfrac{x}{x} + \dfrac{1}{x} = x+1+\dfrac{1}{x}$

87. $x+\dfrac{1}{x-2}$ since

$$
\begin{array}{r}
x \\
x-2 \overline{\smash{\big)}\,x^2-2x+1} \\
\underline{x^2-2x} \\
1
\end{array}
$$

89. $x^2+2x-24$ **91.** $2a^{10}-3a^5-27$

93. $-y-9$ **95.** $6a-6$

97. $w^2+8w+16$ **99.** $16x^2-81$

101. $3y^5-9xy^2$ **103.** $2b-1$

105. x^6-64 **107.** $9w^4-12w^2n+4n^2$ **109.** 1

111. The area is $(x+3)(2x-1) = 2x^2+5x-3$.

113. Divide the volume by the height.

$$
\begin{array}{r}
x^2+5x+6 \\
x+4 \overline{\smash{\big)}\,x^3+9x^2+26x+24} \\
\underline{x^3+4x^2} \\
5x^2+26x \\
\underline{5x^2+20x} \\
6x+24 \\
\underline{6x+24} \\
0
\end{array}
$$

Then the area of the bottom is x^2+5x+6 cm^2.

115. The dimensions of the rectangular habitat are $6-2x$ and $4-2x$. The area is the product of the dimensions which is $4x^2-20x+24$ square miles.

117. a) $0.30 to $0.40

b) $A(.1545)-A(.0618)\approx \$0.34$

119. The fine is $F(25)=500+200(25)=\$5,500$.

121. Let x be the length of a side of the square. The areas of the square and rectangle are x^2 and $(x+10)(x-10)=x^2-100$, respectively. Wilson had 100 ft^2 less area than he thought.

123. Let $x\gg 0$ represent a sufficiently large number. If $x\gg 0$, then the sign of $P(x)$ is the same as the sign of the leading coefficient. If $x<0$, $|x|\gg 0$, and the degree is even, then the sign of $P(x)$ is the same as the sign of the leading coefficient. If $x<0$, $|x|\gg 0$, and the degree is odd, then the sign of $P(x)$ is opposite the sign of the leading coefficient.

For Thought

1. False, since it factors as $(x+8)(x-2)$.

2. True　**3.** True

4. False, since $a^2-1=(a-1)(a+1)$.

5. False, since $a^3-1=(a-1)(a^2+a+1)$.

6. False, since $x^3-y^3=(x-y)(x^2+xy+y^2)$.

7. False, it factors as $(2a+3b)(4a^2-6ab+9b^2)$.

8. False, since $(x+2)(x-6)=x^2-4x-12$.

9. True　**10.** True

P.5 Exercises

1. $6x^2(x-2)$, $-6x^2(-x+2)$

3. $4a(1-2b)$, $-4a(2b-1)$

5. $ax(-x^2+5x-5)$, $-ax(x^2-5x+5)$

7. $1(m-n)$, $-1(n-m)$

9. $x^2(x+2)+5(x+2)=(x^2+5)(x+2)$

11. $y^2(y-1)-3(y-1)=(y^2-3)(y-1)$

13. $ady+d-awy-w=d(ay+1)-w(ay+1)=(d-w)(ay+1)$

15. $x^2y^2-ay^2-(bx^2-ab)=y^2(x^2-a)-b(x^2-a)=(y^2-b)(x^2-a)$

17. $(x+2)(x+8)$　**19.** $(x-6)(x+2)$

21. $(m-2)(m-10)$　**23.** $(t-7)(t+12)$

25. $(2x+1)(x-4)$　**27.** $(4x+1)(2x-3)$

29. $(3y+5)(2y-1)$　**31.** $(t-u)(t+u)$

33. $(t+1)^2$　**35.** $(2w-1)^2$

37. $(y^{2t}-5)(y^{2t}+5)$　**39.** $(3zx+4)^2$

41. $(t-u)(t^2+tu+u^2)$　**43.** $(a-2)(a^2+2a+4)$

45. $(3y+2)(9y^2-6y+4)$

47. $(3xy^2)^3-(2z^3)^3=$ $(3xy^2-2z^3)(9x^2y^4+6xy^2z^3+4z^6)$

49. $(x^n)^3-2^3=(x^n-2)(x^{2n}+2x^n+4)$

51. Replace y^3 by w and y^6 by w^2. Then $w^2+10w+25=(w+5)^2=(y^3+5)^2$.

53. Replace $2a^2b^4$ by w and $4a^4b^8$ by w^2. So, $w^2-4w-5=(w+1)(w-5)=(2a^2b^4+1)(2a^2b^4-5)$.

55. Replace $(2a+1)$ by w and $(2a+1)^2$ by w^2. Then $w^2+2w-24=(w+6)(w-4)=((2a+1)+6)((2a+1)-4)=(2a+7)(2a-3)$.

57. Replace (b^2+2) by w and $(b^2+2)^2$ by w^2. We find $w^2-5w+4=(w-4)(w-1)=((b^2+2)-4)((b^2+2)-1)=(b^2-2)(b^2+1)$.

59. $-3x^3+27x=-3x(x^2-9)=-3x(x-3)(x+3)$

61. $2t(8t^3+27w^3)=2t(2t+3w)(4t^2-6tw+9w^2)$

63. $a^3+a^2-4a-4=a^2(a+1)-4(a+1)=(a^2-4)(a+1)=(a-2)(a+2)(a+1)$

65. $x^4-2x^3-8x+16=x^3(x-2)-8(x-2)=(x-2)(x^3-8)=(x-2)^2(x^2+2x+4)$

67. $-2x(18x^2-9x-2)=-2x(6x+1)(3x-2)$

69. $a^7 - a^6 - 64a + 64 = a^6(a-1) - 64(a-1) =$
$(a^6 - 64)(a-1) = (a^3 - 8)(a^3 + 8)(a-1) =$
$(a-2)(a^2 + 2a + 4)(a+2)(a^2 - 2a + 4)(a-1)$

71. $-(3x + 5)(2x - 3)$

73. Replace $(a^2 + 2)$ by w and $(a^2 + 2)^2$ by w^2.
$w^2 - 4w + 3 = (w - 3)(w - 1) =$
$((a^2 + 2) - 3)((a^2 + 2) - 1) = (a^2 - 1)(a^2 + 1) =$
$(a - 1)(a + 1)(a^2 + 1)$

75. Yes, since

$$
\begin{array}{r}
x^2 + x + 1 \\
x + 3 \overline{\smash{\big)}\, x^3 + 4x^2 + 4x + 3} \\
\underline{x^3 + 3x^2} \\
x^2 + 4x \\
\underline{x^2 + 3x} \\
x + 3 \\
\underline{x + 3} \\
0
\end{array}
$$

and $x^3 + 4x^2 + 4x + 3 = (x + 3)(x^2 + x + 1)$.

77. No, since

$$
\begin{array}{r}
3x^2 + 8x - 4 \\
x - 1 \overline{\smash{\big)}\, 3x^3 + 5x^2 - 12x - 9} \\
\underline{3x^3 - 3x^2} \\
8x^2 - 12x \\
\underline{8x^2 - 8x} \\
-4x - 9 \\
\underline{-4x + 4} \\
-13
\end{array}
$$

and the remainder is not 0.

79.

$$
\begin{array}{r}
x^2 + 5x + 6 \\
x - 1 \overline{\smash{\big)}\, x^3 + 4x^2 + x - 6} \\
\underline{x^3 - x^2} \\
5x^2 + x \\
\underline{5x^2 - 5x} \\
6x - 6 \\
\underline{6x - 6} \\
0
\end{array}
$$

Then $x^3 + 4x^2 + x - 6 = (x - 1)(x^2 + 5x + 6) =$
$(x - 1)(x + 2)(x + 3)$.

81.

$$
\begin{array}{r}
x^2 + 2x + 2 \\
x - 3 \overline{\smash{\big)}\, x^3 - x^2 - 4x - 6} \\
\underline{x^3 - 3x^2} \\
2x^2 - 4x \\
\underline{2x^2 - 6x} \\
2x - 6 \\
\underline{2x - 6} \\
0
\end{array}
$$

So $x^3 - x^2 - 4x - 6 = (x - 3)(x^2 + 2x + 2)$.

83.

$$
\begin{array}{r}
x^3 + 3x^2 - x - 3 \\
x + 2 \overline{\smash{\big)}\, x^4 + 5x^3 + 5x^2 - 5x - 6} \\
\underline{x^4 + 2x^3} \\
3x^3 + 5x^2 \\
\underline{3x^3 + 6x^2} \\
-x^2 - 5x \\
\underline{-x^2 - 2x} \\
-3x - 6 \\
\underline{-3x - 6} \\
0
\end{array}
$$

Then we obtain

$$
\begin{aligned}
x^4 + 5x^3 + 5x^2 - 5x - 6 &= \\
(x + 2)(x^3 + 3x^2 - x - 3) &= \\
(x + 2)(x^2(x + 3) - (x + 3)) &= \\
(x + 2)(x + 3)(x^2 - 1) &= \\
(x + 2)(x + 3)(x - 1)(x + 1).
\end{aligned}
$$

85. The area of the bottom is the volume divided by the height.

$$
\begin{array}{r}
x^2 + x + 1 \\
x - 1 \overline{\smash{\big)}\, x^3 + 0x^2 + 0x - 1} \\
\underline{x^3 - x^2} \\
x^2 + 0x \\
\underline{x^2 - x} \\
x - 1 \\
\underline{x - 1} \\
0
\end{array}
$$

Then $x^2 + x + 1$ ft^2 is the area of the bottom.

87. The volume is $V = x(6 - 2x)(7 - 2x)$.
From the tabulated values,

x	0.5	1	2
Volume	15	20	12

we see $x = 1$ in produces the largest volume.

89. b is not a perfect square trinomial since $(\sqrt{1000}a - b)^2 \neq 1000a^2 - 200b + b^2$. The others are perfect squares.

91. No, since $1^3 + 1^3 = 2 \neq n^3$ for any integer n.

For Thought

1. False, since 2 is not factor. **2.** True

3. False, since the first expression is not defined at $x = 0$.

4. False, since $\dfrac{(a-b)(a+b)}{a-b} = a+b$.

5. False, since the LCD is $x(x+1)$. **6.** True

7. True **8.** True

9. True, since $\dfrac{2(500)+1}{500-3} = \dfrac{1001}{497} \approx 2.014$

10. True, since $\dfrac{5x+1}{x} \approx \dfrac{5x}{x} = 5$ when $|x|$ is large.

P.6 Exercises

1. $\{x \mid x \neq -2\}$

3. $\{x \mid x \neq 4, -2\}$

5. $\{x \mid x \neq \pm 3\}$

7. All real numbers since $x^2 + 3 \neq 0$ for any real number x.

9. $\dfrac{3(x-3)}{(x-3)(x+2)} = \dfrac{3}{x+2}$

11. $\dfrac{10a-8b}{12b-15a} = \dfrac{2(5a-4b)}{-3(5a-4b)} = -\dfrac{2}{3}$

13. $\dfrac{a^3b^6}{a^2b^3 - a^4b^2} = \dfrac{a^3b^6}{a^2b^2(b-a^2)} = \dfrac{ab^4}{b-a^2}$

15. $\dfrac{y^2 z}{x^3}$

17. $\dfrac{a^3-b^3}{a^2-b^2} = \dfrac{(a-b)(a^2+ab+b^2)}{(a-b)(a+b)} = \dfrac{a^2+ab+b^2}{a+b}$

19. $\dfrac{2a}{3b^2} \cdot \dfrac{9b}{14a^2} = \dfrac{1}{b} \cdot \dfrac{3}{7a} = \dfrac{3}{7ab}$

21. $\dfrac{12a}{7} \cdot \dfrac{49}{2a^3} = \dfrac{42}{a^2}$

23. $\dfrac{(a-3)(a+3)}{3(a-2)} \cdot \dfrac{(a-2)(a+2)}{(a-3)(a+2)} = \dfrac{a+3}{3}$

25. $\dfrac{(x-y)(x+y)}{9} \cdot \dfrac{9(2)}{(x+y)^2} = \dfrac{2x-2y}{x+y}$

27. $\dfrac{(x-y)(x+y)}{-3xy} \cdot \dfrac{3xy(2xy^2)}{-2(x-y)} = x^2y^2 + xy^3$

29. $\dfrac{16a}{12a^2}$

31. $\dfrac{x-5}{x+3} \cdot \dfrac{x-3}{x-3} = \dfrac{x^2-8x+15}{x^2-9}$

33. $\dfrac{x}{x+5} \cdot \dfrac{x+1}{x+1} = \dfrac{x^2+x}{x^2+6x+5}$

35. $12a^2b^3$

37. Since $3a+3b = 3(a+b)$ and $2a+2b = 2(a+b)$, the LCD is $6(a+b)$.

39. Since $x^2+5x+6 = (x+3)(x+2)$ and $x^2-x-6 = (x-3)(x+2)$, the LCD is $(x+3)(x-3)(x+2)$.

41. $\dfrac{3(3)}{2x(3)} + \dfrac{x}{6x} = \dfrac{9+x}{6x}$

43. $\dfrac{(x+3)(x+1)}{(x-1)(x+1)} - \dfrac{(x+4)(x-1)}{(x+1)(x-1)} =$ $\dfrac{x^2+4x+3}{(x-1)(x+1)} - \dfrac{x^2+3x-4}{(x+1)(x-1)} =$ $\dfrac{x+7}{(x+1)(x-1)}$

45. $\dfrac{3a}{a} + \dfrac{1}{a} = \dfrac{3a+1}{a}$

47. $\dfrac{(t-1)(t+1)}{t+1} - \dfrac{1}{t+1} = \dfrac{t^2-2}{t+1}$

49. $\dfrac{x}{(x+2)(x+1)} + \dfrac{x-1}{(x+3)(x+2)} =$ $\dfrac{x(x+3)}{(x+2)(x+1)(x+3)} +$ $\dfrac{(x-1)(x+1)}{(x+3)(x+2)(x+1)} =$

$$\frac{2x^2 + 3x - 1}{(x+1)(x+2)(x+3)}$$

51.

$$\frac{1}{x-3} - \frac{5}{-2(x-3)} = \frac{2}{2(x-3)} - \frac{-5}{2(x-3)} =$$

$$\frac{7}{2x-6}$$

53.

$$\frac{y^2}{(x-y)(x^2+xy+y^2)} +$$

$$\frac{(x+y)(x-y)}{(x^2+xy+y^2)(x-y)} =$$

$$\frac{y^2}{(x-y)(x^2+xy+y^2)} +$$

$$\frac{x^2-y^2}{(x^2+xy+y^2)(x-y)} =$$

$$\frac{x^2}{x^3-y^3}$$

55.

$$\frac{(x+1)(x-1)}{x(x+1)(x-1)} + \frac{x(x+1)}{(x-1)(x)(x+1)} -$$

$$\frac{x(x-1)}{(x+1)(x)(x-1)} =$$

$$\frac{x^2-1}{x(x+1)(x-1)} + \frac{x^2+x}{(x-1)(x)(x+1)} -$$

$$\frac{x^2-x}{(x+1)(x)(x-1)} =$$

$$\frac{x^2+2x-1}{x(x^2-1)}$$

57.

$$\frac{\left(\dfrac{4}{a} - \dfrac{3}{b}\right)(ab^2)}{\left(\dfrac{1}{ab} + \dfrac{2}{b^2}\right)(ab^2)} = \frac{4b^2 - 3ab}{b+2a}$$

59.

$$\frac{\dfrac{1}{b^2} - \dfrac{1}{ab^2}}{\dfrac{3}{a^2} + \dfrac{1}{a^2b}} = \frac{\left(\dfrac{1}{b^2} - \dfrac{1}{ab^2}\right)(a^2b^2)}{\left(\dfrac{3}{a^2} + \dfrac{1}{a^2b}\right)(a^2b^2)} = \frac{a^2-a}{3b^2+b}$$

61.

$$\frac{\left(a + \dfrac{4}{a+4}\right)(a+4)}{\left(a - \dfrac{4a+4}{a+4}\right)(a+4)} =$$

$$\frac{a^2 + 4a + 4}{a^2 + 4a - (4a+4)} =$$

$$\frac{(a+2)^2}{(a+2)(a-2)} = \frac{a+2}{a-2}$$

63.

$$\frac{\left(\dfrac{t+2}{t-1} - \dfrac{t-3}{t}\right)((t-1)t)}{\left(\dfrac{t+4}{t} + \dfrac{t-2}{t-1}\right)((t-1)t)} =$$

$$\frac{(t^2+2t) - (t-3)(t-1)}{(t+4)(t-1) + (t^2-2t)} =$$

$$\frac{(t^2+2t) - (t^2-4t+3)}{(t^2+3t-4) + (t^2-2t)} = \frac{6t-3}{2t^2+t-4}$$

65.

$$\frac{(x^{-1}+1)x}{(x^{-1}-1)x} = \frac{1+x}{1-x}$$

67. $\dfrac{(a^2 + a^{-1}b^{-3})\,ab^3}{ab^3} = \dfrac{a^3b^3 + 1}{ab^3}$

69.

$$\frac{(x^2-y^2)xy}{(x^{-1}-y^{-1})xy} = \frac{(x-y)(x+y)xy}{y-x} =$$

$$-(x+y)xy = -x^2y - xy^2$$

71.

$$\left(\frac{1}{m} - \frac{1}{n}\right)^{-2} = \left(\frac{n-m}{mn}\right)^{-2} =$$

$$\left(\frac{mn}{n-m}\right)^2 = \frac{m^2n^2}{n^2 - 2mn + m^2}$$

73. $\dfrac{3}{7}$ **75.** $\dfrac{3}{506}$

77. $S(2) = \dfrac{4-5}{4-9} = \dfrac{1}{5}$

79. $\dfrac{1200 - 5}{1200 - 9} = \dfrac{1195}{1191}$ **81.** $\dfrac{9(16)-1}{3(16)-2} \approx 3.1087$

83. $T(-400) = \dfrac{9(160,000) - 1}{3(160,000) - 2} \approx 3.00001$

85 (a) average cost decreases as capacity increases

(b) The average costs are

n	7	12	22
$A(n)$	\$27.14	\$24.17	\$22.27

87. a) The costs are tabulated below.

p	$C(p)$
50%	$6 million
75%	$18 million
99%	$594 million

b) The cost of cleaning up goes up without bound. A 100% clean up is impossible.

c) domain is $\{p|0 \le p < 100\}$

89. The portion of the invoices Gina and Bert can file in one hour are $\dfrac{1}{4}$ and $\dfrac{1}{6}$, respectively. The part they can file together in one hour is $\dfrac{1}{4} + \dfrac{1}{6} = \dfrac{5}{12}$.

91. Let d be the distance between the restaurant and his home. The number of hours dashing home and returning to the restaurant are $\dfrac{d}{250}$ and $\dfrac{d}{300}$, respectively. His average speed is

$$\frac{2d}{\dfrac{d}{250} + \dfrac{d}{300}} = \frac{2}{\dfrac{1}{250} + \dfrac{1}{300}} \approx 272.7 \text{ mph.}$$

93. If a rational expression simulates an application, then the domain describes the extent to which the simulation applies. This is one reason.

95. (a) Since $1 + \dfrac{1}{2} = \dfrac{3}{2}$, $1 + \dfrac{1}{3/2} = \dfrac{5}{3}$, and so on..., the exact answer is $\dfrac{8}{13}$.

(b) Since $1 - \dfrac{1}{3} = \dfrac{2}{3}$, $1 - \dfrac{1}{2/3} = -\dfrac{1}{2}$, and so on..., the exact answer is $\dfrac{3}{2}$.

Review Exercises

1. False, since $\sqrt{2}$ is an irrational number.

3. False, since -1 is a negative number.

5. False, since terminating decimal numbers are rational numbers.

7. False, since $\{1, 2, 3, ...\}$ is the set of natural numbers.

9. False, since $\dfrac{1}{3} = 0.3333...$

11. False, since the additive inverse of 0.5 is -0.5

13. $-3x - 12 + 20x = 17x - 12$

15. $\dfrac{2x}{10} + \dfrac{x}{10} = \dfrac{3x}{10}$

17. $\dfrac{3(x-2)}{3 \cdot 3} = \dfrac{x-2}{3}$

19. $\dfrac{-6}{8} = -\dfrac{3}{4}$

21. $3 - 5 = -2$ **23.** $|-4| = 4$

25. $8 - 18 \div 3 + 5 = 8 - 6 + 5 = 7$

27. $3 \cdot 3 \div 6 + 3^3 = 9 \div 6 + 27 = 1.5 + 27 = 28.5$

29. 625 **31.** $4 + 40 = 44$

33. $\dfrac{1}{2} + 1 = \dfrac{3}{2}$

35. $\dfrac{-2^3 \cdot 2^1}{3^1} = -\dfrac{16}{3}$

37. $(8^2)^{-1/3} = 64^{-1/3} = \dfrac{1}{4}$ **39.** $5x^2$

41. 11 **43.** $\sqrt{4 \cdot 7s^2 \cdot s} = 2s\sqrt{7s}$

45. $\sqrt[3]{-1000(2)} = -10\sqrt[3]{2}$

47. $\sqrt{\dfrac{5}{2a}} \cdot \sqrt{\dfrac{2a}{2a}} = \dfrac{\sqrt{10a}}{2a}$

49. $\sqrt[3]{\dfrac{2}{5}} \sqrt[3]{\dfrac{25}{25}} = \dfrac{\sqrt[3]{50}}{5}$

51. $3n\sqrt{2n} + 5n\sqrt{2n} = 8n\sqrt{2n}$

53. $\dfrac{2\sqrt{3}}{\sqrt{3}-1} \cdot \dfrac{\sqrt{3}+1}{\sqrt{3}+1} = \dfrac{6 + 2\sqrt{3}}{3 - 1} = 3 + \sqrt{3}$

55. $\dfrac{\sqrt{6}}{2\sqrt{2} + 3\sqrt{2}} = \dfrac{\sqrt{6}}{5\sqrt{2}} = \dfrac{\sqrt{12}}{10} = \dfrac{\sqrt{3}}{5}$

57. $320,000,000$ **59.** 0.000185 **61.** 5.6×10^{-5}

63. 2.34×10^6 **65.** $125 \cdot 10^{18} = 1.25 \times 10^{20}$

67. $\dfrac{(8 \times 10^2)^2 (10^{-5})^{-3}}{(2 \times 10^6)^3 (2 \times 10^{-5})} =$

$\dfrac{(64 \times 10^4)(10^{15})}{(8 \times 10^{18})(2 \times 10^{-5})} =$

$\dfrac{64}{8(2)} \times 10^6 = 4 \times 10^6$

69. $2x^2 + x - 7$

71. $-4x^4 - 3x^3 + x - x^4 + 6x^3 + 2x =$
$-5x^4 + 3x^3 + 3x$

73. $(3a^2 - 2a + 5)a - (3a^2 - 2a + 5) \cdot 2 =$
$3a^3 - 2a^2 + 5a - 6a^2 + 4a - 10 =$
$3a^3 - 8a^2 + 9a - 10$

75. $b^2 - 6by + 9y^2$

77. $3t^2 - 7t - 6$ **79.** $-\dfrac{35y^5}{7y^2} = -5y^3$

81. $9 - 2 = 7$

83. $1 + 2\sqrt{3} + \sqrt{3}^2 = 1 + 2\sqrt{3} + 3 = 4 + 2\sqrt{3}$

85. $2^2\sqrt{5}^2 + 4\sqrt{15} + \sqrt{3}^2 = 20 + 4\sqrt{15} + 3 =$
$23 + 4\sqrt{15}$

87. Quotient $x^2 + 4x - 1$, remainder 1, since

$$\begin{array}{r} x^2 + 4x - 1 \\ x - 2 \enclose{longdiv}{x^3 + 2x^2 - 9x + 3} \\ \underline{x^3 - 2x^2} \\ 4x^2 - 9x \\ \underline{4x^2 - 8x} \\ -x + 3 \\ \underline{-x + 2} \\ 1 \end{array}$$

89. Quotient $3x + 2$, remainder 4, since

$$\begin{array}{r} 3x + 2 \\ 2x - 1 \enclose{longdiv}{6x^2 + x + 2} \\ \underline{6x^2 - 3x} \\ 4x + 2 \\ \underline{4x - 2} \\ 4 \end{array}$$

91. $x - 2 + \dfrac{1}{x + 2}$ since

$$\begin{array}{r} x - 2 \\ x + 2 \enclose{longdiv}{x^2 + 0x - 3} \\ \underline{x^2 + 2x} \\ -2x - 3 \\ \underline{-2x - 4} \\ 1 \end{array}$$

93. $2 + \dfrac{13}{x - 5}$ since

$$\begin{array}{r} 2 \\ x - 5 \enclose{longdiv}{2x + 3} \\ \underline{2x - 10} \\ 13 \end{array}$$

95. $6x(x^2 - 1) = 6x(x - 1)(x + 1)$ **97.** $(3h + 4t)^2$

99. $(t + y)(t^2 - ty + y^2)$

101. $x^2(x + 3) - 9(x + 3) = (x^2 - 9)(x + 3) =$
$(x - 3)(x + 3)(x + 3) = (x - 3)(x + 3)^2$

103. $t^6 - 1 = (t^3 - 1)(t^3 + 1) =$
$(t - 1)(t^2 + t + 1)(t + 1)(t^2 - t + 1)$

105. $(6x + 5)(3x - 4)$

107. $ab(a^2 + 3a - 18) = ab(a + 6)(a - 3)$

109. $x^2(2x + y) - (2x + y) = (x^2 - 1)(2xy + 1) =$
$(x - 1)(x + 1)(2x + y)$

111. $\dfrac{2x + 6}{x + 3} = \dfrac{2(x + 3)}{x + 3} = 2$

113.

$\dfrac{(x - 1)(x + 4)}{(x - 2)(x + 4)} - \dfrac{(x + 3)(x - 2)}{(x + 4)(x - 2)} =$

$\dfrac{x^2 + 3x - 4 - (x^2 + x - 6)}{(x - 2)(x + 4)} = \dfrac{2(x + 1)}{(x + 4)(x - 2)}$

115. $\dfrac{(x - 3)(x + 3)}{x + 3} \cdot \dfrac{1}{-2(x - 3)} = -\dfrac{1}{2}$

117. $\dfrac{c^7}{a^6 b^2} \cdot \dfrac{a^2 b^6 c^{10}}{a^4 b^3} = \dfrac{bc^{17}}{a^8}$

119. $\dfrac{1}{(x - 2)(x + 2)} + \dfrac{3(x + 2)}{(x - 2)(x + 2)} = \dfrac{3x + 7}{x^2 - 4}$

121. $\dfrac{5x}{30x^2} - \dfrac{21}{30x^2} = \dfrac{5x - 21}{30x^2}$

123. $\dfrac{(a-5)(a+5)}{(a-5)(a+1)} \cdot \dfrac{(a-1)(a+1)}{2(a+5)} = \dfrac{a-1}{2}$

125.

$$\dfrac{(x-4)(x+4)}{(x+4)(x+1)} \cdot \dfrac{(x+1)(x^2-x+1)}{-2(x-4)} =$$

$$\dfrac{-x^2+x-1}{2}$$

127.

$$\dfrac{a-2}{(a+5)(a+1)} + \dfrac{2a+1}{(a-1)(a+1)} =$$

$$\dfrac{(a-2)(a-1)}{(a+5)(a+1)(a-1)} +$$

$$\dfrac{(2a+1)(a+5)}{(a-1)(a+1)(a+5)} =$$

$$\dfrac{(a^2-3a+2)+(2a^2+11a+5)}{(a+5)(a+1)(a-1)} =$$

$$\dfrac{3a^2+8a+7}{(a+1)(a-1)(a+5)}$$

129.

$$\dfrac{\left(\dfrac{5}{2x}-\dfrac{3}{4x}\right)(4x)}{\left(\dfrac{1}{2}-\dfrac{2}{x}\right)(4x)} = \dfrac{10-3}{2x-8} = \dfrac{7}{2x-8}$$

131.

$$\dfrac{\left(\dfrac{1}{y^2-2}-3\right)(y^2-2)}{\left(\dfrac{5}{y^2-2}+4\right)(y^2-2)} = \dfrac{1-3(y^2-2)}{5+4(y^2-2)} =$$

$$\dfrac{-3y^2+7}{4y^2-3}$$

133. $\dfrac{(a^{-2}-b^{-3})\,a^2b^3}{(a^{-1}b^{-1})\,a^2b^3} = \dfrac{b^3-a^2}{ab^2}$

135. $(p^{-1}+pq^{-3})\dfrac{pq^3}{pq^3} = \dfrac{q^3+p^2}{pq^3}$

137. $P(2) = (2)^3-3(2)^2+2-9 = 8-3(4)-7 = -11$

139. $P(0) = -9$

141. $R(-1) = \dfrac{3(-1)-1}{2(-1)-9} = \dfrac{-4}{-11} = \dfrac{4}{11}$

143. $R(50) = \dfrac{3(50)-1}{2(50)-9} = \dfrac{149}{91}$

145. a) 300 ft

    b) $\dfrac{130^2}{32} - \dfrac{125^2}{32} \approx 39.8$ feet

147. The number of hydrogen atoms in one kilogram of hydrogen is

$$\dfrac{1000}{1.7 \times 10^{-24}} \approx 5.9 \times 10^{26}.$$

149. The distance is $|-2.35-8.77| = |-11.12| = 11.12$

151. The parts of the lawn Howard and Will can mow in 2 hours are $\dfrac{2}{6}$ and $\dfrac{2}{4}$, respectively. In 2 hours, the portion they will mow together is $\dfrac{2}{6} + \dfrac{2}{4} = \dfrac{5}{6}$.

Chapter P Test

1. All   **2.** $\{-1.22, -1, 0, 2, 10/3\}$

3. $\{-\pi, -\sqrt{3}, \sqrt{5}, 6.020020002...\}$ **4.** $\{0, 2\}$

5. $|6-25| - 6 = |-19| - 6 = 13$ **6.** $8^{1/3} = 2$

7. $-\dfrac{1}{(27^{1/3})^2} = -\dfrac{1}{3^2} = -\dfrac{1}{9}$

8. $\dfrac{9+12+9}{(-5)(6)} = \dfrac{30}{-30} = -1$

9. $6x^5y^7$ **10.** $2x^2 - 2x^2 = 0$

11. $\dfrac{(ab(b+a))^2}{a^2b^2} = \dfrac{a^2b^2(b+a)^2}{a^2b^2} = a^2 + 2ab + b^2$

12. $\dfrac{-8a^{-3}b^{18}}{2^{-2}a^{-6}b^8} = -32a^3b^{10}$

13. $3\sqrt{3} - 2\sqrt{2} + 4\sqrt{2} = 3\sqrt{3} + 2\sqrt{2}$

14.

$$\dfrac{2\sqrt{2}}{\sqrt{6}-\sqrt{2}} \cdot \dfrac{\sqrt{6}+\sqrt{2}}{\sqrt{6}+\sqrt{2}} = \dfrac{2\sqrt{12}+4}{4} =$$

$$\dfrac{2(2\sqrt{3})+4}{4} = \sqrt{3}+1$$

15.

$$\dfrac{1}{x\sqrt[3]{4x}} = \dfrac{\sqrt[3]{2x^2}}{x\sqrt[3]{4x}\sqrt[3]{2x^2}} = \dfrac{\sqrt[3]{2x^2}}{2x^2}$$

16. $\sqrt{4(3)x^2xy^8y \cdot 1} = 2xy^4\sqrt{3xy}$

17. $3x^3 + 3x^2 - 12x$

18. $-x^2 + 3x - 4 - 4x^2 + 6x - 9 = -5x^2 + 9x - 13$

19. $x(x^2 - 2x - 1) + 3(x^2 - 2x - 1) =$
$x^3 - 2x^2 - x + 3x^2 - 6x - 3 = x^3 + x^2 - 7x - 3$

20. $\dfrac{(2h-1)(4h^2 + 2h + 1)}{2h - 1} = 4h^2 + 2h + 1$

21. $x^2 - 6xy - 27y^2$

22.

$$
\begin{array}{r}
x^2 + 2x + 2 \\
x - 2 \overline{) x^3 + 0x^2 - 2x - 4} \\
\underline{x^3 - 2x^2} \\
2x^2 - 2x \\
\underline{2x^2 - 4x} \\
2x - 4 \\
\underline{2x - 4} \\
0
\end{array}
$$

Then $\dfrac{x^3 - 2x - 4}{x - 2} = x^2 + 2x + 2.$

23. $9x^2 - 48x + 64$ **24.** $4t^8 - 1$

25.

$\dfrac{x(x^2 - 5x + 6)}{2x(x - 3)} \cdot \dfrac{4(x^3 + 8)}{2x(x^2 - 4)} =$

$\dfrac{x(x - 3)(x - 2)}{2x(x - 3)} \cdot \dfrac{4(x + 2)(x^2 - 2x + 4)}{2x(x - 2)(x + 2)} =$

$\dfrac{x^2 - 2x + 4}{x}$

26.

$\dfrac{(x + 5)(x + 4)}{(x - 3)(x - 1)(x + 4)} +$

$\dfrac{(x - 1)^2}{(x + 4)(x - 3)(x - 1)} =$

$\dfrac{(x^2 + 9x + 20) + (x^2 - 2x + 1)}{(x + 4)(x - 3)(x - 1)} =$

$\dfrac{2x^2 + 7x + 21}{(x + 4)(x - 1)(x - 3)}$

27.

$\dfrac{a - 1}{(2a - 3)(2a + 3)} + \dfrac{a - 2}{2a - 3} =$

$\dfrac{a - 1}{(2a - 3)(2a + 3)} + \dfrac{(a - 2)(2a + 3)}{(2a - 3)(2a + 3)} =$

$\dfrac{(a - 1) + (2a^2 - a - 6)}{(2a - 3)(2a + 3)} = \dfrac{2a^2 - 7}{4a^2 - 9}$

28.

$\dfrac{\dfrac{1}{2a^2 b} - 2a}{\dfrac{1}{4ab^3} + \dfrac{1}{3b}} \cdot \dfrac{12a^2 b^3}{12a^2 b^3} = \dfrac{6b^2 - 24a^3 b^3}{3a + 4a^2 b^2}$

29. $a(x^2 - 11x + 18) = a(x - 9)(x - 2)$

30. $m(m^4 - 1) = m(m^2 - 1)(m^2 + 1) =$
$m(m - 1)(m + 1)(m^2 + 1)$

31. $(3x - 1)(x + 5)$

32. $bx(x - 3) + w(x - 3) = (bx + w)(x - 3)$

33. The per capita income is $\dfrac{6.102 \times 10^{12}}{2.814 \times 10^8} =$

$\$21,684.43$

34. The annual appreciation rate is
$\left(\dfrac{2.5}{1.3}\right)^{1/3} - 1 \approx 0.24355$ or 24.4%.

35. The altitude is $A(2) = -64 + 240 = 176$ feet.

For Thought

1. True, since $5(1) = 6 - 1$.

2. True, since $x = 3$ is the solution to both equations.

3. False, -2 is not a solution of the first equation since $\sqrt{-2}$ is not a real number.

4. True

5. False, $x = 0$ is the solution.   **6.** True

7. False, since $|x| = -8$ has no solution.

8. False, $\dfrac{x}{x-5}$ is undefined at $x = 5$.

9. False, since we should multiply by $-\dfrac{3}{2}$.

10. False, $0 \cdot x + 1 = 0$ has no solution.

1.1 Exercises

1. No, since $2(3) - 4 = 2 \neq 9$.

3. Yes, since $(-4)^2 = 16$.

5. Since $3x = 5$, the solution set is $\left\{\dfrac{5}{3}\right\}$.

7. Since $-3x = 6$, the solution set is $\{-2\}$.

9. Since $14x = 7$, the solution set is $\left\{\dfrac{1}{2}\right\}$.

11. Since $7 + 3x = 4x - 4$, the solution set is $\{11\}$.

13. Since $x = -\dfrac{4}{3} \cdot 18$, the solution set is $\{-24\}$.

15. Multiplying by 6 we get

$$
\begin{aligned}
3x - 30 &= -72 - 4x \\
7x &= -42.
\end{aligned}
$$

The solution set is $\{-6\}$.

17. Multiply both sides of the equation by 12.

$$
\begin{aligned}
18x + 4 &= 3x - 2 \\
15x &= -6 \\
x &= -\dfrac{2}{5}.
\end{aligned}
$$

The solution set is $\left\{-\dfrac{2}{5}\right\}$.

19. Note, $3(x - 6) = 3x - 18$ is true by the distributive law. It is an identity and the solution set is R.

21. Note, $5x = 4x$ is equivalent to $x = 0$. A conditional equation whose solution set is $\{0\}$.

23. Equivalently, we get $2x + 6 = 3x - 3$ or $9 = x$. A conditional equation whose solution set is $\{9\}$.

25. Using the distributive property, we find

$$
\begin{aligned}
3x - 18 &= 3x + 18 \\
-18 &= 18.
\end{aligned}
$$

The equation is inconsistent and the solution set is $\emptyset$.

27. An identity and the solution set is $\{x \mid x \neq 0\}$.

29. Multiplying $2(w - 1)$, we get

$$
\begin{aligned}
\dfrac{1}{w-1} - \dfrac{1}{2w-2} &= \dfrac{1}{2w-2} \\
2 - 1 &= 1.
\end{aligned}
$$

An identity and the solution set is $\{w \mid w \neq 1\}$

31. Multiply by $6x$.

$$
\begin{aligned}
6 - 2 &= 3 + 1 \\
4 &= 4
\end{aligned}
$$

An identity with solution set $\{x \mid x \neq 0\}$.

33. Multiply by $3(z - 3)$.

$$
\begin{aligned}
3(z + 2) &= -5(z - 3) \\
3z + 6 &= -5z + 15 \\
8z &= 9
\end{aligned}
$$

A conditional equation with solution set $\left\{\dfrac{9}{8}\right\}$.

35. Multiplying by $(x - 3)(x + 3)$.

$$
\begin{aligned}
(x + 3) - (x - 3) &= 6 \\
6 &= 6
\end{aligned}
$$

An identity with solution set $\{x \mid x \neq 3, x \neq -3\}$.

37. Multiply by $(y - 3)$.

$$\begin{aligned} 4(y - 3) + 6 &= 2y \\ 4y - 6 &= 2y \\ y &= 3 \end{aligned}$$

Since division by zero is not allowed, $y = 3$ does not satisfy the original equation. We have an inconsistent equation and so the solution set is $\emptyset$.

39. Multiply by $t + 3$.

$$\begin{aligned} t + 4t + 12 &= 2 \\ 5t &= -10 \end{aligned}$$

A conditional equation with solution set $\{-2\}$.

41. Since $-4.19 = 0.21x$ and $\dfrac{-4.19}{0.21} \approx -19.952$, the solution set is approximately $\{-19.952\}$.

43. Divide by 0.06.

$$\begin{aligned} x - 3.78 &= \frac{1.95}{0.06} \\ x &= 32.5 + 3.78 \\ x &= 36.28 \end{aligned}$$

The solution set is $\{36.28\}$.

45.

$$\begin{aligned} 2a &= -1 - \sqrt{17} \\ a &= \frac{-1 - \sqrt{17}}{2} \\ a &\approx \frac{-1 - 4.1231}{2} \\ a &\approx -2.562 \end{aligned}$$

The solution set is approximately $\{-2.562\}$.

47.

$$\begin{aligned} 0.001 &= 3(y - 0.333) \\ 0.001 &= 3y - 0.999 \\ 1 &= 3y \\ \frac{1}{3} &= y \end{aligned}$$

The solution set is $\left\{\dfrac{1}{3}\right\}$.

49. Factoring x, we get

$$\begin{aligned} x\left(\frac{1}{0.376} + \frac{1}{0.135}\right) &= 2 \\ x(2.6596 + 7.4074) &\approx 2 \\ 10.067x &\approx 2 \\ x &\approx 0.199 \end{aligned}$$

The solution set is approximately $\{0.199\}$.

51.

$$\begin{aligned} x^2 + 6.5x + 3.25^2 &= x^2 - 8.2x + 4.1^2 \\ 14.7x &= 4.1^2 - 3.25^2 \\ 14.7x &= 16.81 - 10.5625 \\ 14.7x &= 6.2475 \\ x &= 0.425 \end{aligned}$$

The solution set is $\{0.425\}$.

53.

$$\begin{aligned} (2.3 \times 10^6)x &= 1.63 \times 10^4 - 8.9 \times 10^5 \\ x &= \frac{1.63 \times 10^4 - 8.9 \times 10^5}{2.3 \times 10^6} \\ x &\approx -0.380 \end{aligned}$$

The solution set is approximately $\{-0.380\}$.

55. Solution set is $\{\pm 9\}$.

57. Since $|5x - 4| = 0$, we get $5x - 4 = 0$. The solution set is $\left\{\dfrac{4}{5}\right\}$.

59. Since $2x - 3 = 7$ or $2x - 3 = -7$, we get $2x = 10$ or $2x = -4$. The solution set is $\{5, -2\}$.

61. Dividing $2|x + 5| = 10$ by 2 we obtain $|x + 5| = 5$. Then $x + 5 = 5$ or $x + 5 = -5$. The solution set is $\{0, -10\}$.

63. Since $3x - 2 = 0$, the solution set is $\left\{\dfrac{2}{3}\right\}$.

65. Solving for $|x|$, we get $|x| = -\dfrac{1}{2}$. The equation has no solution and the solution set is $\emptyset$.

67. Since $0.95x = 190$, the solution set is $\{200\}$.

69.

$$0.1x - 0.05x + 1 = 1.2$$
$$0.05x = 0.2$$

The solution set is $\{4\}$.

71. Simplifying $x^2 + 4x + 4 = x^2 + 4$, we obtain $4x = 0$. The solution set is $\{0\}$.

73. Multiply by 4.

$$2x + 4 = x - 6$$
$$x = -10$$

The solution set is $\{-10\}$.

75. Multiply by 30.

$$15(y - 3) + 6y = 90 - 5(y + 1)$$
$$15y - 45 + 6y = 90 - 5y - 5$$
$$26y = 130$$

The solution set is $\{5\}$.

77. Since $7|x + 6| = 14$, $|x + 6| = 2$. Then $x + 6 = 2$ or $x + 6 = -2$. The solution set is $\{-4, -8\}$.

79. Multiply by $(x - 2)(x + 2)$.

$$3(x + 2) + 4(x - 2) = 7x - 2$$
$$3x + 6 + 4x - 8 = 7x - 2$$
$$7x - 2 = 7x - 2$$

An identity with solution set $\{x \mid x \neq 2, x \neq -2\}$.

81. Multiply $(x + 3)(x - 2)$ to

$$\frac{4}{x + 3} + \frac{3}{x - 2} = \frac{7x + 1}{(x + 3)(x - 2)}.$$

Then we find

$$4(x - 2) + 3(x + 3) = 7x + 1$$
$$4x - 8 + 3x + 9 = 7x + 1$$
$$7x + 1 = 7x + 1.$$

An identity and the solution set is $\{x \mid x \neq 2 \text{ and } x \neq -3\}$.

83. Multiply by $(x - 3)(x - 4)$.

$$(x - 4)(x - 2) = (x - 3)^2$$
$$x^2 - 6x + 8 = x^2 - 6x + 9$$
$$8 = 9$$

An inconsistent equation and so the solution set is $\emptyset$.

85. **a)** About 2005

b) Increasing

c) Let $y = 0.80$. Solving for x, we find

$$0.80 = 0.0102x + 0.644$$
$$\frac{0.80 - 0.644}{0.0102} = x$$
$$15.29 \approx x.$$

In the year 2005 $(= 1990 + 15)$, 80% of mothers will be in the labor force.

87. Since $B = 21,000 - 0.15B$, we obtain $1.15B = 21,000$ and the bonus is

$$B = \frac{21,000}{1.15} = \$18,260.87.$$

89. Rewrite the left-hand side as a sum.

$$10,000 + \frac{500,000,000}{x} = 12,000$$
$$\frac{500,000,000}{x} = 2,000$$
$$500,000,000 = 2000x$$
$$250,000 = x$$

Thus, $250,000$ vehicles must be sold.

For Thought

1. False, $P(1 + rt) = S$ implies $P = \dfrac{S}{1 + rt}$.

2. False, since the perimeter is twice the sum of the length and width.   **3.** False, since $n + 1$ and $n + 3$ are even integers if n is odd.

4. True

5. True, since $x + (-3 - x) = -3$.   **6.** False

7. False, for if the house sells for x dollars then

$$x - 0.09x = 100,000$$
$$0.91x = 100,000$$

and the house sells for $x = \$109,890.11$.

8. True

9. False, a correct equation is $4(x - 2) + 5 = 3x$.

10. False, since 9 and $x + 9$ differ by x.

1.2 Exercises

1. $r = \dfrac{I}{Pt}$

3. Since $F - 32 = \dfrac{9}{5}C$, $C = \dfrac{5}{9}(F - 32)$.

5. Since $2A = bh$, we get $b = \dfrac{2A}{h}$.

7. Since $By = C - Ax$, we obtain $y = \dfrac{C - Ax}{B}$.

9. Multiplying by $RR_1R_2R_3$, we find

$$R_1R_2R_3 = RR_2R_3 + RR_1R_3 + RR_1R_2$$

$$R_1R_2R_3 - RR_1R_3 - RR_1R_2 = RR_2R_3$$
$$R_1(R_2R_3 - RR_3 - RR_2) = RR_2R_3.$$

Then $R_1 = \dfrac{RR_2R_3}{R_2R_3 - RR_3 - RR_2}$.

11. Since $a_n - a_1 = (n-1)d$, we obtain

$$n - 1 = \dfrac{a_n - a_1}{d}$$
$$n = \dfrac{a_n - a_1}{d} + 1$$
$$n = \dfrac{a_n - a_1 + d}{d}.$$

13. Multiplying by 2.37, one obtains

$$2.4(2.37) = L + 2D - F\sqrt{S}$$
$$5.688 - L + F\sqrt{S} = 2D$$

and $D = \dfrac{5.688 - L + F\sqrt{S}}{2}$.

15. By using the formula $I = Prt$, one gets

$$51.30 = 950r \cdot 1$$
$$0.054 = r.$$

The simple interest rate is 5.4%.

17. Since $D = RT$, we find

$$5570 = 2228 \cdot T$$
$$2.5 = T.$$

and the surveillance takes 2.5 hours.

19. Note, $C = \dfrac{5}{9}(F - 32)$. If $F = 23^\circ F$, then

$$C = \frac{5}{9}(23 - 32) = -5^\circ C.$$

21. If x is the cost of the car before taxes, then

$$1.08x = 28,728$$
$$x = \$26,600.$$

23. Let S be the saddle height and let L be the inside measurement.

$$S = 1.09L$$
$$37 = 1.09L$$
$$\frac{37}{1.09} = L$$
$$33.9 \approx L$$

The inside leg measurement is 33.9 inches.

25. Let x be the amount of her game-show winnings.

$$0.14\frac{x}{3} + 0.12\frac{x}{6} = 4000$$
$$6\left(0.14\frac{x}{3} + 0.12\frac{x}{6}\right) = 24000$$
$$0.28x + 0.12x = 24000$$
$$0.40x = 24000$$
$$x = \$60,000.$$

Her winnings is $60,000.

27. If x is the length of the shorter piece in feet, then the length of the longer side is $2x + 2$. Then we obtain

$$x + x + (2x + 2) = 30$$
$$4x = 28$$
$$x = 7.$$

The length of each shorter piece is 7 ft and the longer piece is $2 \cdot 7 + 2$ or 16 ft.

29. If x is the length of the side of the larger square lot then $2x$ is the amount of fencing needed to divide the square lot into four smaller lots. The solution to $4x + 2x = 480$ is $x = 80$. The side of the larger square lot is 80 feet and its area is 6400 ft^2.

31. Note, Bobby will complete the remaining 8 laps in $\dfrac{8}{90}$ of an hour. If Ricky is to finish at the same time as Bobby, then Ricky's average speed s over 10 laps must satisfy $\dfrac{10}{s} = \dfrac{8}{90}$. $\left(\text{Note: } time = \dfrac{distance}{speed}\right)$. The equation is equivalent to $900 = 8s$. Thus, Ricky's average speed must be 112.5 mph.

33. Let d be the halfway distance between San Antonio and El Paso, and let s be the speed in the last half of the trip. Junior took $\dfrac{d}{80}$ hours to get to the halfway point and the last half took $\dfrac{d}{s}$ hours to drive. Since the total distance is $2d$ and $distance = rate \times time$,

$$
\begin{aligned}
2d &= 60\left(\frac{d}{80} + \frac{d}{s}\right) \\
160sd &= 60\left(sd + 80d\right) \\
160sd &= 60sd + 4800d \\
100sd &= 4800d \\
100d(s - 48) &= 0.
\end{aligned}
$$

Since $d \neq 0$, the speed for the last half of the trip was $s = 48$ mph.

35. If x is the part of the start-up capital invested at 5% and $x + 10,000$ is the part invested at 6%, then

$$
\begin{aligned}
0.05x + 0.06(x + 10,000) &= 5880 \\
0.11x + 600 &= 5880 \\
0.11x &= 5280 \\
x &= 48,000.
\end{aligned}
$$

Norma invested \$48,000 at 5% and \$58,000 at 6% for a total start-up capital of \$106,000.

37. Let x and $1500 - x$ be the number of employees from the Northside and Southside, respectively. Then

$$
\begin{aligned}
(0.05)x + 0.80(1500 - x) &= 750 \\
0.05x + 1200 - 0.80x &= 750 \\
450 &= 0.75x \\
600 &= x.
\end{aligned}
$$

There were 600 and 900 employees at the Northside and Southside, respectively.

39. Let x be the number of hours it takes both combines working together to harvest an entire wheat crop.

	rate
old	1/72
new	1/48
combined	1/x

Then $\dfrac{1}{72} + \dfrac{1}{48} = \dfrac{1}{x}$. Multiply both sides by $144x$ and get $2x + 3x = 144$. The solution is $x = 28.8$ hr which is the time it takes both combines to harvest the entire wheat crop.

41. Let t be the number of hours since 8:00 a.m.

	rate	time	work completed
Batman	1/8	$t-2$	$(t-2)/8$
Robin	1/12	t	$t/12$

$$
\begin{aligned}
\frac{t-2}{8} + \frac{t}{12} &= 1 \\
24\left(\frac{t-2}{8} + \frac{t}{12}\right) &= 24 \\
3(t-2) + 2t &= 24 \\
5t - 6 &= 24 \\
t &= 6
\end{aligned}
$$

At 2 p.m., all the crime have been cleaned up.

43. Since there are 5280 feet to a mile and the circumference of a circle is $C = 2\pi r$,

the radius r of the race track is $r = \dfrac{5280}{2\pi}$.

Since the length of a side of the square plot is twice the radius, the area of the plot is

$$\left(2 \cdot \frac{5280}{2\pi}\right)^2 \approx 2,824,677.3 \text{ ft}^2.$$

Dividing this number by $43,560$ results to 64.85 acres which is the acreage of the square lot.

45. The area of a trapezoid is $A = \dfrac{1}{2}h(b_1 + b_2)$.

$$
\begin{aligned}
90,000 &= \frac{1}{2}h(500 + 300) \\
90,000 &= 400h \\
225 &= h
\end{aligned}
$$

Thus, the streets are 225 ft apart.

47. Since the volume of a circular cylinder is

$V = \pi r^2 h$, we have $\dfrac{22,000}{7.5} = \pi 15^2 \cdot h$.

Solving for h, we get $h = 4.15$ ft, the depth of water in the pool.

49. Let r be the radius of the semicircular turns. Since the circumference of a circle is given by $C = 2\pi r$, we have $514 = 2\pi r + 200$. Solving for r, we get

$$r = \frac{157}{\pi} \approx 49.9747 \text{ m}.$$

Note, the width of the rectangular lot is $2r$. Then the dimension of the rectangular lot is 99.9494 m by 199.9494 m; its area is $19,984.82$ m^2, which is equivalent to 1.998 hectares.

51. Let x be Lorinda's taxable income.

$$
\begin{aligned}
14,382 + 0.31(x - 63,550) &= 21,275 \\
0.31x - 5318.50 &= 21,275 \\
0.31x &= 26,593.50 \\
x &\approx 85,785.
\end{aligned}
$$

Lorinda's taxable income is about $85,785.

53. Let x be the amount of water to be added. The volume of the resulting solution is $4 + x$ liters and the amount of pure baneberry in it is $0.05(4)$ liters. Since the resulting solution is a 3% extract, we have

$$
\begin{aligned}
0.03(4 + x) &= 0.05(4) \\
0.12 + 0.03x &= 0.20 \\
x &= \frac{0.08}{0.03} \\
x &= \frac{8}{3}.
\end{aligned}
$$

The amount of water to be added is $\dfrac{8}{3}$ liters.

55. The costs of x pounds of dried apples is $(1.20)4x$ and the cost of $(20 - x)$ pounds of dried apricots is $4(1.80)(20 - x)$. Since the 20 lb-mixture costs $1.68 per quarter-pound, we obtain

$$
\begin{aligned}
4(1.68)(20) &= (1.20)4x + 4(1.80)(20 - x) \\
134.4 &= 4.80x + 144 - 7.20x \\
2.40x &= 9.6 \\
x &= 4.
\end{aligned}
$$

The mix needs 4 lb of dried apples and 16 lb of dried apricots.

57. Let x and $8 - x$ be the number of dimes and nickels, respectively. Since the candy bar costs 55 cents, we have $55 = 10x + 5(8 - x)$. Solving for x, we find $x = 3$. Thus, Dana has 3 dimes and 5 nickels.

59. a) About 1992

   b) Since the revenues are equal, we obtain

$$
\begin{aligned}
13.5n + 190 &= 7.5n + 225 \\
6n &= 35 \\
n &\approx 5.8.
\end{aligned}
$$

In the year 1992 (=1986+6), restaurant revenues and supermarket revenues were the same.

61. If h is the number of hours it will take two hikers to pick a gallon of wild berries, then

$$\frac{1}{2} + \frac{1}{2} = \frac{1}{h}$$

$$1 = \frac{1}{h}$$

$$1 = h.$$

Two hikers can pick a gallon of wild berries in 1 hr.

If m is the number of minutes it will take two mechanics to change the oil of a Saturn, then

$$\frac{1}{6} + \frac{1}{6} = \frac{1}{m}$$

$$\frac{1}{3} = \frac{1}{m}$$

$$m = 3.$$

Two mechanics can change the oil in 3 minutes.

If w is the number of minutes it will take 60 mechanics to change the oil, then

$$60 \cdot \frac{1}{6} = \frac{1}{w}$$

$$w = \frac{1}{10} \text{ min}$$

$$w = 6 \text{ sec.}$$

So, 60 mechanics working together can change the oil in 6 sec (an unreasonable situation and answer).

For Thought

1. False, the point $(2, -3)$ is in Quadrant IV.

2. False, the point $(4, 0)$ does not belong to any quadrant.

3. False, since the distance is $\sqrt{(a-c)^2 + (b-d)^2}$.

4. False, since $Ax + By = C$ is a linear equation.

5. True, since the x-intercept can be obtained by replacing y by 0.

6. False, since $\sqrt{7^2 + 9^2} = \sqrt{130} \approx 11.4$

7. True

8. True

9. True

10. False, it is a circle of radius $\sqrt{5}$.

1.3 Exercises

1. $(4, 1)$, Quadrant I

3. $(1, 0)$, x-axis

5. $(5, -1)$, Quadrant IV

7. $(-4, -2)$, Quadrant III

9. $(-2, 4)$, Quadrant II

11. Distance is $\sqrt{(4-1)^2 + (7-3)^2} = \sqrt{9 + 16} = \sqrt{25} = 5$, midpoint is $(2.5, 5)$

13. Distance is $\sqrt{(-1-1)^2 + (-2-0)^2} = \sqrt{4 + 4} = 2\sqrt{2}$, midpoint is $(0, -1)$

15. Distance is $\sqrt{(-1 + 3\sqrt{3} - (-1))^2 + (4-1)^2} = \sqrt{27 + 9} = 6$, midpoint is $\left(\dfrac{-2 + 3\sqrt{3}}{2}, \dfrac{5}{2} \right)$

17. Distance is $\sqrt{(1.2 + 3.8)^2 + (4.4 + 2.2)^2} = \sqrt{25 + 49} = \sqrt{74}$, midpoint is $(-1.3, 1.3)$

19. Distance is $\sqrt{(a-b)^2 + 0} = |a - b|$, midpoint is $\left(\dfrac{a+b}{2}, 0 \right)$

21. Distance is $\dfrac{\sqrt{\pi^2 + 4}}{2}$, midpoint is $\left(\dfrac{3\pi}{4}, \dfrac{1}{2} \right)$

23. Center$(0, 0)$, radius 4

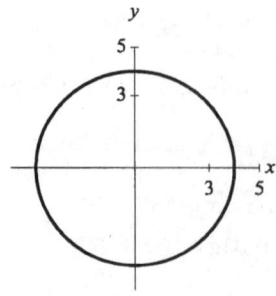

25. Center $(-6, 0)$, radius 6

27. Center $(-1, 0)$, radius 5

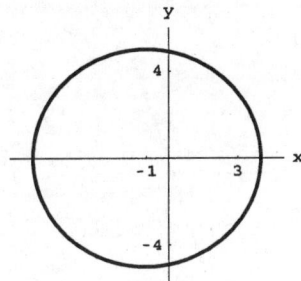

29. Center $(2, -2)$, radius $2\sqrt{2}$

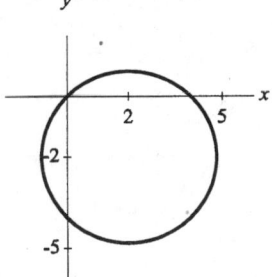

31. $x^2 + y^2 = 49$

33. $(x + 2)^2 + (y - 5)^2 = 1/4$

35. The distance between $(3, 5)$ and the origin is $\sqrt{34}$ which is the radius. The standard equation is $(x - 3)^2 + (y - 5)^2 = 34$.

37. The distance between $(5, -1)$ and $(1, 3)$ is $\sqrt{32}$ which is the radius. The standard equation is $(x - 5)^2 + (y + 1)^2 = 32$.

39. Center $(0, 0)$, radius 3

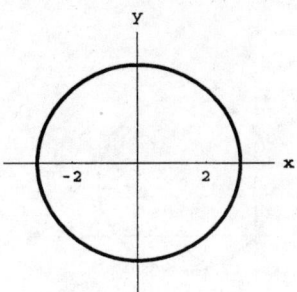

41. Completing the square, we have

$$x^2 + (y^2 + 6y + 9) = 0 + 9$$
$$x^2 + (y + 3)^2 = 9.$$

The center is $(0, -3)$ and the radius is 3.

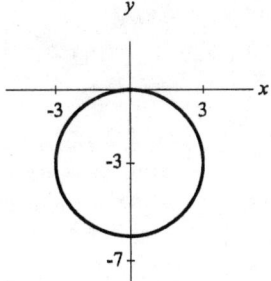

43. Completing the square, we obtain

$$(x^2 - 6x + 9) + (y^2 - 8y + 16) = 9 + 16$$
$$(x - 3)^2 + (y - 4)^2 = 25.$$

The center is $(3, 4)$ and the radius is 5.

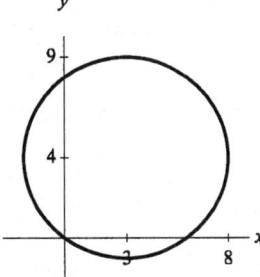

45. Completing the square, we obtain

$$(x^2 - 4x + 4) + \left(y^2 - 3y + \frac{9}{4}\right) = 4 + \frac{9}{4}$$
$$(x - 2)^2 + \left(y - \frac{3}{2}\right)^2 = \frac{25}{4}.$$

The center is $\left(2, \dfrac{3}{2}\right)$ and the radius is $\dfrac{5}{2}$.

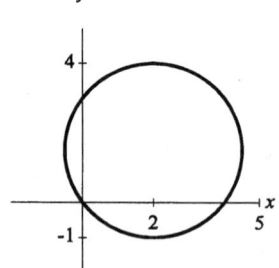

47. Completing the square, we obtain

$$\left(x^2 - \frac{1}{2}x + \frac{1}{16}\right) + \left(y^2 + \frac{1}{3}y + \frac{1}{36}\right) = \frac{1}{36}$$

$$\left(x - \frac{1}{4}\right)^2 + \left(y + \frac{1}{6}\right)^2 = \frac{1}{36}.$$

The center is $\left(\dfrac{1}{4}, -\dfrac{1}{6}\right)$ and the radius is $\dfrac{1}{6}$.

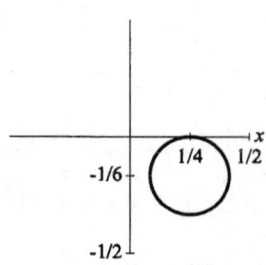

49. Since the center is $(0,0)$ and the radius is 7, the standard equation is $x^2 + y^2 = 49$.

51. The radius, which is the distance between $(1,0)$ and $(3,4)$, is given by

$$\sqrt{(3-1)^2 + (4-0)^2} = \sqrt{20}.$$

Together with the center $(1,0)$, it follows that the standard equation is $(x-1)^2 + y^2 = 20$.

53. The center, by using the midpoint formula, is

$$\left(\frac{3-1}{2}, \frac{5-1}{2}\right) = (1,2).$$ The diameter is

$$\sqrt{(3-(-1))^2 + (5-(-1))^2} = \sqrt{52}.$$ Since the

square of the radius is $\left(\dfrac{1}{2}\sqrt{52}\right)^2 = 13$, the

standard equation is $(x-1)^2 + (y-2)^2 = 13$.

55. $y = 3x - 4$ goes through $(0, -4)$, $\left(\dfrac{4}{3}, 0\right)$.

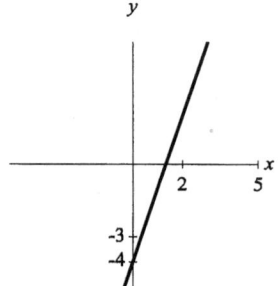

57. $3x - y = 6$ goes through $(0, -6)$, $(2, 0)$.

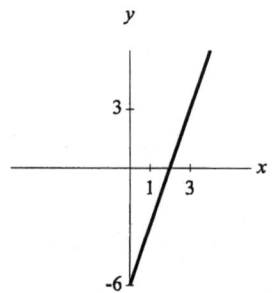

59. $x = 3y - 90$ goes through $(0, 30)$, $(-90, 0)$.

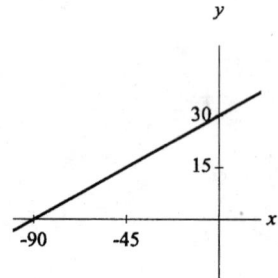

61. $\dfrac{2}{3}y - \dfrac{1}{2}x = 400$ goes through

$(0, 600)$, $(-800, 0)$.

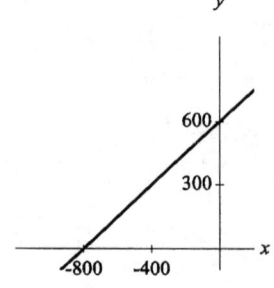

63. Intercepts are $(0, 0.0025), (0.005, 0)$.

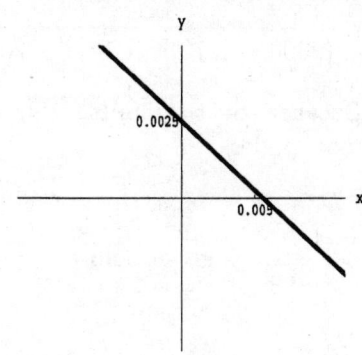

65. Intercepts are $(0, 2500), (5000, 0)$.

67. $x = 5$

69. $y = 4$

71. $x = -4$

73. Solving for y, we have $y = 1$.

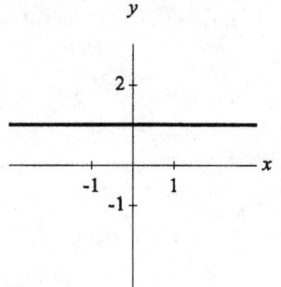

75. Since the x-intercept of $y = 2.4x - 8.64$ is $(3.6, 0)$, the solution set of $2.4x - 8.64 = 0$ is $\{3.6\}$.

77. Since the x-intercept of $y = -\dfrac{3}{7}x + 6$ is $(14, 0)$, the solution set of $-\dfrac{3}{7}x + 6 = 0$ is $\{14\}$.

79. The solution is $x = -\dfrac{3.4}{12} \approx -2.83$.

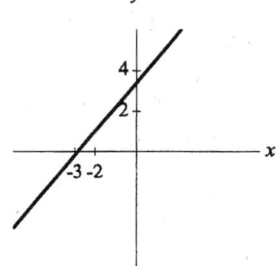

81. The solution is $\dfrac{687}{1.23} \approx 558.54$

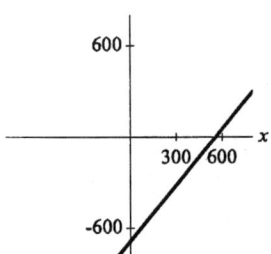

83. The solution is $\dfrac{3497}{0.03} \approx 116,566.67$

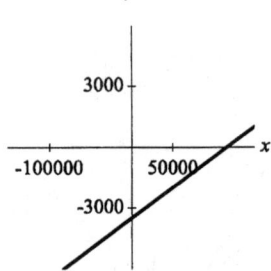

85. Note,

$$
\begin{aligned}
4.3 - 3.1(2.3x) + 3.1(9.9) &= 0 \\
4.3 - 7.13x + 30.69 &= 0 \\
34.99 - 7.13x &= 0 \\
x &= \frac{3499}{713} \\
x &\approx 4.91.
\end{aligned}
$$

The solution is $x \approx 4.91$.

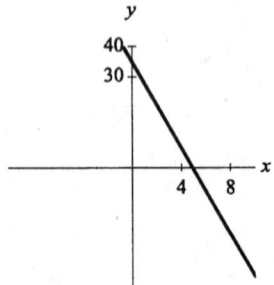

87. a) The midpoint is

$$\left(\frac{1970 + 2000}{2}, \frac{20.8 + 25.1}{2}\right) = (1985, 22.95).$$

The median of age at first marriage in 1985 was 22.95 years.

b) The distance is

$$\sqrt{(2000 - 1970)^2 + (25.1 - 20.8)^2} \approx 30.3$$

Because of the units, the distance is meaningless.

89. Given $D = 22,800$ lbs, the graph of

$C = \dfrac{4B}{\sqrt[3]{22,800}}$ is given below.

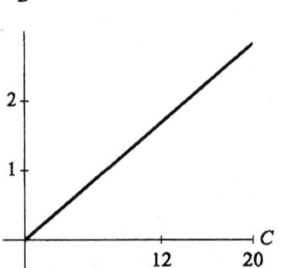

For Island Packet 40, $C = \dfrac{4(12 + \frac{11}{12})}{\sqrt[3]{22,800}} \approx 1.8$.

91. Since $a = \dfrac{a+a}{2} < \dfrac{a+b}{2} < \dfrac{b+b}{2} = b$,

we have $a < \dfrac{a+b}{2} < b$.

93. Note, $AB = \sqrt{(1+4)^2 + (1+5)^2} = \sqrt{25 + 36} = \sqrt{61}$. Similarly, $BC = \sqrt{61}$ and $AC = \sqrt{244} = 2\sqrt{61}$. Since $AB + BC = \sqrt{61} + \sqrt{61} = 2\sqrt{61} = AC$, we conclude A, B, and C are collinear.

95. The distance between $(10, 0)$ and $(0, 0)$ is 10. The distance between $(1, 3)$ and the origin is $\sqrt{10}$. If two points have integer coordinates, then the distance between them is of the form $\sqrt{s^2 + t^2}$ where $s^2, t^2 \in \{0, 1, 2^2, 3^2, 4^2, ...\} = \{0, 1, 4, 9, 16, ...\}$.

Note, there exists no pair s^2 and t^2 in $\{0, 1, 4, 9, 16, ...\}$ satisfying $s^2 + t^2 = 19$. Thus, one cannot find two points with integer coordinates whose distance between them is $\sqrt{19}$.

For Thought

1. False, the slope is $\dfrac{3-2}{3-2} = 1$.

2. False, the slope is $\dfrac{5-1}{-3-(-3)} = \dfrac{4}{0}$
which is undefined.

3. False, slopes of vertical lines are undefined.

4. False, it is a vertical line. **5.** True

6. False, $x = 1$ cannot be written in the
slope-intercept form.

7. False, the slope is -2.

8. True **9.** False **10.** True

1.4 Exercises

1. $\dfrac{5-3}{4+2} = \dfrac{1}{3}$

3. $\dfrac{3+5}{1-3} = -4$

5. $\dfrac{2-2}{5+3} = 0$

7. $\dfrac{1/2 - 1/4}{1/4 - 1/8} = \dfrac{1/4}{1/8} = 2$

9. $\dfrac{3-(-1)}{5-5} = \dfrac{4}{0}$, no slope

11. The slope is $m = \dfrac{4-(-1)}{3-(-1)} = \dfrac{5}{4}$.

Since $y + 1 = \dfrac{5}{4}(x+1)$, we get

$y = \dfrac{5}{4}x + \dfrac{5}{4} - 1$ or $y = \dfrac{5}{4}x + \dfrac{1}{4}$.

13. The slope is $m = \dfrac{-1-6}{4-(-2)} = -\dfrac{7}{6}$.

Since $y + 1 = -\dfrac{7}{6}(x-4)$, we obtain

$y = -\dfrac{7}{6}x + \dfrac{14}{3} - 1$ or $y = -\dfrac{7}{6}x + \dfrac{11}{3}$.

15. The slope is $m = \dfrac{5-5}{-3-3} = 0$.

Since $y - 5 = 0(x-3)$, we get $y = 5$.

17. Since $m = \dfrac{12-(-3)}{4-4} = \dfrac{15}{0}$ is undefined,
the equation of the vertical line is $x = 4$.

19. The slope of the line through $(0, -1)$ and $(3, 1)$
is $m = \dfrac{2}{3}$. Since the y-intercept is $(0, -1)$,

the line is given by $y = \dfrac{2}{3}x - 1$.

21. The slope of the line through $(1, 4)$ and $(-1, 1)$

is $m = \dfrac{5}{2}$. Solving for y in $y - 1 = \dfrac{5}{2}(x+1)$,

we get $y = \dfrac{5}{2}x + \dfrac{3}{2}$.

23. The slope of the line through $(0, 4)$ and $(2, 0)$
is $m = -2$. Since the y-intercept is $(0, 4)$,
the line is given by $y = -2x + 4$.

25. The slope of the line through $(1, 4)$ and

$(-3, -2)$ is $m = \dfrac{3}{2}$. Solving for y in

$y - 4 = \dfrac{3}{2}(x-1)$, we get $y = \dfrac{3}{2}x + \dfrac{5}{2}$.

27. $y = \dfrac{3}{5}x - 2$, slope is $\dfrac{3}{5}$, y-intercept is $(0, -2)$

29. Since $y - 3 = 2x - 8$, $y = 2x - 5$.
The slope is 2 and y-intercept is $(0, -5)$.

31. Since $y + 1 = \dfrac{1}{2}x + \dfrac{3}{2}$, $y = \dfrac{1}{2}x + \dfrac{1}{2}$.

The slope is $\dfrac{1}{2}$ and y-intercept is $\left(0, \dfrac{1}{2}\right)$.

33. Since $y = 4$, the slope is $m = 0$ and the
y-intercept is $(0, 4)$.

35. Since $y - 0.4 = 0.03x - 3$, $y = 0.03x - 2.6$.
The slope is 0.03 and y-intercept is $(0, -2.6)$

37. $y = \dfrac{1}{2}x - 2$ goes through the points

$(0, -2), (2, -1),$ and $(4, 0).$

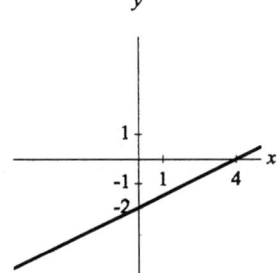

39. $y = -3x + 1$ goes through $(0, 1), (1, -2)$

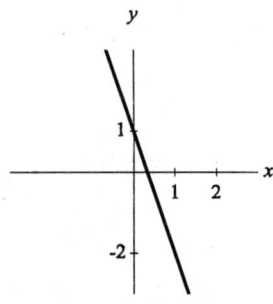

41. $y = -\dfrac{3}{4}x - 1$ goes through $(0, -1), (-4/3, 0)$

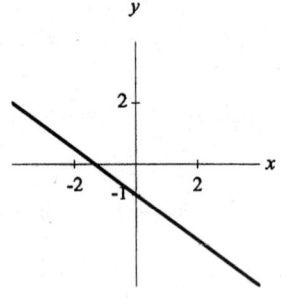

43. $x - y = 3$ goes through $(0, -3), (3, 0)$

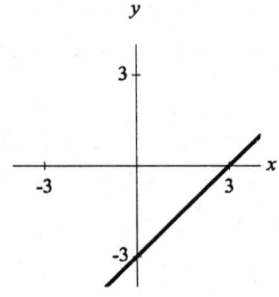

45. $y = 5$ is a horizontal line

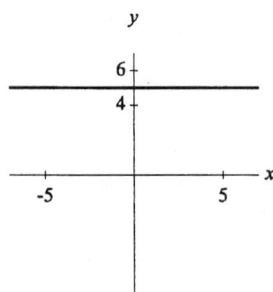

47. Since $m = \dfrac{4}{3}$ and $y - 0 = \dfrac{4}{3}(x - 3),$

we have $4x - 3y = 12.$

49. Since $m = \dfrac{4}{5}$ and $y - 3 = \dfrac{4}{5}(x - 2),$

we obtain $5y - 15 = 4x - 8$ and $4x - 5y = -7.$

51. $x = -4$ is a vertical line.

53. 0.5 **55.** -1 **57.** 0

59. Since $y + 2 = 2(x - 1), 2x - y = 4$

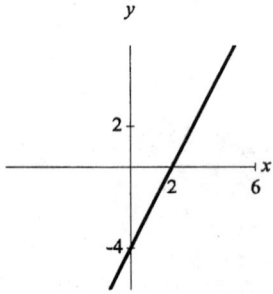

61. Since slope of $y = -3x$ is -3 and
$y - 4 = -3(x - 1),$ we obtain $3x + y = 7.$

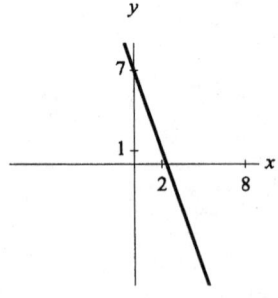

63. Since slope of $y = \frac{1}{2}x - \frac{3}{2}$ is $\frac{1}{2}$ and

$y - 1 = -2(x + 3)$, we find $2x + y = -5$.

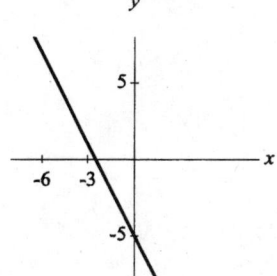

65. Since $x = 4$ is a vertical line, the horizontal line through $(2, 5)$ is $y = 5$.

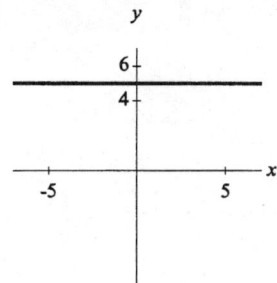

67. Since $\dfrac{5 - 3}{8 + 2} = \dfrac{1}{5} = -\dfrac{1}{a}$, we find $a = -5$.

69. Since $\dfrac{a - 3}{-2 - a} = -\dfrac{1}{2}$, we obtain $2a - 6 = 2 + a$ and $a = 8$.

71. Plot the points $A(-1, 2)$, $B(2, -1)$, $C(3, 3)$, and $D(-2, -2)$, respectively. The slopes of the opposite sides are $m_{AC} = m_{BD} = 1/4$ and $m_{AD} = m_{BC} = 4$. Since the opposite sides are parallel, it is a parallelogram.

73. Plot the points $A(-5, -1)$, $B(-3, -4)$, $C(3, 0)$, and $D(1, 3)$, respectively. The slopes of the opposite sides are $m_{AB} = m_{CD} = -3/2$ and $m_{AD} = m_{BC} = 2/3$. Since the adjacent sides are perpendicular, it is a rectangle.

75. Plot the points $A(-5, 1)$, $B(-2, -3)$, and $C(4, 2)$, respectively. The slopes of the sides are $m_{AB} = -4/3$, $m_{BC} = 5/6$ and $m_{AC} = 1/9$. It is not a right triangle since no two sides are perpendicular.

77. No, they are not parallel since their slopes are not equal, i.e., $\dfrac{1}{3} \neq 0.33$. However, they may appear parallel.

79. Since $x^3 - 8 = (x - 2)(x^2 + 2x + 4)$, we obtain $\dfrac{x^3 - 8}{x^2 + 2x + 4} = x - 2$. We have a linear function for the graph is $y = x - 2$.

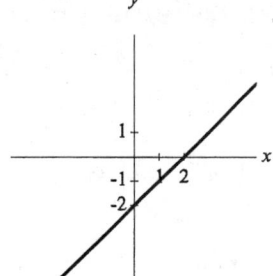

81. The slope is $\dfrac{212 - 32}{100 - 0} = \dfrac{9}{5}$.

Since $F - 32 = \dfrac{9}{5}(C - 0)$, $F = \dfrac{9}{5}C + 32$.

When $C = 150$, $F = \dfrac{9}{5}(150) + 32 = 302^\circ$F.

83. Thee linear function through $(1, 49)$ and $(2, 48)$ is $c = 50 - n$. With $n = 40$ people in a tour she would charge \$10 each and make \$400.

85. The slope is $\dfrac{75 - 95}{4000} = -0.005$.

Since $S - 95 = -0.005(D - 0)$, we obtain $S = -0.005D + 95$.

87. Let c and p be the number of computers and printers, respectively. Since $60000 = 2000c + 1500p$, we have

$$
\begin{aligned}
2000c &= -1500p + 60000 \\
c &= -\frac{3}{4}p + 30.
\end{aligned}
$$

The slope is $-\dfrac{3}{4}$, i.e., if 4 more printers are purchased then 3 fewer computers must be bought.

89. Let $b_1 \neq b_2$. If $y = mx + b_1$ and $y = mx + b_2$ have a point (s, t) in common, then $ms + b_1 = ms + b_2$. After subtracting ms from both sides, we get $b_1 = b_2$; a contradiction. Thus, $y = mx + b_1$ and $y = mx + b_2$ have no points in common if $b_1 \neq b_2$.

For Thought

1. True, a scatter diagram is a graph consisting of ordered pairs.

2. True

3. False, it is possible for the variables to have no relationship.

4. True

5. True, in fact, if $r = 1$, the data is perfectly in line.

6. True. In addition, if $r = -1$ then the data is perfectly in line.

7. False, since $r = 0.002$ is close to zero, we say that there is no positive correlation.

8. False, since $r = -0.001$ is approximatley zero, we say that there is no negative correlation.

9. False, interpolating is making a prediction within the range of the data.

10. False, exterpolating is making a prediction outside the range of the data.

1.5 Exercises

1. Linear relationship

3. No relationship

5. Nonlinear relationship

7. Linear relationship

9. Linear relationship

11. Linear relationship

13. No relationship

15. The missing entries are $(3.3, 160)$ and $(4.0, 193)$.

17. Missing entries are $(132, 34)$ and $(148, 25)$.

19. a) With a graphing calculator, we find

$$y = 0.5148571429x - 1014.510476$$

Approximately, $y = 0.515x - 1014.51$.

b) Let $x = 2003$. Then

$$y = 0.5148571429(2003) - 1014.510476 \approx 16.75.$$

In 2003, there will be 16.75 million cars.

c) Let $y = 0$. The solution to

$$0 = 0.5148571429x - 1014.510476$$

is $x \approx \dfrac{1014.510476}{0.5148571429} \approx 1970.47.$

According to the equation, no cars were produced in the European Union in 1970.

21. a) With a graphing calculator, we find

$$y = 5.057142857x - 10,015.24286$$

Approximately, $y = 5.057x - 10,015.24$.

b) Let $y = 120$. The solution to

$$120 = 5.057142857x - 10,015.24286$$

is $x = \dfrac{10,135.24286}{5.057142857} \approx 2004.$

According to the regression line, in the year 2004, total personal income will be $120 trillion.

c) If $x = 1990$, then

$$5.057142857(1990) - 10,015.24286 \approx 48.5.$$

According to the regression line, in 1990, the total personal income was $48.5 trillion.

23. With a calculator, we get $p = -0.069A + 0.403$. If $A = 4$, then $p = -0.069(4) + 0.403 = 0.127$. Thus, 12.7% of the runners in age group 4 are predicted to be injured.

For Thought

1. False, since $x = 1$ is a solution of the first equation and not of the second equation.

2. False, since $x^2 + 1 = 0$ cannot be factored with real coefficients.

3. False, $\left(x + \dfrac{2}{3}\right)^2 = x^2 + \dfrac{4}{3}x + \dfrac{9}{4}$.

4. False, the solutions to $(x - 3)(2x + 5) = 0$ are $x = 3$ and $x = -\dfrac{5}{2}$.

5. False, $x^2 = 0$ has only $x = 0$ as its solution.

6. True, since $a = 1, b = -3$, and $c = 1$, then by the quadratic formula we obtain

$$x = \frac{3 \pm \sqrt{9 - 4}}{2} = \frac{3 \pm \sqrt{5}}{2}.$$

7. False, the quadratic formula can be used to solve any quadratic equation.

8. False, $x^2 + 1 = 0$ has only imaginary zeros.

9. True, for $b^2 - 4ac = 12^2 - 4(4)(9) = 0$.

10. True, $x^2 - 6x + 9 = (x - 3)^2 = 0$ has only one real solution, namely, $x = 3$.

1.6 Exercises

1. Since $x^2 = 5$, the solution set is $\{\pm\sqrt{5}\}$.

3. Since $x^2 = -\dfrac{2}{3}$, there are no real solutions.

5. Since $x - 3 = \pm 3$, we get $x = 3 \pm 3$.

The solution set is $\{0, 6\}$.

7. By the square root property, we get $3x - 1 = \pm 0 = 0$. Solving for x, we obtain $x = \dfrac{1}{3}$. The solution set is $\left\{\dfrac{1}{3}\right\}$.

9. Since $x - \dfrac{1}{2} = \pm\dfrac{5}{2}$, it follows that $x = \dfrac{1}{2} \pm \dfrac{5}{2}$. The solution set is $\{-2, 3\}$.

11. No real solutions since $\sqrt{-4}$ is not a real number.

13. Since $(x-5)(x+4)=0$, the solution set is $\{5,-4\}$.

15. Since $a^2+3a+2=(a+2)(a+1)=0$, the solution set is $\{-2,-1\}$.

17. Since $(2x+1)(x-3)=0$, the solution set is $\left\{-\dfrac{1}{2},3\right\}$.

19. Since $(2x-1)(3x-2)=0$, the solution set is $\left\{\dfrac{1}{2},\dfrac{2}{3}\right\}$.

21. Note, $y^2+y-12=30$. Subtracting 30 from both sides, one obtains $y^2+y-42=0$ or $(y+7)(y-6)=0$. The solution set is $\{-7,6\}$.

23. $x^2-12x+\left(\dfrac{12}{2}\right)^2=x^2-12x+6^2=x^2-12x+36$

25. $r^2+3r+\left(\dfrac{3}{2}\right)^2=r^2+3r+\dfrac{9}{4}$

27. $w^2+\dfrac{1}{2}w+\left(\dfrac{1}{4}\right)^2=w^2+\dfrac{1}{2}w+\dfrac{1}{16}$

29. By completing the square, we derive

$$\begin{aligned} x^2+6x &= -1 \\ x^2+6x+9 &= -1+9 \\ (x+3)^2 &= 8 \\ x+3 &= \pm 2\sqrt{2}. \end{aligned}$$

The solution set is $\{-3\pm 2\sqrt{2}\}$.

31. By completing the square, we find

$$\begin{aligned} n^2-2n &= 1 \\ n^2-2n+1 &= 1+1 \\ (n-1)^2 &= 2 \\ n-1 &= \pm\sqrt{2}. \end{aligned}$$

The solution set is $\{1\pm\sqrt{2}\}$.

33.

$$\begin{aligned} h^2+3h &= 1 \\ h^2+3h+\dfrac{9}{4} &= 1+\dfrac{9}{4} \\ \left(h+\dfrac{3}{2}\right)^2 &= \dfrac{13}{4} \\ h+\dfrac{3}{2} &= \pm\dfrac{\sqrt{13}}{2} \end{aligned}$$

The solution set is $\left\{\dfrac{-3\pm\sqrt{13}}{2}\right\}$.

35.

$$\begin{aligned} x^2+\dfrac{5}{2}x &= 6 \\ x^2+\dfrac{5}{2}x+\dfrac{25}{16} &= 6+\dfrac{25}{16} \\ \left(x+\dfrac{5}{4}\right)^2 &= \dfrac{121}{16} \\ x &= -\dfrac{5}{4}\pm\dfrac{11}{4} \end{aligned}$$

The solution set is $\left\{\dfrac{3}{2},-4\right\}$.

37.

$$\begin{aligned} x^2+\dfrac{2}{3}x &= -\dfrac{1}{3} \\ x^2+\dfrac{2}{3}x+\dfrac{1}{9} &= -\dfrac{3}{9}+\dfrac{1}{9} \\ \left(x+\dfrac{1}{3}\right)^2 &= -\dfrac{2}{9} \end{aligned}$$

Since $\sqrt{-\dfrac{2}{9}}$ is not a real number, there are no real solutions.

39. Since $a=1,b=3,c=-4$ and
$$x=\dfrac{-3\pm\sqrt{3^2-4(1)(-4)}}{2(1)}=\dfrac{-3\pm\sqrt{25}}{2}=$$
$\dfrac{-3\pm 5}{2}$, the solution set is $\{1,-4\}$.

41. Since $a=2,b=-5,c=-3$ and
$$x=\dfrac{5\pm\sqrt{(-5)^2-4(2)(-3)}}{2(2)}=\dfrac{5\pm\sqrt{49}}{4}=$$
$\dfrac{5\pm 7}{4}$, the solution set is $\left\{-\dfrac{1}{2},3\right\}$.

43. Since $a = 9, b = 6, c = 1$ and

$$x = \frac{-6 \pm \sqrt{6^2 - 4(9)(1)}}{2(9)} = \frac{-6 \pm 0}{18},$$

the solution set is $\left\{-\frac{1}{3}\right\}$.

45. Since $a = 2, b = 0, c = -3$ and

$$x = \frac{0 \pm \sqrt{0^2 - 4(2)(-3)}}{2(2)} = \frac{\pm\sqrt{24}}{4} =$$

$\frac{\pm 2\sqrt{6}}{4}$, the solution set is $\left\{\pm\frac{\sqrt{6}}{2}\right\}$.

47. In $x^2 - 4x + 5 = 0$, $a = 1, b = -4, c = 5$.

Then $x = \dfrac{4 \pm \sqrt{(-4)^2 - 4(1)(5)}}{2(1)} = \dfrac{4 \pm \sqrt{-4}}{2}$.

No real solutions since $\sqrt{-4}$ is not a real number.

49. Since $a = 3.2, b = 7.6$, and $c = -9$,

$$x = \frac{-7.6 \pm \sqrt{(7.6)^2 - 4(3.2)(-9)}}{2(3.2)} \approx$$

$\dfrac{-7.6 \pm \sqrt{172.96}}{6.4} \approx \dfrac{-7.6 \pm 13.151}{6.4}$.

The solution set is $\{0.87, -3.24\}$.

51. Note, $a = 3.25, b = -4.6$, and $c = 20$.

Then $x = \dfrac{4.6 \pm \sqrt{(-4.6)^2 - 4(3.25)(20)}}{2(3.25)}$

$\approx \dfrac{4.6 \pm \sqrt{-238.84}}{6.5}$. No real solutions

since $\sqrt{-238.84}$ is not a real number.

53. The discriminant is $(-30)^2 - 4(9)(25) =$ $900 - 900 = 0$. Only one solution and it is real.

55. The discriminant is $(-6)^2 - 4(5)(2) = 36 - 40 = -4$. There are no real solutions.

57. The discriminant is $12^2 - 4(7)(-1) = 144 + 28 = 172$. There are two distinct real solutions.

59. Note, x-intercepts are $\left(-\frac{2}{3}, 0\right)$ and $\left(\frac{1}{2}, 0\right)$.

The solution set is $\left\{-\frac{2}{3}, \frac{1}{2}\right\}$.

61. Since the x-intercepts are $(-3, 0)$ and $(5, 0)$, the solution set is $\{-3, 5\}$.

63. Note, the graph of $y = 1.44x^2 - 8.4x + 12.25$ has exactly one x-intercept.

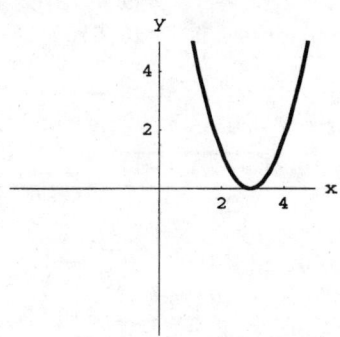

65. Note, the graph of $y = x^2 + 3x + 15$ has no x-intercept.

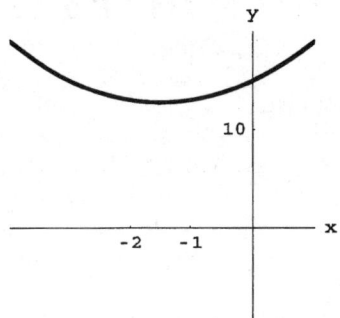

Thus, $x^2 + 3x + 15 = 0$ has no real solution.

67. The graph of $y = x^2 + 3x - 160$ has two x-intercepts.

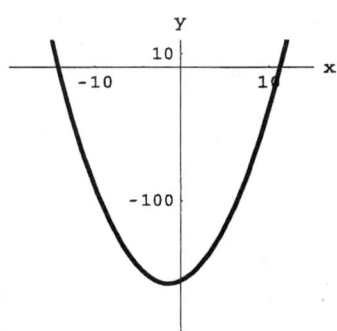

Then $x^2 + 3x - 160 = 0$ has two real solutions.

69. Set the right-hand side to 0.

$$x^2 - \frac{4}{3}x - \frac{5}{9} = 0$$
$$9x^2 - 12x - 5 = 0$$
$$(3x + 1)(3x - 5) = 0$$

The solution set is $\left\{-\dfrac{1}{3}, \dfrac{5}{3}\right\}$.

71. Since $x^2 + 6x - 72 = (x + 12)(x - 6) = 0$, the solution set is $\{-12, 6\}$.

73. Multiply by x to get $x^2 = x + 1$. Then $x^2 - x - 1 = 0$ and by the quadratic formula,
$$x = \frac{1 \pm \sqrt{1 - 4(1)(-1)}}{2} = \frac{1 \pm \sqrt{5}}{2}.$$
The solution set is $\left\{\dfrac{1 \pm \sqrt{5}}{2}\right\}$.

75. Multiplying by $(3 - x)(x + 7)$.
$$
\begin{aligned}
(x - 12)(x + 7) &= (x + 4)(3 - x) \\
x^2 - 5x - 84 &= -x^2 - x + 12 \\
2x^2 - 4x - 96 &= 0 \\
2(x - 8)(x + 6) &= 0
\end{aligned}
$$
The solution set is $\{8, -6\}$.

77. From the revenue function,
$$
\begin{aligned}
x(40 - 0.001x) &= 175{,}000 \\
40x - 0.001x^2 &= 175{,}000
\end{aligned}
$$
By applying the quadratic formula to $0.001x^2 - 40x + 175{,}000 = 0$, we get
$$
\begin{aligned}
x &= \frac{40 \pm \sqrt{(40)^2 - 4(0.001)(175{,}000)}}{0.002} \\
x &= \frac{40 \pm \sqrt{900}}{0.002} = \frac{40 \pm 30}{0.002} \\
x &= 5000 \text{ or } 35{,}000
\end{aligned}
$$
So 5000 or 35,000 units must be produced weekly.

79. The height S (in feet) of the ball from the ground t seconds after it was tossed is given by $S = -16t^2 + 40t + 4$. When the height is 4 feet,
$$
\begin{aligned}
-16t^2 + 40t + 4 &= 4 \\
-16t^2 + 40t &= 0 \\
-8t(2t - 5) &= 0 \\
t &= 0, \frac{5}{2}.
\end{aligned}
$$
The ball returns to a height of 4 ft in 2.5 sec.

81. Let d be the diagonal distance across the field from one goal to the other. By the Pythagorean Theorem, we obtain
$d = \sqrt{300^2 + 160^2} = 340$ ft.

83. Let w and $2w + 2$ be the length and width. From the given area of the court, we obtain
$$
\begin{aligned}
(2w + 2)w &= 312 \\
2w^2 + 2w - 312 &= 0 \\
w^2 + w - 156 &= 0 \\
(w - 12)(w + 13) &= 0
\end{aligned}
$$
Then $w = 12$ yd and the length is 26 yd. The distance between two opposite corners (by the Pythagorean Theorem) is $\sqrt{12^2 + 26^2} = 2\sqrt{205} \approx 28.6$ yd.

85. Substituting the values of S and A, we find that the displacement is
$$
\begin{aligned}
\frac{1}{2^{12}} d^2 (18.8)^3 - 822^3 &= 0 \\
\frac{1}{2^{12}} d^2 (18.8)^3 &= 822^3 \\
d &= \sqrt{\frac{822^3 (2^{12})}{18.8^3}} \\
d &\approx 18{,}503.4 \text{ lbs.}
\end{aligned}
$$

87. Let x be the width of the border. The overall dimensions are $8 + 2x$ and $6 + 2x$. Then
$$
\begin{aligned}
(8 + 2x)(6 + 2x) &= 100 \\
4x^2 + 28x - 52 &= 0 \\
x^2 + 7x - 13 &= 0.
\end{aligned}
$$
By the quadratic formula, $x = \dfrac{-7 \pm \sqrt{101}}{2} \approx$ $1.52, -8.52$. The border should be 1.52 ft wide. The negative value of x is not possible.

89. Let x be the normal speed of the tortoise in ft/hr.

	distance	rate	time
hwy	24	$x+2$	$24/(x+2)$
off hwy	24	x	$24/x$

Since 24 minutes is 2/5 of an hour, we get
$$\frac{2}{5} + \frac{24}{x+2} = \frac{24}{x}$$

$$2x(x+2) + 24(5)x = 24(5)(x+2)$$
$$2x^2 + 4x + 120x = 120x + 240$$
$$x^2 + 2x - 120 = 0$$
$$(x+12)(x-10) = 0$$
$$x = -12, 10.$$

The normal speed of the tortoise is 10 ft/hr.

91. Using $v_1^2 = v_0^2 + 2gS$ with $S = 1.07$ and $v_1 = 0$, we find that

$$v_0^2 + 2(-9.8)(1.07) = 0$$
$$v_0^2 - 20.972 = 0$$
$$v_0 = \pm\sqrt{20.972} \approx \pm 4.58.$$

His initial upward velocity is 4.58 m/sec.

Using $S = \frac{1}{2}gt^2 + v_0t$ with $S = 0$ and $v_o = 4.58$, we find that his time t in the air satisfies

$$\frac{1}{2}(-9.8)t^2 + 4.58t = 0$$
$$t(4.58 - 4.9t) = 0$$
$$t \approx 0, 0.93.$$

Carter is in the air for 0.93 seconds.

93. a) With the aid of a graphing calculator, the regression curve is given by

$$y = -0.57051953x^2 + 2258.14841x - 2233977.7$$

Approximately,

$$y = -0.57x^2 + 2258.1x - 2,233,977.7$$

b) The scatter diagram and regression curve is given below.

c) Set $y = 0$ and solve the quadratic equation

$$-0.57051953x^2 + 2258.14841x - 2233977.7 = 0.$$

The solutions are $x \approx 1949.7$ and $x \approx 2008.4$. By looking at the x-intercept of the regression curve, we conclude 2008 is the year when the production is zero.

For Thought

1. True

2. False, since $-2x < -6$ is equivalent to

$$\frac{-2x}{-2} > \frac{-6}{-2}.$$

3. False, since there is a number between any two distinct real numbers.

4. True, since $|-6-6| = |-12| = 12 > -1$.

5. False, $(-\infty, -3) \cap (-\infty, -2) = (-\infty, -3)$.

6. False, $(5, \infty) \cap (-\infty, -3) = \phi$.

7. False, no real number satisfies $|x - 2| < 0$.

8. False, it is equivalent to $|x| > 3$.

9. False, $|x| + 2 < 5$ is equivalent to $-3 < x < 3$.

10. True

1.7 Exercises

1. $x < 12$ **3.** $x \geq -7$

5. $[-8, \infty)$ **7.** $(-\infty, \pi/2)$

9. Since $3x > 15$ implies $x > 5$, the solution set is $(5, \infty)$ and the graph is

11. Since $10 \leq 5x$ implies $2 \leq x$, the solution set is $[2, \infty)$ and the graph is

13. Multiply 6 to both sides of the inequality.

$$3x - 24 < 2x + 30$$
$$x < 54$$

The solution is the interval $(-\infty, 54)$ and the graph is

15. Multiplying the inequality by 2, we find

$$\begin{aligned} 7 - 3x &\geq -6 \\ 13 &\geq 3x \\ 13/3 &\geq x. \end{aligned}$$

The solution is the interval $(-\infty, 13/3]$ and the graph is

13/3

17. Multiply the inequality by -5 and reverse the direction of the inequality.

$$\begin{aligned} 2x - 3 &\leq 0 \\ 2x &\leq 3 \\ x &\leq \frac{3}{2} \end{aligned}$$

The solution is the interval $(-\infty, 3/2]$ and the graph is

3/2

19. Multiply the left-hand side.

$$\begin{aligned} -6x + 4 &\geq 4 - x \\ 0 &\geq 5x \\ 0 &\geq x. \end{aligned}$$

The solution is the interval $(-\infty, 0]$ and the graph is

0

21. Using the portion of the graph below the x-axis, the solution set is $(-\infty, -3.5)$.

23. Using the part of the graph on or above the x-axis, the solution set is $(-\infty, 1.4]$.

25. By taking the part of the line $y = 2x - 3$ above the horizontal line $y = 5$ and by using $(4, 5)$, the solution set is $(4, \infty)$.

27. Note, the graph of $y = -3x - 7$ is above or on the graph of $y = x + 1$ for $x \leq -2$. Thus, the solution set is $(-\infty, -2]$.

29. $(-3, \infty)$ **31.** $(-3, \infty)$

33. $(-5, -2)$ **35.** ϕ **37.** $(-\infty, 5]$

39. Solve each simple inequality and find the intersection of their solution sets.

$$\begin{aligned} x > 3 \quad &\text{and} \quad 0.5x < 3 \\ x > 3 \quad &\text{and} \quad x < 6 \end{aligned}$$

The intersection of these values of x is the interval $(3, 6)$ and whose graph is

3   6

41. Solve each simple inequality and find the intersection of their solution sets.

$$\begin{aligned} 2x - 5 > -4 \quad &\text{and} \quad 2x + 1 > 0 \\ x > \frac{1}{2} \quad &\text{and} \quad x > -\frac{1}{2} \end{aligned}$$

The intersection of these values of x is the interval $(1/2, \infty)$ and the graph is

1/2

43. Solve each simple inequality and find the union of their solution sets.

$$\begin{aligned} -6 < 2x \quad &\text{or} \quad 3x > -3 \\ -3 < x \quad &\text{or} \quad x > -1 \end{aligned}$$

The union of these values of x is $(-3, \infty)$ and the graph is

-3

45. Solve each simple inequality and find the union of their solution sets.

$$\begin{aligned} x + 1 > 6 \quad &\text{or} \quad x < 7 \\ x > 5 \quad &\text{or} \quad x < 7 \end{aligned}$$

The union of these values of x is $(-\infty, \infty)$ and the graph is

47. Solve each simple inequality and find the intersection of their solution sets.

$$\begin{aligned} 2 - 3x < 8 \quad &\text{and} \quad x - 8 \leq -12 \\ -6 < 3x \quad &\text{and} \quad x \leq -4 \\ -2 < x \quad &\text{and} \quad x \leq -4 \end{aligned}$$

The intersection is empty and there is no solution.

49. Solve an equivalent compound inequality.

$$
\begin{aligned}
-2 < \ & 3x - 1 \ < 2 \\
-1 < \ & \ \ 3x \ \quad < 3 \\
-\frac{1}{3} < \ & \quad x \quad \ < 1
\end{aligned}
$$

The solution set is the interval $(-1/3, 1)$ and the graph is

-1/3   1

51. Solve an equivalent compound inequality.

$$
\begin{aligned}
-1 \leq \ & 5 - 4x \ \leq 1 \\
-6 \leq \ & \ \ -4x \ \ \ \leq -4 \\
\frac{3}{2} \geq \ & \quad x \quad \ \geq 1
\end{aligned}
$$

The solution set is the interval $[1, 3/2]$ and the graph is

1   3/2

53. Solve an equivalent compound inequality.

$$
\begin{aligned}
-5 \leq \ & 4 - x \ \leq 5 \\
-9 \leq \ & \ \ -x \ \ \ \leq 1 \\
9 \geq \ & \quad x \quad \ \geq -1
\end{aligned}
$$

The solution set is the interval $[-1, 9]$ and the graph is

-1   9

55. No solution since an absolute value is never negative.

57. No solution since an absolute value is never negative.

59. Note, $3|x - 2| > 3$ or $|x - 2| > 1$.
We solve an equivalent compound inequality.

$$
\begin{aligned}
x - 2 > 1 \quad &\text{or} \quad x - 2 < -1 \\
x > 3 \quad &\text{or} \quad x < 1
\end{aligned}
$$

The solution set is the interval $(-\infty, 1) \cup (3, \infty)$ and the graph is

1   3

61. Solve an equivalent compound inequality.

$$
\begin{aligned}
\frac{x - 3}{2} > 1 \quad &\text{or} \quad \frac{x - 3}{2} < -1 \\
x - 3 > 2 \quad &\text{or} \quad x - 3 < -2 \\
x > 5 \quad &\text{or} \quad x < 1
\end{aligned}
$$

The solution set is the interval $(-\infty, 1) \cup (5, \infty)$ and the graph is

1   5

63. $|x| < 5$ **65.** $|x| > 3$

67. Since 6 is the midpoint of 4 and 8, the inequality is $|x - 6| < 2$.

69. Since 4 is the midpoint of 3 and 5, the inequality is $|x - 4| > 1$.

71. $|x| \geq 9$

73. Since 7 is the midpoint, the inequality is $|x - 7| \leq 4$.

75. Since 5 is the midpoint, the inequality is $|x - 5| > 2$.

77. Since $x - 2 \geq 0$, the solution set is $[2, \infty)$.

79. Since $2 - x > 0$ is equivalent to $2 > x$, the solution set is $(-\infty, 2)$.

81. Since $|x| \geq 3$ is equivalent to $x \geq 3$ or $x \leq -3$, the solution set is $(-\infty, -3] \cup [3, \infty)$.

83. If x is the price of a car excluding sales tax then it must satisfy $0 \leq 1.1x + 300 \leq 8000$.

This is equivalent to $0 \leq x \leq \dfrac{7700}{1.1} = 7000$.

The price range of Yolanda's car is the interval $[\$0, \$7000]$.

85. Let x be Lucky's score on the final exam.

$$
\begin{aligned}
79 \ &< \ \frac{65 + x}{2} \ < 90 \\
158 \ &< \ 65 + x \ < 180 \\
93 \ &< \ \quad x \quad \ < 115.
\end{aligned}
$$

The final exam score must lie in $(93, 115)$.

87. Let x be Ingrid's final exam score. Since $\dfrac{2x + 65}{3}$ is her weighted average, we obtain

$$
\begin{aligned}
79 \ &< \ \frac{2x + 65}{3} \ < 90 \\
237 \ &< \ 2x + 65 \ < 270 \\
172 \ &< \ \quad 2x \quad \ < 205 \\
86 \ &< \ \quad x \quad \ < 102.5
\end{aligned}
$$

Ingrid's final exam score must lie in $(86, 102.5)$.

89. If h is the height of the box, then

$$
\begin{aligned}
40 + 2(30) + 2h &\leq 130 \\
100 + 2h &\leq 130 \\
2h &\leq 30.
\end{aligned}
$$

The range of the height is $(0 \text{ in}, 15 \text{ in}]$.

91. By substituting $N = 50$ and $w = 27$ into $r = \dfrac{Nw}{n}$ we find $r = \dfrac{1350}{n}$. Moreover if $n = 14$, then $r = \dfrac{1350}{14} = 96.4 \approx 96$. Similarly, the other gear ratios are the following.

n	14	17	20	24	29
r	96	79	67	56	47

Yes, the bicycle has a gear ratio for each of the four types.

93. If x is the price of a BMW 750 iL, then $|x - 66,400| > 25,000$. An equivalent inequality is

$$
\begin{aligned}
x - 66400 > 25000 \quad &\text{or} \quad x - 66400 < -25000 \\
x > 91,400 \quad &\text{or} \quad x < 41,400.
\end{aligned}
$$

The price of a BMW 750 iL is either under $\$41,400$ or over $\$91,400$.

95. If x is the actual temperature, then

$$
\left| \frac{x - 35}{35} \right| < .01
$$

$$
\begin{aligned}
-.35 < x - 35 &< .35 \\
34.65 < x &< 35.35.
\end{aligned}
$$

The actual temperature must lie in the interval $(34.65, 35.35)$.

97. If c is the actual circumference, then $c = \pi d$ and

$$
\begin{aligned}
|\pi d - 7.2| &\leq 0.1 \\
-0.1 \leq \pi d - 7.2 &\leq 0.1 \\
7.1 \leq \pi d &\leq 7.3 \\
2.26 \leq d &\leq 2.32.
\end{aligned}
$$

The actual diameter must lie in the interval $[2.26 \text{ cm}, 2.32 \text{ cm}]$.

99. a) The inequality $|a - 29,676| < 2000$ is equivalent to

$$
\begin{aligned}
-2000 < \quad a - 29,676 \quad &< 2000 \\
27,676 < \quad\quad a \quad\quad &< 31,676.
\end{aligned}
$$

The states within this range are Alaska, Florida, and Hawaii.

b) The inequality $|a - 29,676| > 3000$ is equivalent to

$$
\begin{aligned}
x - 29,676 > 3000 \quad &\text{or} \quad x - 29,676 < -3000 \\
x > 32,676 \quad &\text{or} \quad x < 26,676.
\end{aligned}
$$

The states satisfying the inequality are Alabama, Maine, Colorado, Maryland, New Jersey, and Connecticut.

Chapter 1 Review Exercises

1. Since $3x = 2$, the solution set is $\{2/3\}$.

3. Multiply by 60 to get $30y - 20 = 15y + 12$, or $15y = 32$. The solution set is $\{32/15\}$.

5. Multiply by $x(x - 1)$ to get $2x - 2 = 3x$. The solution set is $\{-2\}$.

7. Multiply by $(x + 1)(x - 3)$ and get $-2x - 3 = x - 2$. Then $-1 = 3x$. The solution set is $\{-1/3\}$.

9. The distance is $\sqrt{(-3 - 2)^2 + (5 - (-6))^2} = \sqrt{(-5)^2 + 11^2} = \sqrt{25 + 121} = \sqrt{146}$. The midpoint is $\left(\dfrac{-3 + 2}{2}, \dfrac{5 - 6}{2} \right) = \left(\dfrac{-1}{2}, \dfrac{-1}{2} \right)$.

11. Distance is $\sqrt{\left(\dfrac{1}{2} - \dfrac{1}{4} \right)^2 + \left(\dfrac{1}{3} - 1 \right)^2} = \sqrt{\left(\dfrac{1}{4} \right)^2 + \left(-\dfrac{2}{3} \right)^2} = \sqrt{\dfrac{1}{16} + \dfrac{4}{9}} = \sqrt{\dfrac{73}{144}} = \dfrac{\sqrt{73}}{12}$. Midpoint is $\left(\dfrac{1/2 + 1/4}{2}, \dfrac{1/3 + 1}{2} \right) = \left(\dfrac{3/4}{2}, \dfrac{4/3}{2} \right) = \left(\dfrac{3}{8}, \dfrac{2}{3} \right)$.

13. Circle with radius 5 and center at the origin.

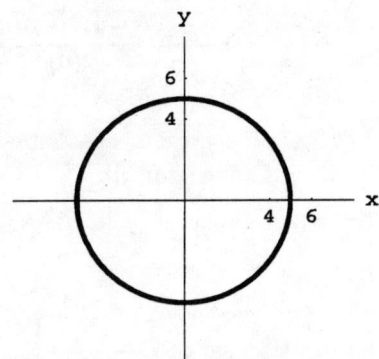

15. Equivalently, by using the method of completing the square, the circle is given by $(x+2)^2 + y^2 = 4$. It has radius 2 and center $(-2, 0)$.

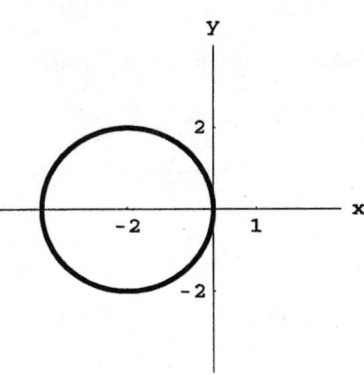

17. The line $y = -x + 25$ has intercepts $(0, 25)$, $(25, 0)$.

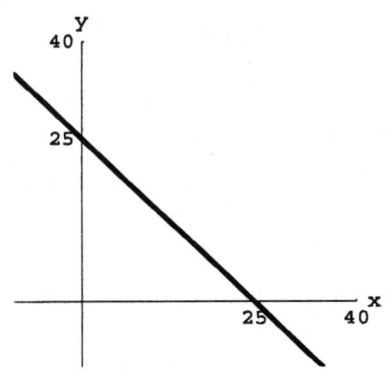

19. The line $y = 3x - 4$ has intercepts $(0, -4)$, $(4/3, 0)$.

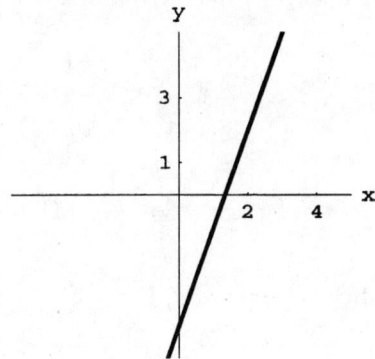

21. Vertical line $x = 5$ has intercept $(5, 0)$.

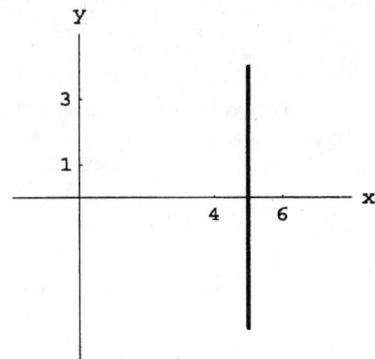

23. Simplify $(x - (-3))^2 + (y - 5)^2 = \left(\sqrt{3}\right)^2$. The standard equation is $(x+3)^2 + (y-5)^2 = 3$.

25. Substitute $y = 0$ in $3x - 4y = 12$. Then $3x = 12$ or $x = 4$. The x-intercept is $(4, 0)$. Substitute $x = 0$ in $3x - 4y = 12$ to get $-4y = 12$ or $y = -3$. The y-intercept is $(0, -3)$.

27. $\dfrac{2 - (-6)}{-1 - 3} = \dfrac{8}{-4} = -2$

29. Note, $m = \dfrac{-1 - 3}{5 - (-2)} = -\dfrac{4}{7}$. Solving for y in

$y - 3 = -\dfrac{4}{7}(x + 2)$, we obtain $y = -\dfrac{4}{7}x + \dfrac{13}{7}$.

31. Note, the slope of $3x + y = -5$ is -3.
The standard form for the line through $(2, -4)$
with slope $\dfrac{1}{3}$ is derived below.

$$
\begin{aligned}
y + 4 &= \frac{1}{3}(x - 2) \\
3y + 12 &= x - 2 \\
-x + 3y &= -14 \\
x - 3y &= 14
\end{aligned}
$$

33. Since $2x - 6 = 3y$, $y = \dfrac{2}{3}x - 2$.

35. Note, $y(x - 3) = 1$. Then $y = \dfrac{1}{x - 3}$.

37. Note, $by = -ax + c$. Then $y = -\dfrac{a}{b}x + \dfrac{c}{b}$
provided $b \neq 0$.

39. The discriminant of $x^2 - 4x + 2$ is $(-4)^2 - 4(2) = 8$. There are two distinct real solutions.

41. The discriminant is $(-20)^2 - 4(4)(25) = 0$. Only one real solution.

43. Since $x^2 = 5$, the solution set is $\{\pm\sqrt{5}\}$.

45. Since $x - 2 = \pm\sqrt{17}$, we get $x = 2 \pm \sqrt{17}$.
The solution set is $\left\{2 \pm \sqrt{17}\right\}$.

47. Since $(x + 3)(x - 4) = 0$, the solution set is $\{-3, 4\}$.

49. The discriminant of $b^2 - 6b + 10$ is

$$(-6)^2 - 4(1)(10) = -4.$$

Since the discriminant is negative, the equation $b^2 + 10 = 6b$ has no real solution.

51. We apply the method of completing the square.

$$
\begin{aligned}
s^2 - 4s &= -1 \\
s^2 - 4s + 4 &= -1 + 4 \\
(s - 2)^2 &= 3 \\
s - 2 &= \pm\sqrt{3}
\end{aligned}
$$

The solution set is $\left\{2 \pm \sqrt{3}\right\}$.

53. Use the quadratic formula to solve $4x^2 - 4x - 5 = 0$.

$$
\begin{aligned}
x &= \frac{4 \pm \sqrt{(-4)^2 - 4(4)(-5)}}{2(4)} \\
&= \frac{4 \pm \sqrt{96}}{8} \\
&= \frac{4 \pm 4\sqrt{6}}{8} \\
&= \frac{1 \pm \sqrt{6}}{2}
\end{aligned}
$$

The solution set is $\left\{\dfrac{1 \pm \sqrt{6}}{2}\right\}$.

55. Multiplying by $2x(x - 1)$, one obtains

$$
\begin{aligned}
2x - 2 + 2x &= 3x^2 - 3x \\
0 &= 3x^2 - 7x + 2 \\
0 &= (3x - 1)(x - 2).
\end{aligned}
$$

The solution set is $\{1/3, 2\}$.

57. Solve an equivalent statement

$$
\begin{aligned}
3q - 4 = 2 \quad &\text{or} \quad 3q - 4 = -2 \\
3q = 6 \quad &\text{or} \quad 3q = 2.
\end{aligned}
$$

The solution set is $\{2/3, 2\}$.

59. We obtain

$$
\begin{aligned}
|2h - 3| &= 0 \\
2h - 3 &= 0 \\
h &= \frac{3}{2}.
\end{aligned}
$$

The solution set is $\left\{\dfrac{3}{2}\right\}$.

61. No solution since absolute values are nonnegative.

63. The solution set of $x > 3$ is the interval $(3, \infty)$ and the graph is

65. The solution set of $8 > 2x$ is the interval $(-\infty, 4)$ and the graph is

67. Since $-\dfrac{7}{3} > \dfrac{1}{2}x$, the solution set is $(-\infty, -14/3)$ and the graph is

69. After multiplying the inequality by 2 we have

$$-4 < x - 3 \leq 10$$
$$-1 < x \leq 13.$$

The solution set is the interval $(-1, 13]$ and the graph is

-1   13

71. The solution set of $\frac{1}{2} < x$ and $x < 1$ is the interval $(1/2, 1)$ and the graph is

1/2   1

73. The solution set of $x > -4$ or $x > -1$ is the interval $(-4, \infty)$ and the graph is

-4

75. Solving an equivalent statement, we get

$$x - 3 > 2 \quad \text{or} \quad x - 3 < -2$$
$$x > 5 \quad \text{or} \quad x < 1.$$

The solution set is $(-\infty, 1) \cup (5, \infty)$ and the

graph is

1   5

77. Since an absolute value is nonnegative, $2x - 7 = 0$. The solution set is $\{7/2\}$ and the graph is

7/2

79. Since absolute values are nonnegative, the solution set is $(-\infty, \infty)$ and

the graph is

81. The solution set is $\{10\}$ since the x-intercept is $(10, 0)$.

83. Since the x-intercept is $(8, 0)$ and the y-values are negative in quadrants 3 and 4, the solution set is $(-\infty, 8)$.

85. Let x be the length of one side of the square. Since dimensions of the base are $8 - 2x$ and $11 - 2x$, we obtain

$$(11 - 2x)(8 - 2x) = 50$$
$$4x^2 - 38x + 38 = 0$$
$$2x^2 - 19x + 19 = 0$$
$$x = \frac{19 \pm \sqrt{209}}{4} \approx 8.36, 1.14.$$

But $x = 8.36$ is too big and so $x = 1.14$ inch.

87. Let x be the number of hours it takes Lisa or Taro to drive to the restaurant. Since the sum of the driving distances is 300, we obtain $300 = 50x + 60x$. Thus, $x = \dfrac{300}{110} \approx 2.7272$ and Lisa drove $50(2.7272) \approx 136.4$ miles.

89. Let x and $8000 - x$ be the number of fish in Homer Lake and Mirror lake, respectively. Then

$$0.2x + 0.3(8000 - x) = 0.28(8000)$$
$$-0.1x + 2400 = 2240$$
$$1600 = x.$$

There were originally 1600 fish in Homer Lake.

91. Let x be the distance she hiked in the northern direction. Then she hiked $32 - x$ miles in the eastern direction. By the Pythagorean Theorem, we obtain

$$x^2 + (32 - x)^2 = (4\sqrt{34})^2$$
$$2x^2 - 64x + 480 = 0$$
$$2 \cdot (x - 20)(x - 12) = 0$$
$$x = 20, 12.$$

Since the eastern direction was the shorter leg of the journey, the northern direction was 20 miles.

93. Let x and $x+50$ be the cost of a haircut at Joe's and Renee's, respectively. Since 5 haircuts at Joe's is less than one haircut at Renee's, we have

$$5x < x + 50.$$

Thus, the price range of a haircut at Joe's is $x < \$12.50$ or $(0, \$12.50)$.

95. Let x and $x + 2$ be the length and width of a picture frame in inches, respectively. Since there are between 32 and 50 inches of molding, we get

$$32 < 2x + 2(x + 2) < 50$$
$$32 < 4x + 4 < 50$$
$$28 < 4x < 46$$
$$7 < x < 11.5.$$

The set of possible widths is $(7 \text{ in}, 11.5 \text{ in})$.

97. If the average gas mileage is increased from 29.5 mpg to 31.5 mpg, then the amount of gas saved is

$$\frac{10^{12}}{29.5} - \frac{10^{12}}{31.5} \approx 2.15 \times 10^9 \text{ gallons.}$$

Suppose the mileage is increased to x from 29.5 mpg. Then x must satisfy

$$\frac{10^{12}}{29.5} - \frac{10^{12}}{x} = \frac{10^{12}}{27.5} - \frac{10^{12}}{29.5}$$
$$\frac{1}{29.5} - \frac{1}{x} = \frac{1}{27.5} - \frac{1}{29.5}$$
$$-\frac{1}{x} \approx -0.031433$$
$$x \approx 31.8.$$

The mileage must be increased to 31.8 mpg.

99. Let x be the number of years since 1985. The equation of the line passing through $(0, 21.5)$ and $(15, 160.5)$ is

$$y = \frac{139}{15}x + 21.5.$$

In 1993, when $x = 8$, there were

$$\frac{139}{15}(8) + 21.5 \approx 95.6 \text{ million computers in use.}$$

101. Let a be the age in years and p be the percentage. The equation of the line passing through $(20, 0.23)$ and $(50, 0.47)$ is $p = 0.008a + 0.07$. If $a = 65$, then $p = 0.008(65) + 0.07 \approx 0.59$. Thus, the percentage of body fat in a 65-year old woman is 59%.

103. Let C be the cost and let y be the year.

a) The regression line is
$C = 163.8y - 323,094.5$ and its graph is given in the next page.

b) Substituting $y = 2001$, one finds
$C = 163.8(2001) - 323,094.5 \approx \$4669.3$.

Chapter 1 Test

1. Since $2x - x = -6 - 1$, the solution set is $\{-7\}$.

2. Multiplying the original equation by 6, we get $3x - 2x = 1$. The solution set is $\{1\}$.

3. Since $x^2 = \frac{2}{3}$, one obtains $x = \pm\frac{\sqrt{2}}{\sqrt{3}} = \pm\frac{\sqrt{6}}{3}$. The solution set is $\left\{\pm\frac{\sqrt{6}}{3}\right\}$.

4. By completing the square, we obtain

$$x^2 - 6x = -1$$
$$(x-3)^2 = -1 + 9$$
$$(x-3)^2 = 8$$
$$x - 3 = \pm\sqrt{8}.$$

The solution set is $\{3 \pm 2\sqrt{2}\}$.

5. Since $x^2 - 9x + 14 = (x-2)(x-7) = 0$, the solution set is $\{2, 7\}$.

6. After cross-multiplying, we get

$$(x-1)(x-6) = (x+3)(x+2)$$
$$x^2 - 7x + 6 = x^2 + 5x + 6$$
$$-7x = 5x$$
$$0 = 12x$$

The solution set is $\{0\}$.

7. We use the method of completing the square.

$$x^2 - 2x = -5$$
$$(x-1)^2 = -5 + 1$$
$$(x-1)^2 = -4$$

Since the square of a real number is **never negative**, we conclude that there are **no real solutions**.

8. The line $3x - 4y = 120$ passes through $(0, -30)$ and $(40, 0)$.

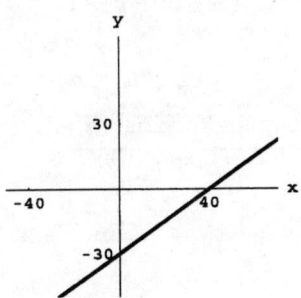

9. Circle with center $(0, 0)$ and radius 20.

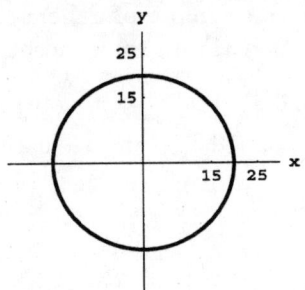

10. By using the method of completing the square, we obtain $x^2 + (y+2)^2 = 4$. A circle with center $(0, -2)$ and radius 2.

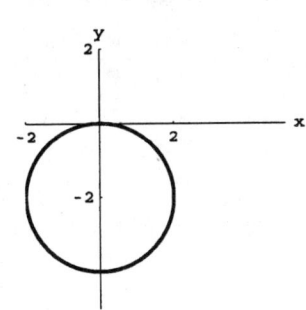

11. The line $y = -\dfrac{2}{3}x + 4$ passes through $(0, 4)$ and $(6, 0)$.

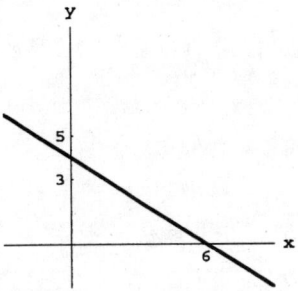

12. The horizontal line $y = 4$.

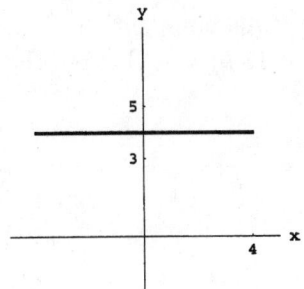

13. The vertical line $x = -2$.

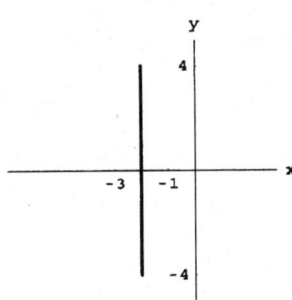

14. Since $y = \dfrac{3}{5}x - \dfrac{8}{5}$, the slope is $\dfrac{3}{5}$.

15. $\dfrac{-4 - 6}{5 - (-3)} = \dfrac{-10}{8} = -\dfrac{5}{4}$

16. We rewrite $2x - 3y = 6$ as $y = \dfrac{2}{3}x - 2$. Note, the slope of $y = \dfrac{2}{3}x - 2$ is $\dfrac{2}{3}$. Then we use $m = -\dfrac{3}{2}$ and the point $(1, -2)$.

$$y + 2 \;=\; -\dfrac{3}{2}(x - 1)$$

$$y = -\frac{3}{2}x + \frac{3}{2} - 2.$$

The perpendicular line is $y = -\frac{3}{2}x - \frac{1}{2}$.

17. The slope is $\dfrac{-1-2}{3} = -1$. We use the point $(3, -4)$.

$$y + 4 = -(x - 3)$$
$$y = -x + 3 - 4.$$

The parallel line is $y = -x - 1$.

18. $\sqrt{(-3-2)^2 + (1-4)^2} = \sqrt{25 + 9} = \sqrt{34}$

19. $\left(\dfrac{-1+1}{2}, \dfrac{1+0}{2}\right) = \left(0, \dfrac{1}{2}\right)$

20. Since the discriminant is negative, namely, $(-5)^2 - 4(1)(9) = -11$, then there are no real solutions.

21. We solve for y.

$$5 - 4 = 3xy + 2y$$
$$1 = y(3x + 2)$$
$$\frac{1}{3x + 2} = y$$

22. Since $-4 > 2x$, the solution set is $(-\infty, -2)$ and the graph is

23. The solution set to $x > 6$ and $x > 5$ is the interval $(6, \infty)$ and the graph is

24. Solving an equivalent statement, we obtain

$$-3 \leq 2x - 1 \leq 3$$
$$-2 \leq 2x \leq 4$$
$$-1 \leq x \leq 2.$$

The solution set is the interval $[-1, 2]$ and the graph is

25. We rewrite $|x - 3| > 2$ without any absolute values. Then

$$x - 3 > 2 \quad \text{or} \quad x - 3 < -2$$
$$x > 5 \quad \text{or} \quad x < 1.$$

The solution set is $(-\infty, 1) \cup (5, \infty)$ and the graph is

26. If x is the original length of one side of the square, then

$$(x + 20)(x + 10) = 999$$
$$x^2 + 30x + 200 = 999$$
$$x^2 + 30x - 799 = 0$$
$$\frac{-30 \pm \sqrt{900 + 4(799)}}{2} = x$$
$$\frac{-30 \pm 64}{2} = x$$
$$17, -47 = x.$$

Thus, $x = 17$ and the original area is $17^2 = 289$ ft^2.

27. Let x be the number of gallons of the 20% solution. From the concentrations,

$$0.3(10 + x) = 0.5(10) + 0.2x$$
$$3 + 0.3x = 5 + 0.2x$$
$$0.1x = 2$$
$$x = 20.$$

Then 20 gallons of the 20% solution are needed.

28. Let n be the number of years since 1994 and let P be the median price of a new home in Springfield. Since the slope is given by

$$m = \frac{92,800 - 88,000}{3 - 0} = 1600$$

and the P-intercept is $(0, 88000)$, an equation of the line is $P = 1600n + 88,000$.

Using this equation, the median price in the year 2003 is $P = 1600(9) + 88,000 = \$102,400$.

29. a) Let y be the year and let x be the number of cattle. The linear regression line is $y = 0.0461077844x + 7.16247505$ or approximately

$$y = 0.046x + 7.16.$$

While the equation of the quadratic regression curve is
$y = -0.1855488576x^2 + 740.6027085x - 738911.4892$
or about

$$y = -0.186x^2 + 740.6x - 738,911.5.$$

b) Let $x = 2004$. Then by using the lienar regression line we obtain
$$y = 0.0461077844(2004) + 7.16247505 \approx 99.6 \text{ million.}$$
Substituting $x = 2004$ into the quadratic regression equation, we get
$$y = -0.1855488576(2004)^2 + 740.6027085(2004) - 738911.4892 \approx 89.2$$
million. The difference in the extrapolations is 10.4 million cattle.

Tying It All Together

1. $7x$ **2.** $30x^2$ **3.** $\dfrac{2}{2x} + \dfrac{1}{2x} = \dfrac{3}{2x}$

4. $x^2 + 6x + 9$ **5.** $6x^2 + x - 2$

6. $\dfrac{x^2 + 2xh + h^2 - x^2}{h} = \dfrac{2xh + h^2}{h} = 2x + h$

7. $\dfrac{x+1}{(x-1)(x+1)} + \dfrac{x-1}{(x+1)(x-1)} = \dfrac{2x}{x^2 - 1}$

8. $x^2 + 3x + \dfrac{9}{4}$

9. An identity and the solution set is $(-\infty, \infty)$.

10. Since $30x^2 - 11x = x(30x - 11) = 0$, the solution set is $\left\{0, \dfrac{11}{30}\right\}$.

11. Since $\dfrac{3}{2x} = \dfrac{3}{2x}$, the solution set is $(-\infty, 0) \cup (0, \infty)$.

12. Subtract $x^2 + 9$ from $x^2 + 6x + 9 = x^2 + 9$ and get $6x = 0$. The solution set is $\{0\}$.

13. Since $(2x - 1)(3x + 2) = 0$, the solution set is $\left\{\dfrac{1}{2}, -\dfrac{2}{3}\right\}$.

14. Multiply the equation by $8(x+1)(x-1)$.
$$\begin{aligned} 8x + 8 + 8x - 8 &= 5(x^2 - 1) \\ 0 &= 5x^2 - 16x - 5 \\ x &= \frac{16 \pm \sqrt{356}}{10} \\ x &= \frac{16 \pm 2\sqrt{89}}{10} \end{aligned}$$
The solution set is $\left\{\dfrac{8 \pm \sqrt{89}}{5}\right\}$.

15. Since $7x - 7x^2 = 7x(1 - x) = 0$, the solution set is $\{0, 1\}$.

16. Since $7x - 7 = 7(x - 1) = 0$, the solution set is $\{1\}$.

17. 0 **18.** $-1 - 3 + 2 = -2$

19. $-\dfrac{1}{8} - 3\left(\dfrac{2}{8}\right) + \dfrac{16}{8} = \dfrac{9}{8}$

20. $-\dfrac{1}{27} - 3\left(\dfrac{3}{27}\right) + \dfrac{54}{27} = \dfrac{44}{27}$

21. $-1 - 3 - 4 = -8$

22. $-4 + 6 - 4 = -2$ **23.** -4

24. $-0.25 + 1.5 - 4 = -2.75$

For Thought

1. False, since $\{(1,2),(1,3)\}$ is not a function.

2. False, since $f(5)$ is not defined.   **3.**   True

4. False, since a student's exam grade is a function of the student's preparation. If two classmates had the same IQ and only one prepared then the one who prepared will most likely achieve a higher grade.

5. False, since $(x+h)^2 = x^2 + 2xh + h^2$

6. False, since the domain is all real numbers.

7. True   **8.**   True   **9.**   True

10. False, since $\left(\frac{3}{8}, 8\right)$ and $\left(\frac{3}{8}, 5\right)$ are two ordered pairs with the same first coordinate and different second coordinates.

2.1 Exercises

1. Note, $b = 2\pi a$ is equivalent to $a = \dfrac{b}{2\pi}$.

Thus, a is a function of b, and b is a function of a.

3. a is a function of b since a given denomination has a unique length. Since a dollar bill and a five-dollar bill have the same length, then b is not a function of a.

5. Since an item has only one price, b is a function of a. Since two items may have the same price, a is not a function of b.

7. a is not a function of b since it is possible that two different students can obtain the same final exam score but the times spent on studying are different.

b is not a function of a since it is possible that two different students can spend the same time studying but obtain different final exam scores.

9. Since 1 in ≈ 2.54 cm, a is a function of b and b is a function of a.

11. No   **13.**   Yes   **15.**   Yes

17. Function

19. Not a function since 25 has two different second coordinates.

21. Not a function since 3 has two different second coordinates.

23. Function

25. Since the ordered pairs in the graph of $y = 3x - 8$ are $(x, 3x - 8)$, there are no two ordered pairs with the same first coordinate and different second coordinates. We have a function.

27. Since $y = (x + 9)/3$, the ordered pairs are $(x, (x+9)/3)$. Thus, there are no two ordered pairs with the same first coordinate and different second coordinates. We have a function.

29. Since $y = \pm x$, the ordered pairs are $(x, \pm x)$. Thus, there are two ordered pairs with the same first coordinate and different second coordinates. We do not have a function.

31. Since $y = x^2$, the ordered pairs are (x, x^2). Thus, there are no two ordered pairs with the same first coordinate and different second coordinates. We have a function.

33. Since $y = |x| - 2$, the ordered pairs are $(x, |x| - 2)$. Thus, there are no two ordered pairs with the same first coordinate and different second coordinates. We have a function.

35. Since $(2, 1)$ and $(2, -1)$ are two ordered pairs with the same first coordinate and different second coordinates, the equation does not define a function.

37. domain $\{-3, 4, 5\}$, range $\{1, 2, 6\}$

39. Domain $(-\infty, \infty)$, range $\{4\}$

41. Domain $(-\infty, \infty)$; since $|x| \geq 0$, the range of $y = |x| + 5$ is $[5, \infty)$

43. since $|y| - 3 \geq -3$, the domain of $x = |y| - 3$ is $[-3, \infty)$; range $(-\infty, \infty)$

45. Since $\sqrt{x-4}$ is a real number whenever $x \geq 4$, the domain of $y = \sqrt{x-4}$ is $[4, \infty)$.

Since $y = \sqrt{x-4} \geq 0$ for $x \geq 4$, the range is $[0, \infty)$.

47. Since $x = -y^2 \le 0$, the domain of $x = -y^2$ is $(-\infty, 0]$; range is $(-\infty, \infty)$;

49. 6   **51.** $g(2) = 3(2) + 5 = 11$

53. Since $(3, 8)$ is the ordered pair, one obtains $f(3) = 8$. Then $x = 3$.

55. Solving for $3x + 5 = 26$, we find $x = 7$.

57. $f(4) + g(4) = 5 + 17 = 22$

59. $3a^2 - a$ **61.** $4(a + 2) - 2 = 4a + 6$

63. $3(x^2 + 2x + 1) - (x + 1) = 3x^2 + 5x + 2$

65. $4(x + h) - 2 = 4x + 4h - 2$

67. $(3x^2 + 6xh + 3h^2 - x - h) - 3x^2 + x = 3h^2 + 6xh - h$

69. The average rate of change is

$$\frac{4,000 - 16,000}{5} = -\$2,400 \text{ per year.}$$

71. The average rate of change on $[0, 2]$ is

$$\frac{h(2) - h(0)}{2 - 0} = \frac{0 - 64}{2 - 0} = -32 \text{ ft/sec.}$$

The average rate of change on $[1, 2]$ is

$$\frac{h(2) - h(1)}{2 - 1} = \frac{0 - 48}{2 - 1} = -48 \text{ ft/sec.}$$

The average rate of change on $[1.9, 2]$ is

$$\frac{h(2) - h(1.9)}{2 - 1.9} = \frac{0 - 6.24}{0.1} = -62.4 \text{ ft/sec.}$$

The average rate of change on $[1.99, 2]$ is

$$\frac{h(2) - h(1.99)}{2 - 1.99} = \frac{0 - 0.6384}{0.01} = -63.84 \text{ ft/sec.}$$

The average rate of change on $[1.999, 2]$ is

$$\frac{h(2) - h(1.999)}{2 - 1.999} = \frac{0 - 0.063984}{0.001} = -63.984$$
ft/sec.

73. The average rate of change is $\dfrac{958 - 1056}{2000 - 1984} =$
-6.125 million hectares per year.

75.

$$\frac{f(x + h) - f(x)}{h} = \frac{4(x + h) - 4x}{h}$$
$$= \frac{4h}{h}$$
$$= 4$$

77.

$$\frac{f(x + h) - f(x)}{h} = \frac{3(x + h) + 5 - 3x - 5}{h}$$
$$= \frac{3h}{h}$$
$$= 3$$

79. Let $g(x) = x^2 + x$. Then we obtain

$$\frac{g(x + h) - g(x)}{h} =$$
$$\frac{(x + h)^2 + (x + h) - x^2 - x}{h} =$$
$$\frac{2xh + h^2 + h}{h} =$$
$$2x + h + 1.$$

81. Difference quotient is

$$= \frac{-(x + h)^2 + (x + h) - 2 + x^2 - x + 2}{h}$$
$$= \frac{-2xh - h^2 + h}{h}$$
$$= -2x - h + 1$$

83. Difference quotient is

$$= \frac{3\sqrt{x + h} - 3\sqrt{x}}{h} \cdot \frac{3\sqrt{x + h} + 3\sqrt{x}}{3\sqrt{x + h} + 3\sqrt{x}}$$
$$= \frac{9(x + h) - 9x}{h(3\sqrt{x + h} + 3\sqrt{x})}$$
$$= \frac{9h}{h(3\sqrt{x + h} + 3\sqrt{x})}$$
$$= \frac{3}{\sqrt{x + h} + \sqrt{x}}$$

85. Difference quotient is

$$= \frac{\sqrt{x + h + 2} - \sqrt{x + 2}}{h} \cdot \frac{\sqrt{x + h + 2} + \sqrt{x + 2}}{\sqrt{x + h + 2} + \sqrt{x + 2}}$$
$$= \frac{(x + h + 2) - (x + 2)}{h(\sqrt{x + h + 2} + \sqrt{x + 2})}$$
$$= \frac{h}{h(\sqrt{x + h + 2} + \sqrt{x + 2})}$$
$$= \frac{1}{\sqrt{x + h + 2} + \sqrt{x + 2}}$$

87. Difference quotient is

$$= \frac{\dfrac{1}{x+h} - \dfrac{1}{x}}{h} \cdot \frac{x(x+h)}{x(x+h)}$$

$$= \frac{x - (x+h)}{xh(x+h)}$$

$$= \frac{-h}{xh(x+h)}$$

$$= \frac{-1}{x(x+h)}$$

89. Difference quotient is

$$= \frac{\dfrac{3}{x+h+2} - \dfrac{3}{x+2}}{h} \cdot \frac{(x+h+2)(x+2)}{(x+h+2)(x+2)}$$

$$= \frac{3(x+2) - 3(x+h+2)}{h(x+h+2)(x+2)}$$

$$= \frac{-3h}{h(x+h+2)(x+2)}$$

$$= \frac{-3}{(x+h+2)(x+2)}$$

91. a) $A = s^2$ b) $s = \sqrt{A}$ c) $s = \dfrac{\sqrt{2}d}{2}$

   d) $d = \sqrt{2}s$ e) $P = 4s$ f) $s = P/4$

   g) $A = P^2/16$ h) $d = \sqrt{2A}$

93. $C = 35n + 50$

95.

   (a) In 1999, the combined payments by Medicare and Medicaid is $E(9) + P(9) = (15.6(9)+99.8)+(12.6(9)+81.3) = \$434.9$ billion.

   (b) By solving $12.6n + 81.3 = 250$, we obtain

   $$n = \frac{250 - 81.3}{12.6} \approx 13.4$$

   Medicaid payments will reach \$250 billion in 2003

For Thought

1. True, since the graph is a parabola opening down with vertex at the origin.

2. False, the graph is decreasing.

3. True

4. True, since $f(-4.5) = [-1.5] = -2$.

5. False, since the range is $\{\pm 1\}$.

6. True   7.   True   8.   True

9. False, since the range is the interval $[0, 4]$.

10. True

2.2 Exercises

1. Function $y = 2x$ includes the points $(0, 0)$, $(1, 2)$, domain and range are both R

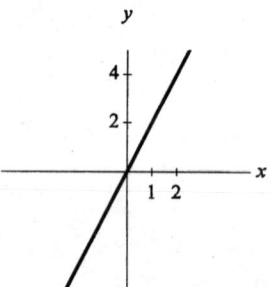

3. Function $x - y = 0$ includes the points $(-1, -1)$, $(0, 0)$, $(1, 1)$, domain and range are both R

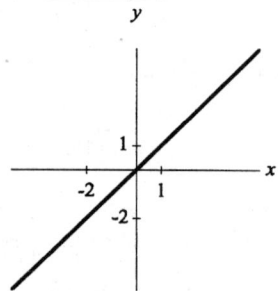

5. Function $y = 5$ includes the points $(0,5), (\pm2,5)$, domain is R, range is $\{5\}$

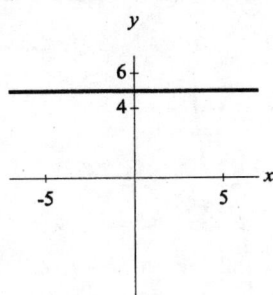

7. Function $y = 2x^2$ includes the points $(0,0)$, $(\pm1,2)$, domain is R, range is $[0,\infty)$

9. Function $y = 1 - x^2$ includes the points $(0,1)$, $(\pm1,0)$, domain is R, range is $(-\infty, 1]$

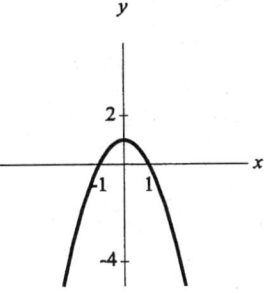

11. Function $y = 1 + \sqrt{x}$ includes the points $(0,1)$, $(1,2), (4,3)$, domain is $[0,\infty)$, range is $[1,\infty)$

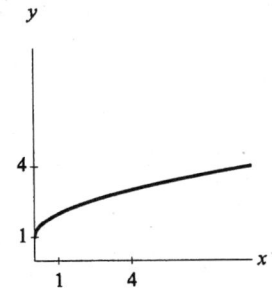

13. $x = y^2 + 1$ is not a function and includes the points $(1,0), (2,\pm1)$, domain is $[1,\infty)$, range is R

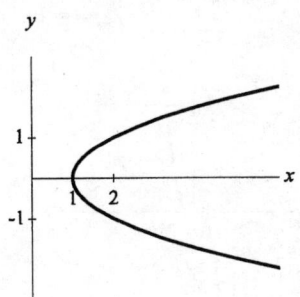

15. Function $x = \sqrt{y}$ goes through $(0,0), (2,4), (3,9)$, domain and range is $[0,\infty)$

17. Function $y = \sqrt[3]{x} + 1$ goes through $(-1,0), (1,2), (8,3)$, domain $(-\infty,\infty)$, and range $(-\infty,\infty)$

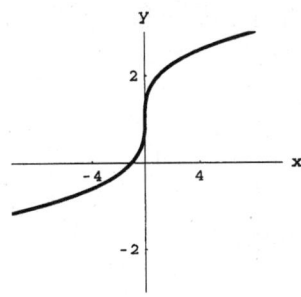

19. Function, $x = \sqrt[3]{y}$ goes through $(0,0), (1,1), (2,8)$, domain $(-\infty, \infty)$, and range $(-\infty, \infty)$

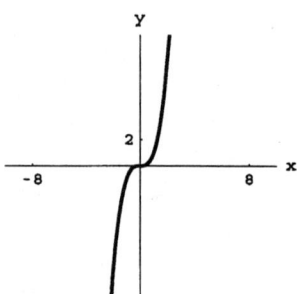

21. Not a function, $y^2 = 1 - x^2$ goes through $(1,0), (0,1), (-1,0)$, domain $[-1,1]$, and range $[-1,1]$

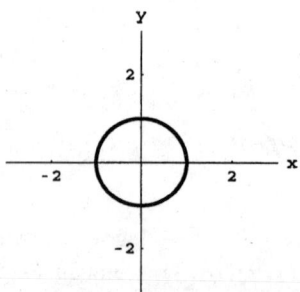

23. Function, $y = \sqrt{1 - x^2}$ goes through $(\pm 1, 0), (0,1)$, domain $[-1,1]$, and range $[0,1]$

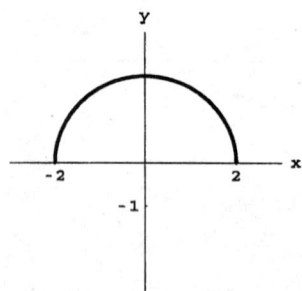

25. Function $y = x^3$ includes the points $(0,0), (1,1), (2,8)$, domain and range are both R

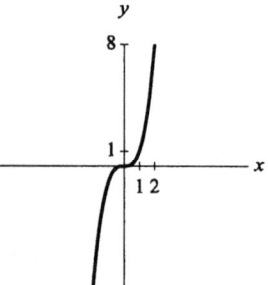

27. Function $y = 2|x|$ includes the points $(0,0), (\pm 1, 2)$, domain is R, range is $[0, \infty)$

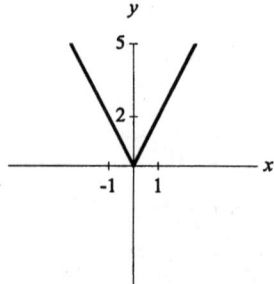

29. Function $y = -|x|$ includes the points $(0,0), (\pm 1, -1)$, domain is R, range is $(-\infty, 0]$

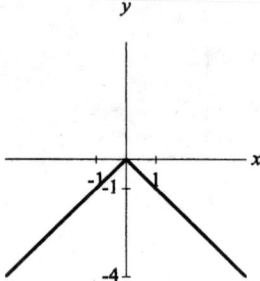

31. Not a function, graph of $x = |y|$ includes the points $(0,0), (2,2), (2,-2)$, domain is $[0, \infty)$, range is R

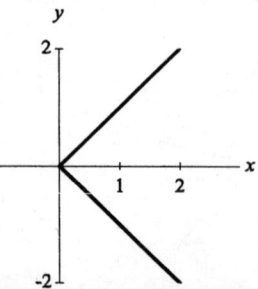

33. Domain is $(-\infty, \infty)$, range is $\{\pm 2\}$, some points are $(-3, -2)$, $(1, -2)$

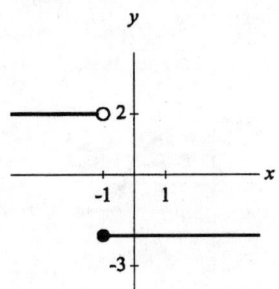

35. Domain is $(-\infty, \infty)$, range is $(-\infty, -2] \cup (2, \infty)$, some points are $(2, 3)$, $(1, -2)$

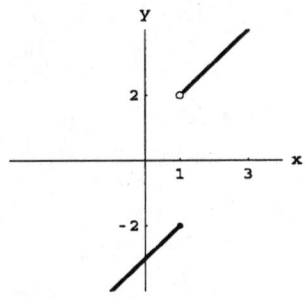

37. Domain is $[-2, \infty)$, range is $(-\infty, 2]$, some points are $(2, 2)$, $(-2, 0)$, $(3, 1)$

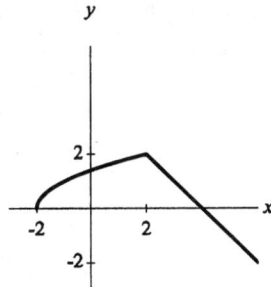

39. Domain is R, range is $[0, \infty)$, some points are $(-1, 1)$, $(-4, 2)$, $(4, 2)$

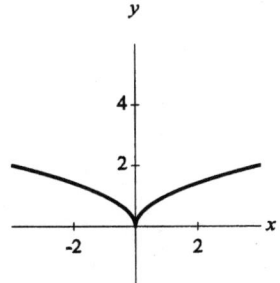

41. Domain is $(-\infty, \infty)$, range is $(-\infty, \infty)$, some points are $(-2, 4)$, $(1, -1)$

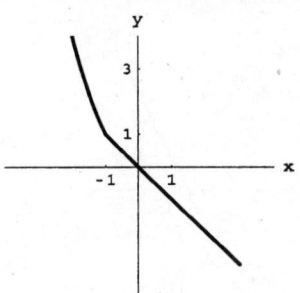

43. Domain is R, range is the set of integers, some points are $(0, 1)$, $(1, 2)$, $(1.5, 2)$

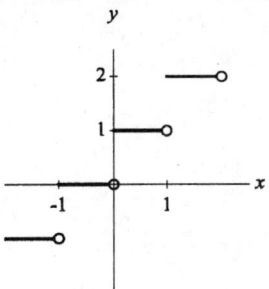

45. Domain $[0, 4)$, range is $\{2, 3, 4, 5\}$, some points are $(0, 2)$, $(1, 3)$, $(1.5, 3)$

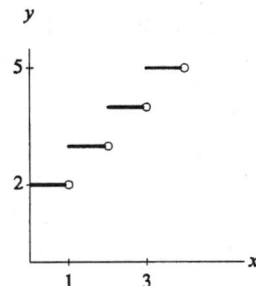

47. Domain and range are both $(-\infty, \infty)$, decreasing on $(-\infty, \infty)$

49. Domain is $(-\infty, \infty)$, range is $(-\infty, 4]$ increasing on $(-\infty, 0)$, decreasing on $(0, \infty)$

51. Domain $(-\infty, 2]$, range $(-\infty, 3]$, increasing on $(-\infty, -2)$, constant on $(-2, 2)$

53. Domain is $[-4, 4]$, range is $[0, 4]$ increasing on $(-4, 0)$, decreasing on $(0, 4)$

55. Domain and range are both $(-\infty, \infty)$ increasing on $(-\infty, \infty)$, some points are $(0, 1)$, $(1, 3)$

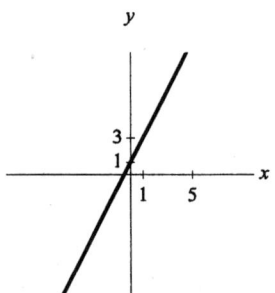

57. Domain is $(-\infty, \infty)$, range is $[0, \infty)$, increasing on $(1, \infty)$, decreasing on $(-\infty, 1)$, some points are $(0, 1)$, $(1, 0)$

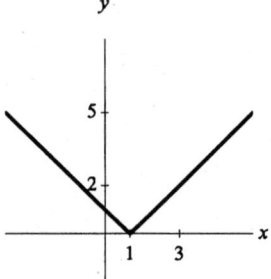

59. Domain is $(-\infty, 0) \cup (0, \infty)$, range is $\{\pm 1\}$, constant on $(-\infty, 0)$ and $(0, \infty)$, some points are $(1, 1)$, $(-1, -1)$

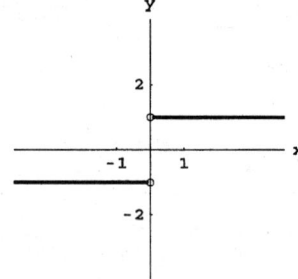

61. Domain is $[-3, 3]$, range is $[0, 3]$, increasing on $(-3, 0)$, decreasing on $(0, 3)$, some points are $(\pm 3, 0)$, $(0, 3)$

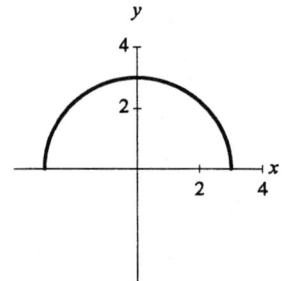

63. Domain and range are both $(-\infty, \infty)$, increasing on $(-\infty, 3)$ and $(3, \infty)$, some points are $(4, 5)$, $(0, 2)$

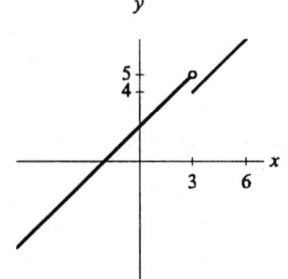

65. Domain is $(-\infty, \infty)$, range is $(-\infty, 2]$, increasing on $(-\infty, -2)$ and $(-2, 0)$, decreasing on $(0, 2)$ and $(2, \infty)$, some points are $(-3, 0)$, $(0, 2)$, $(4, -1)$

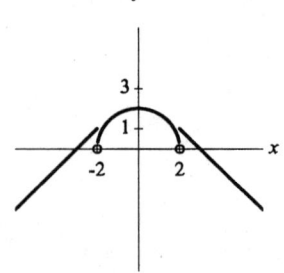

67. $f(x) = \begin{cases} 2 & \text{for } x > -1 \\ -1 & \text{for } x \le -1 \end{cases}$

69. The line joining $(-1, 1)$ and $(-3, 3)$ is $y = -x$, and the line joining $(-1, -2)$ and $(3, 2)$ is $y = x - 1$. The piecewise function is

$$f(x) = \begin{cases} x - 1 & \text{for } x \ge -1 \\ -x & \text{for } x < -1. \end{cases}$$

71. The line joining $(0, -2)$ and $(2, 2)$ is $y = 2x - 2$, and the line joining $(0, -2)$ and $(-3, 1)$ is $y = -x - 2$. The piecewise function is

$$f(x) = \begin{cases} 2x - 2 & \text{for } x \ge 0 \\ -x - 2 & \text{for } x < 0. \end{cases}$$

73. increasing on the interval $(0.83, \infty)$, decreasing on $(-\infty, 0.83)$

75. increasing on $(-\infty, -1)$ and $(1, \infty)$, decreasing on $(-1, 1)$

77. increasing on $(-1.73, 0)$ and $(1.73, \infty)$, decreasing on $(-\infty, -1.73)$ and $(0, 1.73)$

79. c, graph was increasing at first, then suddenly dropped and became constant, then increased slightly

81. d, graph was decreasing at first, then fluctuated between increases and decreases, then the market increased

83. The independent variable is time t where t is the number of minutes after 7:45 and the dependent variable is distance D from the holodeck.

D is increasing on the intervals $(0, 3)$ and $(6, 15)$, decreasing on $(3, 6)$ and $(30, 39)$, and constant on $(15, 30)$.

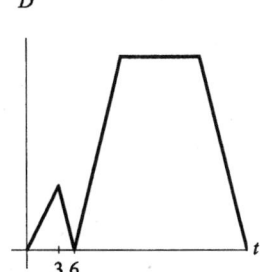

85. Independent variable is time t in years, dependent variable is savings s in dollars

s is increasing on the interval $(0, 2)$; s is constant on $(2, 2.5)$; s is decreasing on $(2.5, 4.5)$.

87. In 1988, there were $M(18) = 565$ million cars. In 2006, it is projected that there will be $M(36) = 760$ million cars.
The average rate of change from 1984 to 1994 is $\dfrac{M(24) - M(14)}{10} = 14.5$ million cars per year.

89. Constant on $(0, 10000)$, increasing on $(10000, \infty)$

91. The cost is over \$235 for $t \ge 5$.

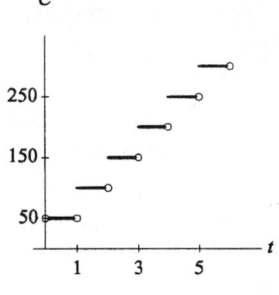

93. $f(x) = \begin{cases} 150 & \text{if } 0 < x < 3 \\ 50x & \text{if } 3 \le x \le 10 \end{cases}$

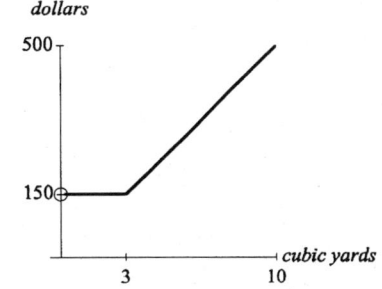

95. A mechanic's fee is \$20 for each half-hour of work with any fraction of a half-hour charged as a half-hour. If y is the fee in dollars and x is the number of hours, then $y = -20[-2x]$.

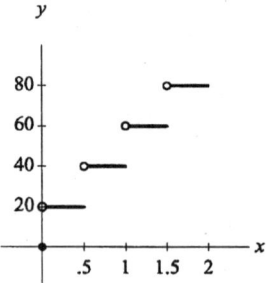

97.

(a) Let x be the number of days. For the first 55 days, the sales y (in dollars) was increasing at an increasing rate. At $x = 55$, sales dropped suddenly due to a product shortage. When $55 < x \le 75$, y was increasing at a constant rate. Another product shortage occured at $x = 75$. For $75 < x \le 100$, the sales were the same for each day.

(b) Suppose an antibiotic is being taken by a patient every 7 hours. The graph given can represent the amount, y, of antibiotic in the body as a function of the number of hours, x, that has passed since the antibiotic was first taken.

(c) The price of a stock was decreasing in 1990-95, with the price bottoming out in 1995, and then its price has been increasing and is still projected to increase faster.

For Thought

1. False, it is a reflection in the y-axis.

2. True   **3.** False, rather it is a left translation.

4. True   **5.** True

6. False, the down shift should come after the reflection.   **7.** True

8. False, since their domains are different.

9. True   **10.**   True

2.3 Exercises

1. $f(x) = |x|, g(x) = |x| - 4$

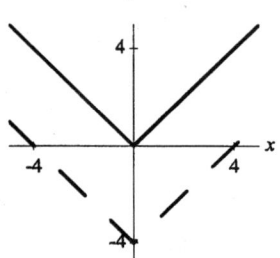

3. $f(x) = x, g(x) = x + 3$

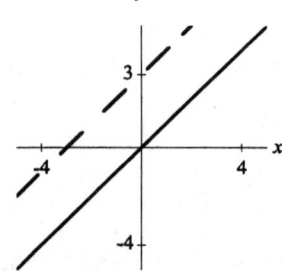

5. $y = x^2, y = (x-3)^2$

7. $y = \sqrt{x}, y = \sqrt{x+9}$

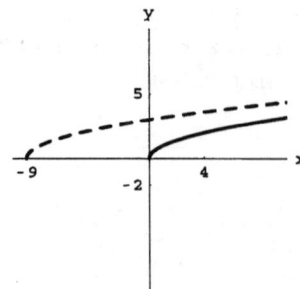

9. $f(x) = \sqrt{x}, g(x) = -\sqrt{x}$

11. $y = \sqrt{x}, y = 3\sqrt{x}$

13. $y = x^2, y = \frac{1}{4}x^2$

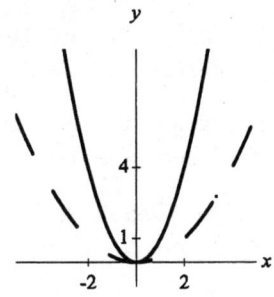

15. $y = \sqrt{4-x^2}, y = -\sqrt{4-x^2}$

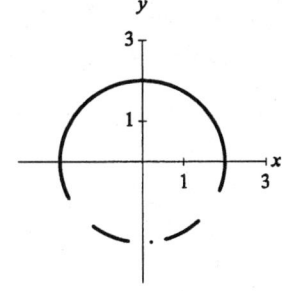

17. g **19.** b

21. c **23.** f

25. $y = \sqrt{x} + 2$ **27.** $y = (x-5)^2$

29. $y = (x-10)^2 + 4$

31. $y = -(3\sqrt{x} + 5)$ or $y = -3\sqrt{x} - 5$

33. $y = -3|x-7| + 9$

35. $y = (x-1)^2 + 2$; right by 1, up by 2

37. $y = |x-1| + 3$; right by 1, up by 3

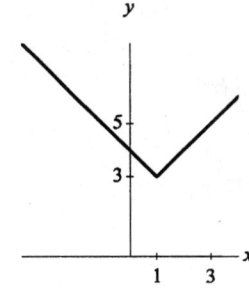

39. $y = 3x - 40$

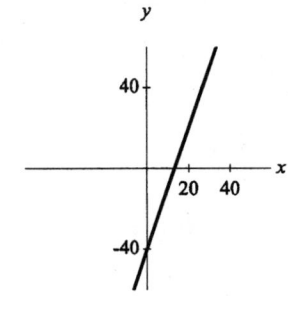

41. $y = \dfrac{1}{2}x - 20$

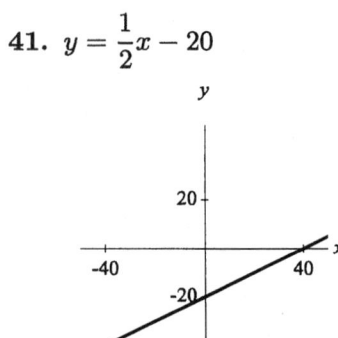

43. $y = -\dfrac{1}{2}|x| + 40$, shrink by 1/2,

reflect about x-axis, up by 40

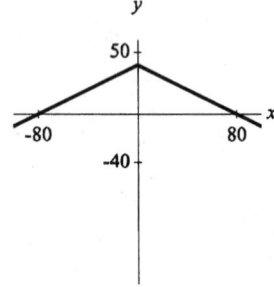

45. $y = -\dfrac{1}{2}|x + 4|$, left by 4,

reflect about x-axis, shrink by 1/2

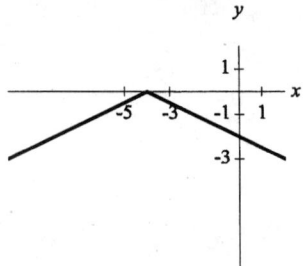

47. $y = -\sqrt{x - 3} + 1$, right by 3,
reflect about x-axis, up by 1

49. $y = -2\sqrt{x + 3} + 2$, left by 3, stretch by 2,
reflect about x-axis, up by 2

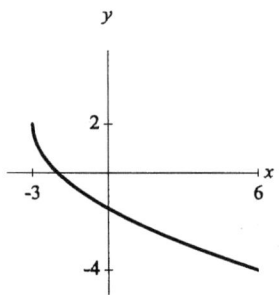

51. Symmetric about y-axis, even function
since $f(-x) = f(x)$

53. No symmetry, neither even nor odd
since $f(-x) \neq f(x)$ and $f(-x) \neq -f(x)$

55. Symmetric about $x = -3$, neither even nor
odd since $f(-x) \neq f(x)$ and $f(-x) \neq -f(x)$

57. Symmetry about $x = 2$, not an even or odd
function since $f(-x) \neq f(x)$ and
$f(-x) \neq -f(x)$

59. Symmetric about the origin, odd function
since $f(-x) = -f(x)$

61. No symmetry, not an even or odd function
since $f(-x) \neq f(x)$ and $f(-x) \neq -f(x)$

63. No symmetry, not an even or odd function
since $f(-x) \neq f(x)$ and $f(-x) \neq -f(x)$

65. Symmetric about the y-axis, even function
since $f(-x) = f(x)$

67. No symmetry, not an even or odd function
since $f(-x) = -f(x)$ and $f(-x) \neq -f(x)$

69. Symmetric about the y-axis, even function
since $f(-x) = f(x)$

71. e   **73.**   g

75. b   **77.**   c

79. $(-\infty, -1] \cup [1, \infty)$ **81.** $(-\infty, -1) \cup (5, \infty)$

83. Graph of $y = (x-1)^2 - 9$ shows solution is $(-2, 4)$

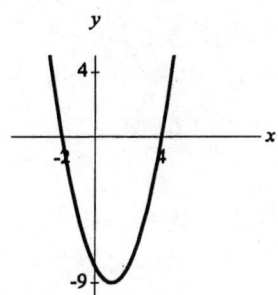

85. Graph of $y = 5 - \sqrt{x}$ shows solution is $[0, 25]$

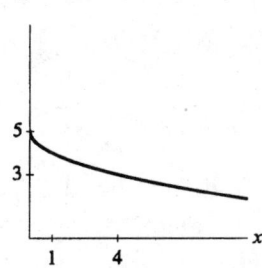

87. Note, the points of intersection of $y = 3$ and $y = (x-2)^2$ are $(2 \pm \sqrt{3}, 3)$. The solution set of $(x-2)^2 > 3$ is $\left(-\infty, 2-\sqrt{3}\right) \cup \left(2+\sqrt{3}, \infty\right)$.

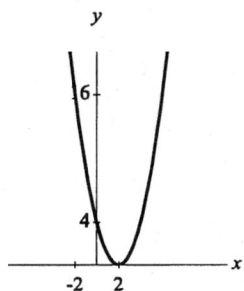

89. Graph of $y = \sqrt{25 - x^2}$ shows solution is $(-5, 5)$

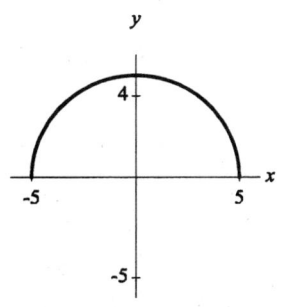

91. From the graph of $y = \sqrt{3}x^2 + \pi x - 9$,

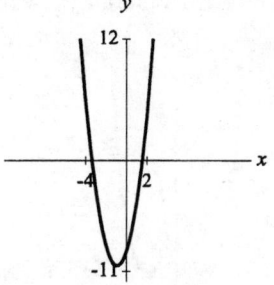

the solution set of $\sqrt{3}x^2 + \pi x - 9 < 0$ is $(-3.36, 1.55)$.

93. $N(x) = x + 2000$

95. If inflation rate is less than 50%, then $1 - \sqrt{x} < \dfrac{1}{2}$. This simplifies to $\dfrac{1}{2} < \sqrt{x}$. After squaring we have $\dfrac{1}{4} < x$ and so $x > 25\%$.

97.

(a) Both functions are even functions and the graphs are identical

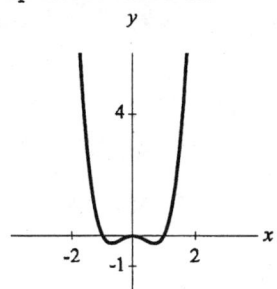

(b) One graph is a reflection of the other about the y-axis. Both functions are odd.

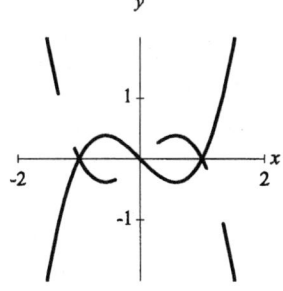

(c) The second graph is obtained by shifting the first one to the left by 1 unit.

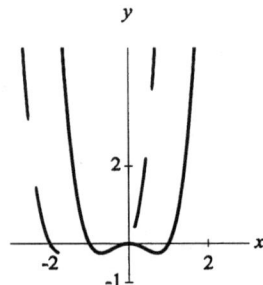

(d) The second graph is obtained by translating the first one to the right by 2 units and 3 units up.

For Thought

1. False, since $f + g$ has an empty domain.

2. True   **3.**   True   **4.**   True

5. True, since $A = P^2/16$.   **6.**   True

7. False, since $(f \circ g)(x) = \sqrt{x} - 2$ **8.**   True

9. False, since $(h \circ g)(x) = x^2 - 9$.

10. True, since x belongs to the domain if $\sqrt{x-2}$ is a real number, i.e., if $x \geq 2$.

2.4 Exercises

1. $-1 + 2 = 1$

3. $-5 - 6 = -11$

5. $(-4) \cdot 2 = -8$

7. $1/12$

9. $(a - 3) + (a^2 - a) = a^2 - 3$

11. $(a - 3)(a^2 - a) = a^3 - 4a^2 + 3a$

13. $f + g = \{(-3, 1 + 2), (2, 0 + 6)\} = \{(-3, 3), (2, 6)\}$, domain $\{-3, 2\}$

15. $f - g = \{(-3, 1 - 2), (2, 0 - 6)\} = \{(-3, -1), (2, -6)\}$, domain $\{-3, 2\}$

17. $f \cdot g = \{(-3, 1 \cdot 2), (2, 0 \cdot 6)\} = \{(-3, 2), (2, 0)\}$, domain $\{-3, 2\}$

19. $g/f = \{(-3, 2/1)\} = \{(-3, 2)\}$, domain $\{-3\}$

21. $(f + g)(x) = \sqrt{x} + x - 4$, domain is $[0, \infty)$

23. $(f - h)(x) = \sqrt{x} - \dfrac{1}{x - 2}$, domain is $[0, 2) \cup (2, \infty)$

25. $(g \cdot h)(x) = \dfrac{x - 4}{x - 2}$, domain is $(-\infty, 2) \cup (2, \infty)$

27. $\left(\dfrac{g}{f}\right)(x) = \dfrac{x - 4}{\sqrt{x}}$, domain is $(0, \infty)$

29. $\{(-3, 0), (1, 0), (4, 4)\}$ **31.** $\{(1, 4)\}$

33. $\{(-3, 4), (1, 4)\}$

35. $f(2) = 5$ **37.** $f(2) = 5$

39. $f(20.2721) = 59.8163$

41. $(g \circ h \circ f)(2) = (g \circ h)(5) = g(2) = 5$

43. $(f \circ g)(x) = \sqrt{x} - 2$, domain $[0, \infty)$

45. $(f \circ h)(x) = \dfrac{1}{x} - 2$, domain $(-\infty, 0) \cup (0, \infty)$

47. $(h \circ g)(x) = \dfrac{1}{\sqrt{x}}$, domain $(0, \infty)$

49. $(f \circ f)(x) = (x - 2) - 2 = x - 4$, domain $(-\infty, \infty)$

51. $(h \circ g \circ f)(x) = h(\sqrt{x} - 2) = \dfrac{1}{\sqrt{x} - 2}$, domain $(2, \infty)$

53. $(h \circ f \circ g)(x) = h(\sqrt{x} - 2) = \dfrac{1}{\sqrt{x} - 2}$, domain $(0, 4) \cup (4, \infty)$

55. $F = g \circ h$ **57.** $H = h \circ g$

59. $N = h \circ g \circ f$ **61.** $P = g \circ f \circ g$

63. $S = g \circ g$

65. $y = 2(3x + 1) - 3 = 6x - 1$

67. $y = (x^2 + 6x + 9) - 2 = x^2 + 6x + 7$

69. $y = 3 \cdot \dfrac{x+1}{3} - 1 = x + 1 - 1 = x$

71. Since $m = n - 4$ and $y = m^2$, $y = (n - 4)^2$.

73. Since $w = x + 16$, $z = \sqrt{w}$, and $y = \dfrac{z}{8}$,

we obtain $y = \dfrac{\sqrt{x + 16}}{8}$.

75. After multiplying y by $\dfrac{x+1}{x+1}$ we have

$$y = \frac{\dfrac{x-1}{x+1} + 1}{\dfrac{x-1}{x+1} - 1} = \frac{(x-1) + (x+1)}{(x-1) - (x+1)} = -x$$

The domain of the original function is $(-\infty, -1) \cup (-1, \infty)$ while the domain of the simplified function is $(-\infty, \infty)$.
The two functions are not the same.

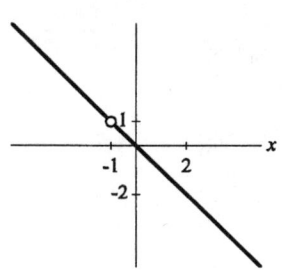

77. Domain $[-1, \infty)$, range $[-7, \infty)$

79. Domain $[1, \infty)$, range $[0, \infty)$

81. Domain $[0, \infty)$, range $[4, \infty)$

83. $P(x) = (3000x - 20x^2) - (600x + 4000) = -20x^2 + 2400x - 4000$.

85. $A = d^2/2$

87. $(f \circ f)(x) = 0.899x$ and $(f \circ f \circ f)(x) = 0.852x$ are the amounts of forest land at the start of 2002 and 2003, respectively.

89. Total cost is $(T \circ C)(x) = 1.05(1.20x) = 1.26x$.

91. Note, $D = \dfrac{d/2240}{x} = \dfrac{d/2240}{L^3/100^3} = \dfrac{100^3 d}{2240 L^3}$

$= \dfrac{100^3 (26000)}{2240 L^3} = \dfrac{100^4 (26)}{224 L^3} = \dfrac{100^4 (13)}{112 L^3}$.

Expressing D as a function of L, we write $D = \dfrac{(13)100^4}{112 L^3}$ or $D = \dfrac{1.16 \times 10^7}{L^3}$.

93. The area of a semicircle with radius $s/2$ is $(1/2)\pi(s/2)^2 = \pi s^2/8$. The area of the square is s^2. The area of the window is

$$W = \frac{\pi s^2}{8} + s^2 = \frac{s^2(\pi + 8)}{8}.$$

95. Form a right triangle with two sides of length s and a hypotenuse of length d. By the Pythagorean Theorem, we obtain

$$d^2 = s^2 + s^2.$$

Solving for s, we have $s = \dfrac{\sqrt{2}d}{2}$.

97. If a coat is on sale at 25% off and there is an additional 10% off, then the coat will cost $0.90(.75x) = 0.675x$ where x is the regular price. Thus, the discount sale is 32.5% off and not 35% off.

For Thought

1. False, since the inverse function is $\{(3,2),(5,5)\}$.

2. False, since it is not one-to-one.

3. False, $g^{-1}(x)$ does not exist since g is not one-to-one.

4. True

5. False, a function that fails the horizontal line test has no inverse.

6. False, since it fails the horizontal line test.

7. False, since $f^{-1}(x) = \left(\dfrac{x}{3}\right)^2 + 2$ where $x \geq 0$.

8. False, $f^{-1}(x)$ does not exist since f is not one-to-one.

9. False, since $y = |x|$ is V-shaped and the horizontal line test fails.

10. True

2.5 Exercises

1. Not invertible, there can be two different items with the same price.

3. Invertible, since the playing time is a function of the length of the VCR tape.

5. Invertible, assuming that cost is simply a multiple of the number of days. If cost includes extra charges, then the function may not be invertible.

7. Invertible, $\{(3,9),(2,2)\}$

9. Not invertible

11. Invertible, $\{(3,3),(2,2),(4,4),(7,7)\}$

13. Not invertible

15. $f^{-1} = \{(1,2),(5,3)\}$, $f^{-1}(5) = 3$, $(f^{-1} \circ f)(2) = 2$

17. $f^{-1} = \{(-3,-3),(5,0),(-7,2)\}$, $f^{-1}(5) = 0$, $(f^{-1} \circ f)(2) = 2$

19. Not one-to-one

21. One-to-one

23. Not one-to-one

25. One-to-one; since the graph of $y = 2x-3$ shows $y = 2x - 3$ is an increasing function, the Horizontal Line Test implies $y = 2x - 3$ is one-to-one.

27. One-to-one; for if $q(x_1) = q(x_2)$ then

$$
\begin{aligned}
\frac{1 - x_1}{x_1 - 5} &= \frac{1 - x_2}{x_2 - 5} \\
(1 - x_1)(x_2 - 5) &= (1 - x_2)(x_1 - 5) \\
x_2 - 5 - x_1 x_2 + 5x_1 &= x_1 - 5 - x_2 x_1 + 5x_2 \\
x_2 + 5x_1 &= x_1 + 5x_2 \\
4(x_1 - x_2) &= 0 \\
x_1 - x_2 &= 0.
\end{aligned}
$$

Thus, if $q(x_1) = q(x_2)$ then $x_1 = x_2$. Hence, q is one-to-one.

29. Not one-to-one for $p(-2) = p(0) = 1$.

31. Not one-to-one for $w(1) = w(-1) = 4$.

33. One-to-one; for if $k(x_1) = k(x_2)$ then

$$
\begin{aligned}
\sqrt[3]{x_1 + 9} &= \sqrt[3]{x_2 + 9} \\
(\sqrt[3]{x_1 + 9})^3 &= (\sqrt[3]{x_2 + 9})^3 \\
x_1 + 9 &= x_2 + 9 \\
x_1 &= x_2.
\end{aligned}
$$

Thus, if $k(x_1) = k(x_2)$ then $x_1 = x_2$. Hence, k is one-to-one.

35. Not invertible since it fails the Horizontal Line Test.

37. Not invertible since it fails the Horizontal Line Test.

39. Interchange x and y then solve for y.

$$
\begin{aligned}
x &= 3y - 7 \\
\frac{x+7}{3} &= y \\
\frac{x+7}{3} &= f^{-1}(x)
\end{aligned}
$$

41. Interchange x and y then solve for y.

$$
\begin{aligned}
x &= 2 + \sqrt{y-3} \quad \text{for } x \ge 2 \\
(x-2)^2 &= y - 3 \quad \text{for } x \ge 2 \\
f^{-1}(x) &= (x-2)^2 + 3 \quad \text{for } x \ge 2
\end{aligned}
$$

43. Interchange x and y then solve for y.

$$
\begin{aligned}
x &= -y - 9 \\
y &= -x - 9 \\
f^{-1}(x) &= -x - 9
\end{aligned}
$$

45. Interchange x and y then solve for y.

$$
\begin{aligned}
x &= \frac{y+3}{y-5} \\
xy - 5x &= y + 3
\end{aligned}
$$

$$
\begin{aligned}
xy - y &= 5x + 3 \\
y(x - 1) &= 5x + 3 \\
f^{-1}(x) &= \frac{5x+3}{x-1}
\end{aligned}
$$

47. Interchange x and y then solve for y.

$$
\begin{aligned}
x &= -\frac{1}{y} \\
xy &= -1 \\
f^{-1}(x) &= -\frac{1}{x}
\end{aligned}
$$

49. Interchange x and y then solve for y.

$$
\begin{aligned}
x &= \sqrt[3]{y-9} + 5 \\
x - 5 &= \sqrt[3]{y-9} \\
(x-5)^3 &= y - 9 \\
f^{-1}(x) &= (x-5)^3 + 9
\end{aligned}
$$

51. Interchange x and y then solve for y.

$$
\begin{aligned}
x &= (y-2)^2 \quad x \ge 0 \\
\sqrt{x} &= y - 2 \\
f^{-1}(x) &= \sqrt{x} + 2
\end{aligned}
$$

53. Since $(g \circ f)(x) = 0.25(4x+4) - 1 = x$ and $(f \circ g)(x) = 4(0.25x - 1) + 4 = x$, g and f are inverse functions of each other.

55. Since $(f \circ g)(x) = \left(\sqrt{x-1}\right)^2 + 1 = x$ and and $(g \circ f)(x) = \sqrt{x^2 + 1 - 1} = \sqrt{x^2} = |x|$, g and f are not inverse functions of each other.

57. We find

$$
\begin{aligned}
(f \circ g)(x) &= \frac{1}{1/(x-3)} + 3 \\
&= x - 3 + 3 \\
(f \circ g)(x) &= x
\end{aligned}
$$

and

$$
\begin{aligned}
(g \circ f)(x) &= \frac{1}{\left(\dfrac{1}{x} + 3\right) - 3} \\
&= \frac{1}{1/x} \\
(g \circ f)(x) &= x.
\end{aligned}
$$

Then g and f are inverse functions of each other.

59. We obtain

$$(f \circ g)(x) = \sqrt[3]{\frac{5x^3 + 2 - 2}{5}}$$

$$= \sqrt[3]{\frac{5x^3}{5}}$$

$$= \sqrt[3]{x^3}$$

$$(f \circ g)(x) = x$$

and

$$(g \circ f)(x) = 5\left(\sqrt[3]{\frac{x-2}{5}}\right)^3 + 2$$

$$= 5\left(\frac{x-2}{5}\right) + 2$$

$$= (x - 2) + 2$$

$$(g \circ f)(x) = x.$$

Thus, g and f are inverse functions of each other.

61. y_1 and y_2 are inverse functions of each other and $y_3 = y_2 \circ y_1$.

63. No, since they fail the Horizontal Line Test.

65. Yes, since the graphs are symmetric about the line $y = x$.

67. Graph of f^{-1}

69. Graph of f^{-1}

71. $f^{-1}(x) = \dfrac{x-2}{3}$

73. $f^{-1}(x) = \sqrt{x+4}$

75. $f^{-1}(x) = \sqrt[3]{x}$

77. $f^{-1}(x) = (x+3)^2$ for $x \geq -3$

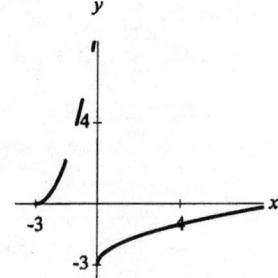

79. a) $f^{-1}(x) = x/5$

b) $f^{-1}(x) = x + 88$

c) $f^{-1}(x) = (x+7)/3$

d) $f^{-1}(x) = \dfrac{x-4}{-3}$

e) $f^{-1}(x) = 2(x+9) = 2x + 18$

f) $f^{-1}(x) = -x$

g) $f^{-1}(x) = (x+9)^3$

h) $f^{-1}(x) = \sqrt[3]{\dfrac{x+7}{3}}$

81. $C = 1.08P$ expresses the total cost as a function of the purchase price; and $P = C/1.08$ is the purchase price as a function of the total cost.

83. The graph of t as a function of r satisfies the Horizontal Line Test and is invertible. Solving for r we find,

$$\begin{aligned} t - 7.89 &= -0.39r \\ r &= \dfrac{t - 7.89}{-0.39} \end{aligned}$$

and the inverse function is $r = \dfrac{7.89 - t}{0.39}$.

If $t = 5.55$ min., then $r = \dfrac{7.89 - 5.55}{0.39} = 6$ rowers.

85. Solving for w, we obtain

$$\begin{aligned} 1.496w &= V^2 \\ w &= \dfrac{V^2}{1.496} \end{aligned}$$

and the inverse function is $w = \dfrac{V^2}{1.496}$. If $V = 115$ ft./sec., then $w = \dfrac{115^2}{1.496} \approx 8,840.2$ lbs.

87. a) Let $V = \$28,000$. The depreciation rate is

$$r = 1 - \left(\dfrac{28,000}{50,000}\right)^{1/5} \approx 0.109$$

or $r \approx 10.9\%$.

b) Writing V as a function of r we find

$$\begin{aligned} 1 - r &= \left(\dfrac{V}{50,000}\right)^{1/5} \\ (1-r)^5 &= \dfrac{V}{50,000} \end{aligned}$$

and $V = 50,000(1-r)^5$.

89. Since $g^{-1}(x) = \dfrac{x+5}{3}$ and $f^{-1}(x) = \dfrac{x-1}{2}$, we have

$$g^{-1} \circ f^{-1}(x) = \dfrac{\dfrac{x-1}{2} + 5}{3} = \dfrac{x+9}{6}.$$

Likewise, since $(f \circ g)(x) = 6x - 9$, we get

$$(f \circ g)^{-1}(x) = \dfrac{x+9}{6}.$$

Hence, $(f \circ g)^{-1} = g^{-1} \circ f^{-1}$.

For Thought

1. False

2. False, since cost varies directly with the number of pounds purchased.

3. True   **4.** True

5. True, since the area of a circle varies directly with the square of its radius.

6. False, since $y = k/x$ is undefined when $x = 0$.

7. True   **8.** True   **9.** True

10. False, the surface area is not equal to $k \cdot length \cdot width \cdot height$ for any constant k.

2.6 Exercises

1. $G = kn$ **3.** $V = k/P$

5. $C = khr$ **7.** $Y = \dfrac{kx}{\sqrt{z}}$

9. A varies directly with the square of r

11. y varies inversely with x

13. Not a variation expression

15. a varies jointly with z and w

17. H varies directly with the square root of t and inversely with s

19. D varies jointly with L and J and inversely with W

21. Since $y = kx$ and $5 = k \cdot 9$, $k = 5/9$. Then $y = 5x/9$.

23. Since $T = k/y$ and $-30 = k/5$, $k = -150$. Thus, $T = -150/y$.

25. Since $m = kt^2$ and $54 = k \cdot 18$, $k = 3$. Thus, $m = 3t^2$.

27. Since $y = kx/\sqrt{z}$ and $2.192 = k(2.4)/\sqrt{2.25}$, we obtain $k = 1.37$. Hence, $y = 1.37x/\sqrt{z}$.

29. Since $y = kx$ and $9 = k(2)$, we obtain
$$y = \frac{9}{2} \cdot (-3) = -27/2.$$

31. Since $P = k/w$ and $2/3 = \dfrac{k}{1/4}$, we find
$$k = \frac{2}{3} \cdot \cdot \frac{1}{4} = \frac{1}{6}. \text{ Thus, } P = \frac{1/6}{1/6} = 1.$$

33. Since $A = kLW$ and $30 = k(3)(5\sqrt{2})$, we obtain $A = \sqrt{2}(2\sqrt{3})\dfrac{1}{2} = \sqrt{6}$.

35. Since $y = ku/v^2$ and $7 = k \cdot 9/36$,

we find $y = 28 \cdot 4/64 = 7/4$.

37. Let L_i and L_f be the length in inches and feet, respectively. Then $L_i = 12L_f$ is a direct variation.

39. Let P and n be the cost per person and the number of persons, respectively. Then $P = 20/n$ is an inverse variation.

41. Let S_m and S_k be the speeds of the car in mph and kph, respectively. Then $S_m \approx S_k/1.6 \approx 0.6 \cdot S_k$ is a direct variation.

43. Not a variation

45. Let A and W be the area and width, respectively. Then $A = 30W$ is a direct variation.

47. Let n and p be the number of gallons and price per gallon, respectively. Since $np = 5$, we obtain that $n = \dfrac{5}{p}$ is an inverse variation.

49. If p is the pressure at depth d, then $p = kd$. Since $4.34 = k(10)$, $k = 0.434$. At $d = 6000$ ft, the pressure is $p = 0.434(6000) = 2604$ lb per square inch.

51. If h is the number of hours, p is the number of pounds, and w is the number of workers then $h = kp/w$. Since $8 = k(3000)/6$, $k = 0.016$. Five workers can process 4000 pounds in $h = (0.016)(4000)/5 = 12.8$ hours.

53. Since $I = kPt$ and $20.80 = k(4000)(16)$, we find $k = 0.000325$. The interest from a deposit of \$6500 for 24 days is
$$I = (0.000325)(6500)(24) \approx \$50.70.$$

55. Since $C = kDL$ and $18.60 = k(6)(20)$, we obtain $k = 0.155$. The cost of a 16 ft pipe with a diameter of 8 inches is
$$C = 0.155(8)(16) = \$19.84.$$

57. Since $w = khd^2$ and $14.5 = k(4)(6^2)$, we find $k = \dfrac{14.5}{144}$. Then a 5-inch high can with a diameter of 6 inches has weight
$$w = \frac{14.5}{144}(5)(6^2) = 18.125 \text{ ounces.}$$

59. Since $V = kh/l$ and $10 = k(50)/(200)$, we get $k = 40$. The velocity, if the head is 60 ft and the length is 300 ft, is $V = (40)(60)/(300) = 8$ ft/year.

61. No, it is not directly proportional otherwise the following ratios $\dfrac{42,506}{1.34} \approx 31,720$,

$\dfrac{59,085}{0.295} \approx 200,288$, and

$\dfrac{738,781}{0.958} \approx 771,170$ would be

the same but they are not.

63. Since $g = ks/p$ and $76 = k(12)/(10)$, $k = \dfrac{190}{3}$.
If Calvin studies for 9 hours and plays for 15 hours, then his score is

$$g = \frac{190}{3} \cdot \frac{s}{p} = \frac{190}{3} \cdot \frac{9}{15} = 38.$$

65. Since $h = kv^2$ and $16 = k(32)^2$, we get $k = \dfrac{1}{64}$.
To reach a height of $20'2.5''$, the velocity v must satisfy

$$20 + \frac{2.5}{12} = \frac{1}{64}v^2.$$

Solving for v, we find $v \approx 35.96$ ft/sec.

Review Exercises

1. Function, domain and range are both $\{0, 1, -2\}$

3. $y = 3 - x$ is a function, domain and range are both $(-\infty, \infty)$

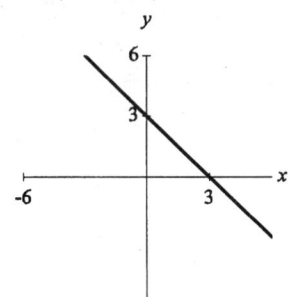

5. Not a function, domain is $\{2\}$, range is $(-\infty, \infty)$

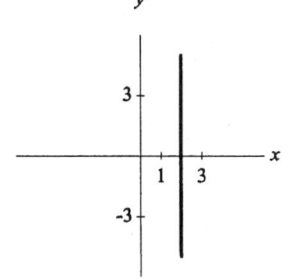

7. $x^2 + y^2 = 0.01$ is not a function, domain and range are both $[-0.1, 0.1]$

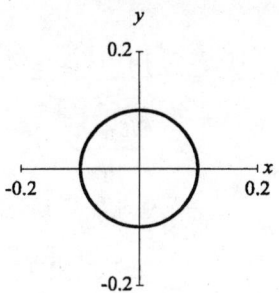

9. $x = y^2 + 1$ is not a function, domain is $[1, \infty)$, range is $(-\infty, \infty)$

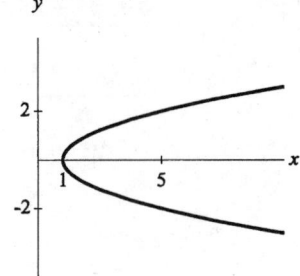

11. $y = \sqrt{x} - 3$ is a function, domain is $[0, \infty)$, range is $[-3, \infty)$

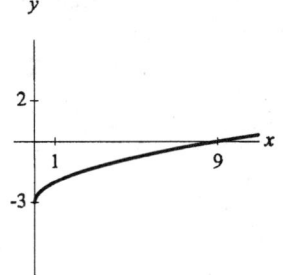

13. $9 + 3 = 12$ **15.** $24 - 7 = 17$

17. If $x^2 + 3 = 19$, then $x^2 = 16$ or $x = \pm 4$.

19. $g(12) = 17$ **21.** $7 + (-3) = 4$

23. $(4)(-9) = -36$ **25.** $f(-3) = 12$

27.

$$\begin{aligned} f(g(x)) &= f(2x - 7) \\ &= (2x - 7)^2 + 3 \\ &= 4x^2 - 28x + 52 \end{aligned}$$

29. $(x^2 + 3)^2 + 3 = x^4 + 6x^2 + 12$

31. $(a+1)^2 + 3 = a^2 + 2a + 4$

33.

$$\frac{f(3+h) - f(3)}{h} = \frac{(9 + 6h + h^2) + 3 - 12}{h}$$

$$= \frac{6h + h^2}{h}$$

$$= 6 + h$$

35.

$$\frac{f(x+h) - f(x)}{h} =$$

$$\frac{(x^2 + 2xh + h^2) + 3 - x^2 - 3}{h} =$$

$$\frac{2xh + h^2}{h} =$$

$$2x + h =$$

37. $g\left(\dfrac{x+7}{2}\right) = (x+7) - 7 = x$

39. $g^{-1}(x) = \dfrac{x+7}{2}$

41. $f(x) = \sqrt{x}, g(x) = 2\sqrt{x+3}$; left by 3, stretch by 2

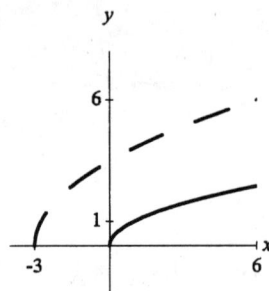

43. $f(x) = |x|, g(x) = -2|x+2| + 4$; left by 2, stretch by 2, reflect about x-axis, up by 4

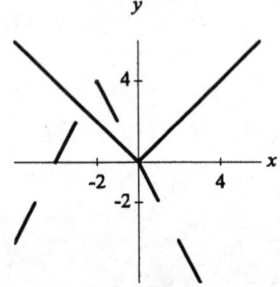

45. $f(x) = x^2, g(x) = \dfrac{1}{2}(x-2)^2 + 1$; right by 2, stretch by $\dfrac{1}{2}$, up by 1

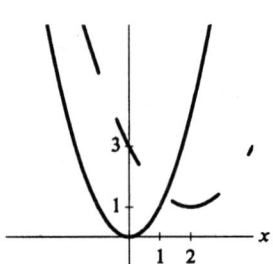

47. $F = f \circ g$ **49.** $H = f \circ h \circ g \circ j$

51. $N = h \circ f \circ j$ **53.** $R = g \circ h \circ j$

55.

$$\frac{f(x+h) - f(x)}{h} = \frac{-5(x+h) + 9 + 5x - 9}{h}$$

$$= \frac{-5h}{h}$$

$$= -5$$

57.

$$\frac{f(x+h) - f(x)}{h} =$$

$$= \frac{\dfrac{1}{2x+2h} - \dfrac{1}{2x}}{h} \cdot \frac{(2x+2h)(2x)}{(2x+2h)(2x)}$$

$$= \frac{(2x) - (2x + 2h)}{h(2x+2h)(2x)}$$

$$= \frac{-2}{(2x+2h)(2x)}$$

$$= \frac{-1}{(x+h)(2x)}$$

59. Domain is $[-10, 10]$, range is $[0, 10]$, increasing on $(-10, 0)$, decreasing on $(0, 10)$

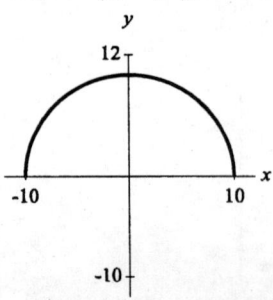

61. Domain and range are both $(-\infty, \infty)$, increasing on $(-\infty, \infty)$

63. Domain is $(-\infty, \infty)$, range is $[-2, \infty)$, increasing on $(-2, 0)$ and $(2, \infty)$, decreasing on $(-\infty, -2)$ and $(0, 2)$

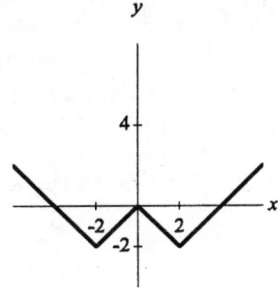

65. $y = |x| - 3$, domain is $(-\infty, \infty)$, range is $[-3, \infty)$

67. $y = -2|x| + 4$, domain is $(-\infty, \infty)$, range is $(-\infty, 4]$

69. $y = |x + 2| + 1$, domain is $(-\infty, \infty)$, range is $[1, \infty)$

71. Symmetry: y-axis

73. Symmetric about origin

75. Neither symmetry

77. Symmetric about y-axis

79. From the graph of $y = |x - 3| - 1$, the solution set is $(-\infty, 2] \cup [4, \infty)$

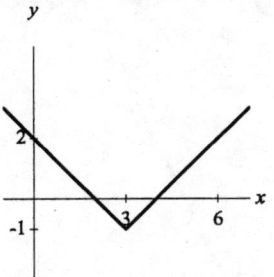

81. From the graph of $y = -2x^2 + 4$, the solution set is $(-\sqrt{2}, \sqrt{2})$

83. No solution since $-\sqrt{x + 1} - 2 \le -2$

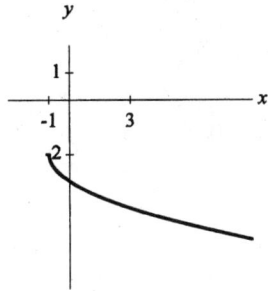

85. inverse functions,
$$f(x) = \sqrt{x + 3}, g(x) = x^2 - 3 \text{ for } x \ge 0$$

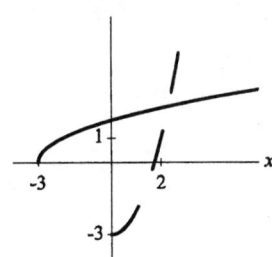

87. inverse functions,

$$f(x) = 2x - 4, g(x) = \frac{1}{2}x + 2$$

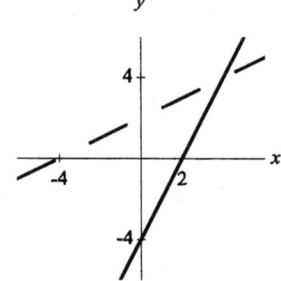

89. Not invertible

91. Inverse is $f^{-1}(x) = \dfrac{x + 21}{3}$ with domain and range both $(-\infty, \infty)$

93. Not invertible

95. Inverse is $f^{-1}(x) = x^2 + 9$ for $x \geq 0$ with domain $[0, \infty)$ and range $[9, \infty)$

97. Inverse is $f^{-1}(x) = \dfrac{5x + 7}{1 - x}$ with domain $\{x | x \neq 1\}$ and range $\{y | y \neq -5\}$

99. Inverse is $f^{-1}(x) = -\sqrt{x - 1}$ with domain $[1, \infty)$ and range $(-\infty, 0]$

101. Let x be the number of roses. The cost function is $C(x) = 40 + 1.20x$, the revenue function is $R(x) = 2x$, and the profit function is $P(x) = R(x) - C(x)$ or $P(x) = 0.80x - 40$. Since $P(50) = 0$, to make a profit she must sell at least 51 roses.

103. Since $h(0) = 64$ and $h(2) = 0$, the range of $h = -16t^2 + 64$ is the interval $[0, 64]$. Then the domain of the inverse function is $[0, 64]$. Solving for t, we obtain

$$\begin{aligned} 16t^2 &= 64 - h \\ t^2 &= \frac{64 - h}{16} \\ t &= \frac{\sqrt{64 - h}}{4}. \end{aligned}$$

The inverse function is $t = \dfrac{\sqrt{64 - h}}{4}$.

105. Since $A = \pi \left(\dfrac{d}{2}\right)^2$, $d = 2\sqrt{\dfrac{A}{\pi}}$.

107. The average rate of change is
$$\frac{8 - 6}{4} = 0.5 \text{ inch/lb.}$$

109. Since $D = kW$ and $9 = k \cdot 25$, we obtain
$$D = \frac{9}{25}100 = 36.$$

111. Since $V = k\sqrt{h}$ and $45 = k\sqrt{1.5}$, the velocity of a Triceratops is $V = \dfrac{45}{\sqrt{1.5}} \cdot \sqrt{2.8} \approx 61$ kph.

113. Since $C = kd^2$ and $4.32 = k \cdot 36$, a 16-inch diameter globe costs $C = \dfrac{4.32}{36} \cdot 16^2 = \$30.72$.

Chapter 2 Test

1. No, since $(0, 5)$ and $(0, -5)$ are two ordered pairs with the same first coordinate and different second coordinates.

2. Yes, since to each x-coordinate there is exactly one y-coordinate, namely, $y = \dfrac{3x - 20}{5}$.

3. No, since $(1, -1)$ and $(1, -3)$ are two ordered pairs with the same first coordinate and different second coordinates.

4. Yes, since to each x-coordinate there is exactly one y-coordinate, namely, $y = x^3 - 3x^2 + 2x - 1$.

5. Domain is $\{2, 5\}$, range is $\{-3, -4, 7\}$

6. Domain is $[9, \infty)$, range is $[0, \infty)$

7. Domain is $[0, \infty)$, range is $(-\infty, \infty)$

8. Graph of $3x - 4y = 12$ includes the points $(4, 0), (0, -3)$

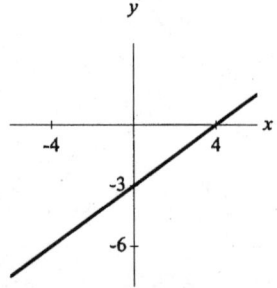

9. Graph of $y = 2x - 3$ includes the points $(3/2, 0), (0, -3)$

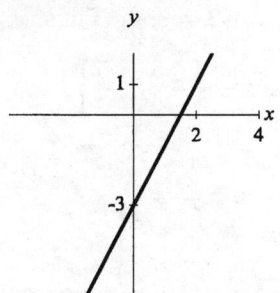

10. $y = \sqrt{25 - x^2}$ is a semicircle with radius 5

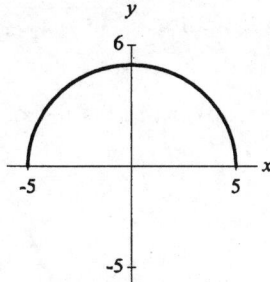

11. $y = -(x - 2)^2 + 5$ is a parabola with vertex $(2, 5)$

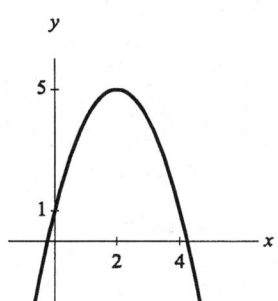

12. $y = 2|x| - 4$ includes the points $(0, -4), (\pm 3, 2)$

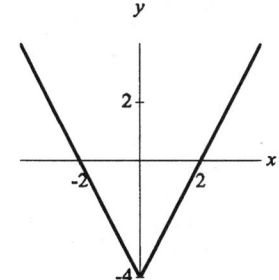

13. $y = \sqrt{x + 3} - 5$ includes the points $(1, -3), (6, -2)$

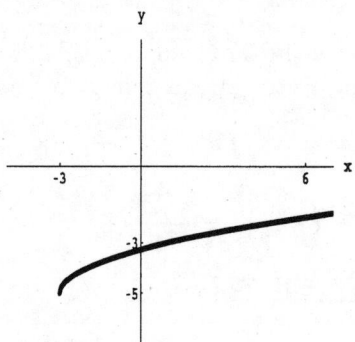

14. Graph includes the points $(-2, -2), (0, 2), (3, 2)$

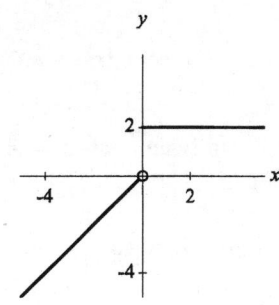

15. $\sqrt{9} = 3$ **16.** $f(5) = \sqrt{7}$

17. $f(3x - 1) = \sqrt{(3x - 1) + 2} = \sqrt{3x + 1}$.

18. $g^{-1}(x) = \dfrac{x + 1}{3}$ **19.** $\sqrt{16} + 41 = 45$

20.

$$\frac{g(x + h) - g(x)}{h} = \frac{3(x + h) - 1 - 3x + 1}{h}$$

$$= \frac{3h}{h}$$

$$= 3$$

21. Increasing on $(3, \infty)$, decreasing on $(-\infty, 3)$

22. Symmetric about y-axis

23. Add 1 to $-3 < x - 1 < 3$ to obtain $-2 < x < 4$. Thus, the solution set $(-2, 4)$.

24. $g^{-1}(x) = (x - 3)^3 + 2$

25. $\dfrac{60 - 35}{200} = \$0.125$ per envelope

26. Since $I = k/d^2$ and $300 = k/4$, we get $k = 1200$. If $d = 10$, then $I = 1200/100 = 12$ candlepower.

27. Let s be the length of one side of the cube. By the Pythagorean Theorem we have

$s^2 + s^2 = d^2$. Then $s = \dfrac{d}{\sqrt{2}}$ and the volume

is $V = \left(\dfrac{d}{\sqrt{2}}\right)^3 = \dfrac{\sqrt{2}d^3}{4}$.

Tying It All Together

1. Add 3 to both sides of $2x - 3 = 0$ to get $2x = 3$.
Then $x = \dfrac{3}{2}$ and the solution set is $\left\{\dfrac{3}{2}\right\}$.

2. Add $2x$ to both sides of $-2x + 6 = 0$ to get $6 = 2x$. Then $3 = x$ and the solution set is $\{3\}$.

3. Note, $|x| = 100$ is equivalent to $x = \pm 100$. The solution set is $\{\pm 100\}$.

4. The equation is equivalent to $\dfrac{1}{2} = |x + 90|$.

Then $x + 90 = \pm\dfrac{1}{2}$ and $x = \pm\dfrac{1}{2} - 90$.

The solution set is $\{-90.5, -89.5\}$.

5. Note, $3 = 2\sqrt{x + 30}$. If we divide by 2 and square both sides, we get $x + 30 = \dfrac{9}{4}$.

Since $x = \dfrac{9}{4} - 30 = -27.75$, the solution set

is $\{-27.75\}$.

6. Note, $\sqrt{x - 3}$ is not a real number if $x < 3$. Since $\sqrt{x - 3}$ is nonnegative for $x \geq 3$, it follows that $\sqrt{x - 3} + 15$ is at least 15. In particular, $\sqrt{x - 3} + 15 = 0$ has no real solution.

7. Rewriting, we obtain $(x - 2)^2 = \dfrac{1}{2}$. By the

square root property, $x - 2 = \pm\dfrac{\sqrt{2}}{2}$.

Thus, $x = 2 \pm \dfrac{\sqrt{2}}{2}$. The solution set

is $\left\{\dfrac{4 \pm \sqrt{2}}{2}\right\}$.

8. Rewriting, we get $(x + 2)^2 = \dfrac{1}{4}$. By the square

root property, $x + 2 = \pm\dfrac{1}{2}$. Thus, $x = -2 \pm \dfrac{1}{2}$.

The solution set is $\left\{-\dfrac{3}{2}, -\dfrac{5}{2}\right\}$.

9. Squaring both sides of $\sqrt{9 - x^2} = 2$, we get $9 - x^2 = 4$. Then $5 = x^2$.
The solution set is $\left\{\pm\sqrt{5}\right\}$.

10. Note, $\sqrt{49 - x^2}$ is not a real number if $49 - x^2 < 0$. Since $\sqrt{49 - x^2}$ is nonnegative for $49 - x^2 \geq 0$, we get that $\sqrt{49 - x^2} + 3$ is at least 3. In particular, $\sqrt{49 - x^2} + 3 = 0$ has no real solution.

11. Domain is $(-\infty, \infty)$, range is $(-\infty, \infty)$, x-intercept $(3/2, 0)$

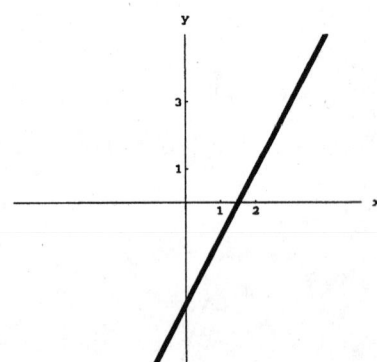

12. Domain is $(-\infty, \infty)$, range is $(-\infty, \infty)$, x-intercept $(3, 0)$

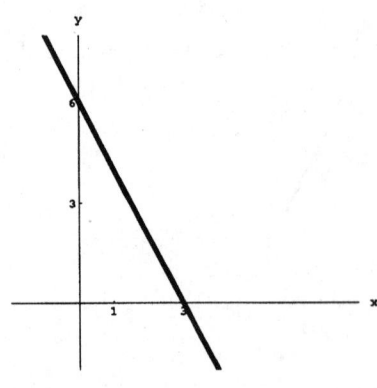

13. Domain is $(-\infty, \infty)$, range is $[-100, \infty)$, x-intercepts $(\pm 100, 0)$

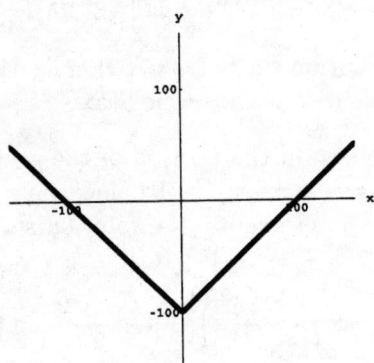

14. Domain is $(-\infty, \infty)$, range is $(-\infty, 1]$, x-intercepts $(-90.5, 0)$ and $(-89.5, 0)$.

15. Domain is $[-30, \infty)$ since we need to require $x + 30 \geq 0$, range is $(-\infty, 3]$, x-intercept is $(-27.75, 0)$ since the solution to $3 - 2\sqrt{x + 30} = 0$ is -27.75

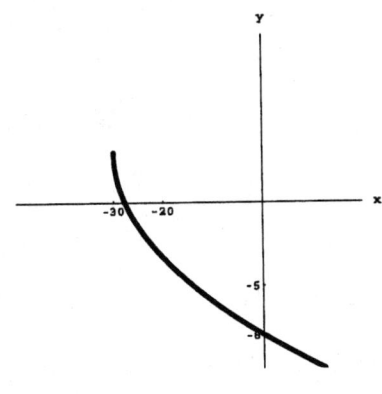

16. Domain is $[3, \infty)$ since we need to require $x - 3 \geq 0$, range is $[15, \infty)$, no x-intercept

17. Domain is $(-\infty, \infty)$, range is $(-\infty, 1]$ since the vertex is $(2, 1)$, and the x-intercepts are $\left(\dfrac{4 \pm \sqrt{2}}{2}, 0\right)$ since the solutions of $0 = -2(x - 2)^2 + 1$ are $x = \dfrac{4 \pm \sqrt{2}}{2}$.

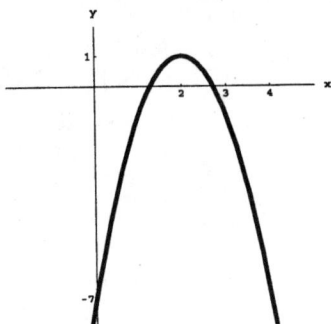

18. Domain is $(-\infty, \infty)$, range is $[-1, \infty)$ since the vertex is $(-2, -1)$, and the x-intercepts are $(-5/2, 0)$ and $(-3/2, 0)$ for the solutions to $0 = 4(x + 2)^2 - 1$ are $x = -5/2, -3/2$.

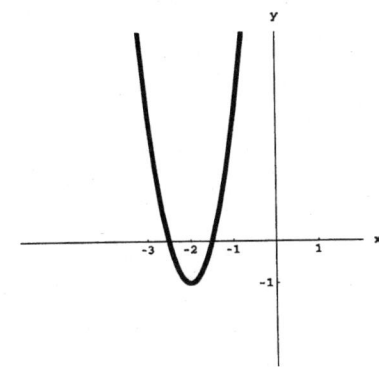

19. Domain is $[-3, 3]$, range is $[-2, 1]$ since the graph is a semi-circle with center $(0, -2)$ and radius 3. x-intercepts are $(\pm\sqrt{5}, 0)$ for the solutions to $0 = \sqrt{9 - x^2} - 2$ are $x = \pm\sqrt{5}$. The graph is shown below.

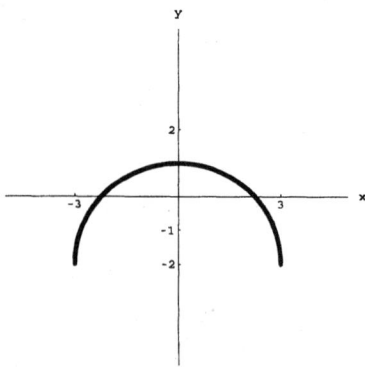

20. Domain is $[-7, 7]$, range is $[3, 10]$ since the graph is a semi-circle with center $(0, 3)$ and radius 7, there are no x-intercepts as seen from the graph

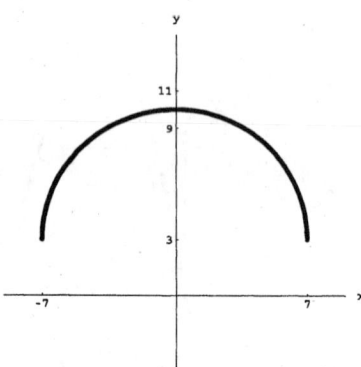

21. Since $2x > 3$ or $x > 3/2$, the solution set is $(3/2, \infty)$.

22. Since $-2x \leq -6$ or $x \geq 3$, the solution set is $[3, \infty)$.

23. Based on the portion of the graph of $y = |x| - 100$ above the x-axis, and its x-intercepts, the solution set of $|x| - 100 \geq 0$ is $(-\infty, -100] \cup [100, \infty)$.

24. Based on the part of the graph of $y = 1 - 2|x + 90|$ above the x-axis, and its x-intercepts, the solution set of $1 - 2|x + 90| > 0$ is $(-90.5, -89.5)$.

25. Based on the part of the graph of $y = 3 - 2\sqrt{x + 30}$ below the x-axis, and its x-intercepts, the solution set of $3 - 2\sqrt{x + 30} \leq 0$ is $[-27.75, \infty)$.

26. Solution set is $(-\infty, \infty)$ since the graph is entirely above the x-axis

27. Based on the portion of the graph of $y = -2(x - 2)^2 + 1$ below the x-axis, and its x-intercepts, the solution set of $-2(x - 2)^2 + 1 < 0$ is
$$\left(-\infty, \frac{4 - \sqrt{2}}{2}\right) \cup \left(\frac{4 + \sqrt{2}}{2}, \infty\right).$$

28. Based on the part of the graph of $y = 4(x + 2)^2 - 1$ above the x-axis, and its x-intercepts, the solution set of $4(x + 2)^2 - 1 < 0$ is $(-\infty, -5/2] \cup [-3/2, \infty)$.

29. Based on the part of the graph of $y = \sqrt{9 - x^2} - 2$ above the x-axis, and its x-intercepts, the solution set of $\sqrt{9 - x^2} - 2 \geq 0$ is $[-\sqrt{5}, \sqrt{5}]$.

30. Since the graph of $y = \sqrt{49 - x^2} + 3$ is entirely above the line $y = 3$, the solution to $\sqrt{49 - x^2} + 3 \leq 0$ is the empty set $\emptyset$.

31. $f(2) = 3(2 + 1)^3 - 24 = 3(27) - 24 = 57$

32. The graph of $f(x) = 3(x + 1)^3 - 24$ is given.

33. $f = K \circ F \circ H \circ G$

34. Solving for x, we get

$$
\begin{aligned}
3(x+1)^3 &= 24 \\
(x+1)^3 &= 8 \\
x+1 &= 2 \\
x &= 1.
\end{aligned}
$$

The solution set is $\{1\}$.

35. Based on the part of the graph of $y = 3(x+1)^3 - 24$ above the x-axis, and its x-intercept $(1,0)$, the solution set of $3(x+1)^3 - 24 \geq 0$ is $[1, \infty)$.

36. Solving for x, we obtain

$$
\begin{aligned}
y + 24 &= 3(x+1)^3 \\
\frac{y+24}{3} &= (x+1)^3 \\
\sqrt[3]{\frac{y+24}{3}} &= x+1 \\
\sqrt[3]{\frac{y+24}{3}} - 1 &= x.
\end{aligned}
$$

37. Based on the answer from number 36, the inverse is $f^{-1}(x) = \sqrt[3]{\dfrac{x+24}{3}} - 1$

38. The graph of f^{-1} is given.

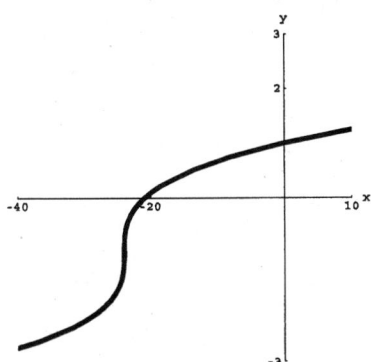

39. Based on the part of the graph of f^{-1} above the x-axis, and its x-intercept $(-21, 0)$, the solution set of $f^{-1}(x) > 0$ is $(-21, \infty)$.

40. Since $f = K \circ F \circ H \circ G$, we get

$$
f^{-1} = G^{-1} \circ H^{-1} \circ F^{-1} \circ K^{-1}.
$$

For Thought

1. False, the range of $y = x^2$ is $[0, \infty)$.

2. False, the vertex is the point $(3, -1)$.

3. True **4.** True **5.** True, since $\dfrac{-b}{2a} = \dfrac{6}{2 \cdot 3} = 1$.

6. True, the x-intercept of $y = (3x + 2)^2$ is the vertex $(-2/3, 0)$ and the y-intercept is $(0, 4)$.

7. True

8. True, since $(x - \sqrt{3})^2$ is always nonnegative.

9. True, since if x and $\dfrac{p - 2x}{2}$ are the length and the width, respectively, of a rectangle with perimeter p, then the area is $y = x \cdot \dfrac{p - 2x}{2}$. This is a parabola opening down with vertex $\left(\dfrac{p}{4}, \dfrac{p^2}{16} \right)$. Thus, the maximum area is $\dfrac{p^2}{16}$.

10. False

3.1 Exercises

1. Completing the square, we get

$$
\begin{aligned}
y &= \left(x^2 + 4x + \left(\frac{4}{2}\right)^2 \right) - \left(\frac{4}{2}\right)^2 \\
y &= \left(x^2 + 4x + 4 \right) - 4 \\
y &= (x + 2)^2 - 4.
\end{aligned}
$$

3. $y = \left(x^2 - 3x + \dfrac{9}{4} \right) - \dfrac{9}{4} = \left(x - \dfrac{3}{2} \right)^2 - \dfrac{9}{4}$

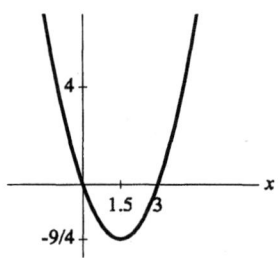

5. Completing the square, we get

$$
\begin{aligned}
y &= 2\left(x^2 - 6x + \left(\frac{6}{2}\right)^2 \right) - 2\left(\frac{6}{2}\right)^2 + 22 \\
y &= 2(x - 3)^2 + 4.
\end{aligned}
$$

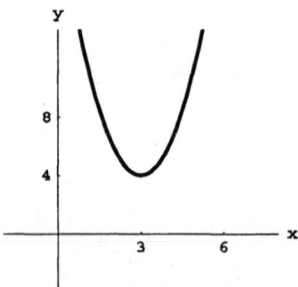

7. Completing the square, we find

$$
\begin{aligned}
y &= -3\left(x^2 - 2x + \left(\frac{2}{2}\right)^2 \right) + 3\left(\frac{2}{2}\right)^2 - 3 \\
y &= -3(x - 1)^2.
\end{aligned}
$$

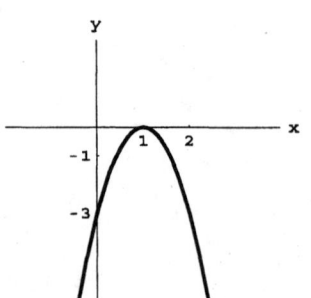

9. $y = \left(x^2 + 3x + \dfrac{9}{4}\right) - \dfrac{9}{4} + \dfrac{5}{2} = \left(x + \dfrac{3}{2}\right)^2 + \dfrac{1}{4}$

11. $y = -2\left(x^2 - \dfrac{3}{2}x + \dfrac{9}{16}\right) + \dfrac{9}{8} - 1 =$

$-2\left(x - \dfrac{3}{4}\right)^2 + \dfrac{1}{8}$

13. Since $\dfrac{-b}{2a} = \dfrac{12}{6} = 2$ and $f(2) = 12 - 24 + 1 =$

-11, the vertex is $(2, -11)$.

15. Vertex: $(4, 1)$

17. Since $\dfrac{-b}{2a} = \dfrac{1/3}{-1} = -\dfrac{1}{3}$ and $f(-1/3) =$

$-\dfrac{1}{18} + \dfrac{1}{9} = \dfrac{1}{18}$, the vertex is $\left(-\dfrac{1}{3}, \dfrac{1}{18}\right)$.

19. Up, vertex $(1, -4)$, axis of symmetry $x = 1$, range $[-4, \infty)$, minimum value -4, decreasing on $(-\infty, 1)$, inreasing on $(1, \infty)$.

21. Since it opens down with vertex $(0, 3)$, range is $(-\infty, 3]$, maximum value is 3, decreasing on $(0, \infty)$, and increasing on $(-\infty, 0)$.

23. Since it opens up with vertex $(1, -1)$, range is $[-1, \infty)$, minimum value is -1, decreasing on $(-\infty, 1)$, and increasing on $(1, \infty)$.

25. Since it opens up with vertex $(-4, -18)$, range is $[-18, \infty)$, minimum value is -18, decreasing on $(-\infty, -4)$, and increasing on $(-4, \infty)$.

27. Since it opens up with vertex is $(3, 4)$, range is $[4, \infty)$, minimum value is 4, decreasing on $(-\infty, 3)$, and increasing on $(3, \infty)$.

29. Since it opens down with vertex $(3/2, 27/2)$, range is $(-\infty, 27/2]$, maximum value is $27/2$, decreasing on $(3/2, \infty)$, and increasing on $(-\infty, 3/2)$.

31. Since it opens down with vertex is $(1/2, 9)$, range is $(-\infty, 9]$, maximum value is 9, decreasing on $(1/2, \infty)$, and increasing on $(-\infty, 1/2)$.

33. Vertex $(0, -3)$, axis $x = 0$, y-intercept $(0, -3)$, x-intercepts $(\pm\sqrt{3}, 0)$, opening up

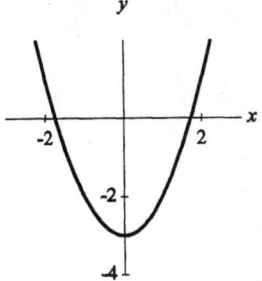

35. Vertex $(1/2, -1/4)$, axis $x = 1/2$, y-intercept $(0, 0)$, x-intercepts $(0, 0)$, $(1, 0)$, opening up

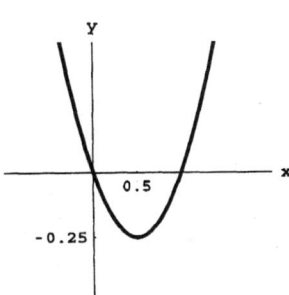

37. Vertex $(-3, 0)$, axis $x = -3$, y-intercept $(0, 9)$, x-intercept $(-3, 0)$, opening up

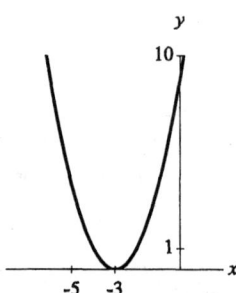

39. Vertex $(3, -4)$, axis $x = 3$, y-intercept $(0, 5)$, x-intercepts $(1, 0), (5, 0)$, opening up

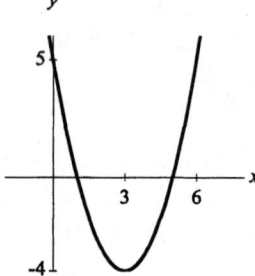

41. Vertex $(2, 12)$, axis $x = 2$, y-intercept $(0, 0)$, x-intercepts $(0, 0), (4, 0)$, opening down

43. Vertex $(1, 3)$, axis $x = 1$, y-intercept $(0, 1)$, x-intercepts $\left(1 \pm \dfrac{\sqrt{6}}{2}, 0\right)$, opening down

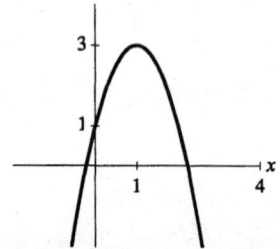

45. The sign graph of $(2x - 3)(x + 1) < 0$ is

$$- - - - - - - - \ 0 \ + + + +$$
$$- - - - \ 0 + + + + + + + +$$

$$\xleftarrow{\hspace{1cm}} \quad \xrightarrow{\hspace{1cm}}$$
$$\qquad -1 \qquad\quad 3/2$$

The solution set is the interval $(-1, 3/2)$

and the graph is $\xleftarrow{\quad \overset{-1 \quad\; 3/2}{(\!=\!=\!=\!)} \quad}\to$

47. The sign graph of $(x + 3)(x - 5) > 0$ is

$$- - - - - - - - \ 0 \ + + + +$$
$$- - - - \ 0 + + + + + + + +$$

$$\xleftarrow{\hspace{1cm}} \quad \xrightarrow{\hspace{1cm}}$$
$$\qquad -3 \qquad\quad 5$$

The solution set is $(-\infty, -3) \cup (5, \infty)$

and the graph is $\xleftarrow{\overset{-3 \ \; 5}{=\!=\!)\;(\!=\!=}}\to$

49. The sign graph of $(w + 2)(w - 6) \geq 0$ is

$$- - - - - - - - \ 0 \ + + + +$$
$$- - - - \ 0 + + + + + + + +$$

$$\xleftarrow{\hspace{1cm}} \quad \xrightarrow{\hspace{1cm}}$$
$$\qquad -2 \qquad\quad 6$$

The solution set is $(-\infty, -2] \cup [6, \infty)$

and the graph is $\xleftarrow{\overset{-2 \ \; 6}{=\!=\!]\;[\!=\!=}}\to$

51. The sign graph of $(t - 4)(t + 4) \leq 0$ is

$$- - - - - - - - \ 0 \ + + + +$$
$$- - - - \ 0 + + + + + + + +$$

$$\xleftarrow{\hspace{1cm}} \quad \xrightarrow{\hspace{1cm}}$$
$$\qquad -4 \qquad\quad 4$$

The solution set is $[-4, 4]$ and the

graph is $\xleftarrow{\overset{-4 \quad\; 4}{[\!=\!=\!=\!]}}\to$

53. The sign graph of $(a + 3)^2 \leq 0$ is

$$+ + + + \ 0 \ + + + + + + + +$$

$$\xleftarrow{\hspace{1cm}} \quad \xrightarrow{\hspace{1cm}}$$
$$\qquad -3$$

The solution set is $\{-3\}$ and the

graph is $\xleftarrow{\overset{-3}{\bullet}}\to$

55. The sign graph of $(2z - 3)^2 > 0$ is

$$+ + + + \ 0 \ + + + + + + + +$$

$$\xleftarrow{\hspace{1cm}} \quad \xrightarrow{\hspace{1cm}}$$
$$\qquad 3/2$$

The solution set is $(-\infty, 3/2) \cup (3/2, \infty)$

and the graph is

57. The roots of $x^2 - 4x + 2 = 0$ are $x_1 = 2 - \sqrt{2}$ and $x_2 = 2 + \sqrt{2}$.

If $x = -5$, then $(-5)^2 - 4(-5) + 2 > 0$.
If $x = 2$, then $(2)^2 - 4(2) + 2 < 0$.
If $x = 5$, then $(5)^2 - 4(5) + 2 > 0$.

The solution set of $x^2 - 4x + 2 < 0$

is $(2 - \sqrt{2}, 2 + \sqrt{2})$ and its graph follows.

59. The roots of $x^2 - 10 = 0$ are $x_1 = -\sqrt{10}$ and $x_2 = \sqrt{10}$.

If $x = -4$, then $(-4)^2 - 10 > 0$.
If $x = 0$, then $(0)^2 - 10 < 0$.
If $x = 4$, then $(4)^2 - 10 > 0$.

The solution set of $x^2 - 9 \geq 1$

is $(-\infty, -\sqrt{10}] \cup [\sqrt{10}, \infty)$ and its

graph is

61. The roots of $y^2 - 10y + 18 = 0$ are $y_1 = 5 - \sqrt{7}$ and $y_2 = 5 + \sqrt{7}$.

If $y = 2$, then $(2)^2 - 10(2) + 18 > 0$.
If $y = 5$, then $(5)^2 - 10(5) + 18 < 0$.
If $y = 8$, then $(8)^2 - 10(8) + 18 > 0$.

The solution set of $y^2 - 10y + 18 > 0$

is $(-\infty, 5 - \sqrt{7}) \cup (5 + \sqrt{7}, \infty)$ and its

graph is

63. Note, $p^2 + 9 = 0$ has no real roots. If $p = 0$, then $(0)^2 + 9 > 0$. The signs of $p^2 + 9$ are shown below.

The solution set of $p^2 + 9 > 0$ is $(-\infty, \infty)$

and its graph follows.

65. Note, $a^2 - 8a + 20 = 0$ has no real roots. If $a = 0$, then $(0)^2 - 8(0) + 20 > 0$. The signs of $a^2 - 8a + 20$ are shown below.

The solution set of $a^2 - 8a + 20 \leq 0$ is $\emptyset$.

67. Note, $2w^2 - 5w + 6 = 0$ has no real roots. If $w = 0$, then $2(0)^2 - 5(0) + 6 > 0$. The signs of $2w^2 - 5w + 6$ are shown below.

The solution set of $2w^2 - 5w + 6 > 0$ is $(-\infty, \infty)$

and its graph follows.

69. $(-\infty, -1] \cup [3, \infty)$

71. $(-3, 1)$

73. $[-3, 1]$

75. a) Since $x^2 - 3x - 10 = (x - 5)(x + 2) = 0$, the solution set is $\{5, -2\}$.

b) Since $x^2 - 3x - 10 = -10$, we get $x^2 - 3x = 0$ or $x(x - 3) = 0$. The solution set is $\{0, 3\}$.

c) If $x = -3$, then $(-3)^2 - 3(-3) - 10 > 0$.
If $x = 0$, then $(0)^2 - 3(0) - 10 < 0$.
If $x = 6$, then $(6)^2 - 3(6) - 10 > 0$.
The signs of $x^2 - 3x - 10$ are shown below.

The solution set of $x^2 - 3x - 10 > 0$.

is $(-\infty, -2) \cup (5, \infty)$

d) Using the sign graph of $x^2 - 3x - 10$ given in part c), the solution set of $x^2 - 3x - 10 \leq 0$ is $[-2, 5]$.

e) By using the method of completing the square, one obtains

$$x^2 - 3x - 10 = \left(x - \frac{3}{2}\right)^2 - 10 - \frac{9}{4}$$

$$x^2 - 3x - 10 = \left(x - \frac{3}{2}\right)^2 - \frac{49}{4}.$$

The graph of f is obtained from the graph of $y = x^2$ by shifting to the right by $\frac{3}{2}$ unit, and down by $\frac{49}{4}$.

f) Domain is $(-\infty, \infty)$, range is $\left[-\frac{49}{4}, \infty\right)$, minimum y-value is $-\frac{49}{4}$

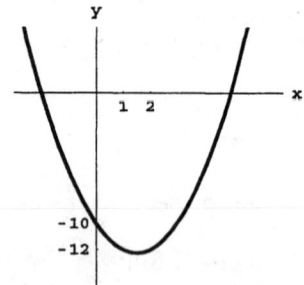

g) The solution to $f(x) > 0$ can be obtained by considering the part of the parabola above the x-axis. While, the solution to $f(x) \leq 0$ can be obtained by considering the part of the parabola on and below the x-axis.

h) x-intercepts are $(5, 0)$ and $(-2, 0)$, y-intercept $(0, -10)$, axis of symmetry $x = \frac{3}{2}$, vertex $\left(\frac{3}{2}, -\frac{49}{4}\right)$, opens up, increasing on $\left(\frac{3}{2}, \infty\right)$, decreasing on $\left(-\infty, \frac{3}{2}\right)$

77. Since $\dfrac{-b}{2a} = \dfrac{-128}{-2(-16)} = 4$, the maximum height is $h(4) = 261$ ft.

79. a) About 100 mph

b) The value of A that would maximize M is
$$A = \frac{-b}{2a} = \frac{-0.127}{-0.001306} \approx 97.24 \text{ mph.}$$

c) Lindbergh flying at 97 mph would use $\dfrac{97}{1.2} \approx 80.83$ lbs. of fuel or

$$\frac{80.83 \text{ lbs}}{6.12 \text{ lbs per gal}} \approx 13.2 \text{ gallons per hour.}$$

81. Let x and y be the length and width, respectively. Since $2x + 2y = 200$, we find $y = 100 - x$. The area as a function of x is

$$f(x) = x(100 - x) = 100x - x^2.$$

The graph of f is a parabola and its vertex is $(50, 2500)$. Thus, the maximum area is 2500 yd^2. Using $x = 50$ from the vertex, we get $y = 100 - x = 100 - 50 = 50$. The dimensions are 50 yd by 50 yd.

83. Let the length of the sides be x, x, and $30 - 2x$. The area of rectangular enclosure is $A(x) = x(30 - 2x) = 30x - 2x^2$. This is a parabola opening down. Since $-\dfrac{b}{2a} = 7.5$ and $y = 30 - 2(7.5) = 15$, the optimal dimensions are 15 ft by 7.5 ft.

85. Let x be the length of a folded side. The area of the cross-section is $A = x(10 - 2x)$. This is a parabola opening down with $-b/2a = 2.5$. The dimensions of the cross-section are 2.5 in. high and 5 in. wide.

87. Let n and p be the number of persons and the price of a tour per person, respectively. The function expressing p as a function of n is $p = 50 - n$. The revenue is $R = (50 - n)n$. This is a parabola opening down with vertex $(25, 625)$. Therefore, 25 persons will give her the maximum revenue of $625.

89. Since $v = 50p - 50p^2$ is a parabola opening down, to maximize v choose

$$p = \frac{-b}{2a} = \frac{50}{100} = 1/2.$$

91. (a) A graph of atmospheric pressure versus altitude is given below.

(b) From the graph, one finds the atmospheric pressure is decreasing as they went to $h = 29,029$ feet.

(c) Since $\dfrac{-b}{2a} = \dfrac{3.48 \times 10^{-5}}{2(3.89 \times 10^{-10})} \approx 44,730$ feet, the function is decreasing on the interval $(0, 44730)$ and increasing on $(44,730, \infty)$.

(d) No, it does not make sense to speak of atmospheric pressure at heights that lie in $(44,730, \infty)$ since it is higher than the summit of 29,029 ft.

(e) It is valid for altitudes less than $30,000$ feet which is less than the height of the summit.

93. (a) Using a graphing calculator, we obtain that the equation of the linear regression is $y = -1550.916667x + 18790.13889$ or approximately $y = -1551x + 18,790$.

The equation of the quadratic regression is $y = 102.4729437x^2 - 2575.646104x + 20668.80952$ or approximately $y = 102x^2 - 2576x + 20,669$.

(b) Judging from the graph, the quadratic function appears to be a better fit.

(c) We substitute $x = 11$ into the regression equations in part (a), in particular, into the equations with more precision. Using the linear function, the price of an eleven year old car is \$1,730. Using the quadratic equation, we obtain \$4736. The \$4736 amount seems to be a more appropriate price of an eleven year old car.

For Thought

1. True, since $i \cdot (-i) = 1$.

2. True, since $\overline{0 + i} = 0 - i = -i$.

3. False, the set of real numbers is a subset of the complex numbers.

4. True, $(\sqrt{3} - i\sqrt{2})(\sqrt{3} + i\sqrt{2}) = 3 + 2 = 5$.

5. False, since $(2 + 5i)(2 + 5i) = 4 + 20i + 25i^2 = 4 + 20i - 25 = -21 + 20i$.

6. False, $5 - \sqrt{-9} = 5 - 3i$.

7. True, since $P(3i) = (3i)^2 + 9 = -9 + 9 = 0$.

8. True, since $(3i)^2 + 9 = -9 + 9 = 0$.

9. True, since $i^4 = i^2 \cdot i^2 = (-1)(-1) = 1$.

10. False, $i^{18} = (i^4)^4 i^2 = (1)^4(-1) = -1$.

3.2 Exercises

1. $0 + 6i$, imaginary

3. $\dfrac{1}{3} + \dfrac{1}{3}i$, imaginary

5. $\sqrt{7} + 0i$, real

7. $\dfrac{\pi}{2} + 0i$, real

9. $7 + 2i$

11. $1 - i - 3 - 2i = -2 - 3i$

13. $-18i + 12i^2 = -12 - 18i$

15. $8 + 12i - 12i - 18i^2 = 26 + 0i$

17. $(5 - 2i)(5 + 2i) = 25 - 4i^2 = 25 - 4(-1) = 29$

19. $(\sqrt{3} - i)(\sqrt{3} + i) = 3 - i^2 = 3 - (-1) = 4$

21. $9 + 24i + 16i^2 = -7 + 24i$

23. $5 - 4i\sqrt{5} + 4i^2 = 1 - 4i\sqrt{5}$

25. $(i^4)^4 \cdot i = (1)^4 \cdot i = i$

27. $(i^4)^{24} i^2 = 1^{24}(-1) = -1$

29. $(i^4)^{-1} = 1^{-1} = 1$

31. Since $i^4 = 1$, we get $i^{-1} = i^{-1}i^4 = i^3 = -i$.

33. $(3 - 9i)(3 + 9i) = 9 - 81i^2 = 90$

35. $\left(\dfrac{1}{2} + 2i\right)\left(\dfrac{1}{2} - 2i\right) = \dfrac{1}{4} - 4i^2 = \dfrac{1}{4} + 4 = \dfrac{17}{4}$

37. $i(-i) = -i^2 = 1$

39. $(3 - i\sqrt{3})(3 + i\sqrt{3}) = 9 - 3i^2 = 9 - 3(-1) = 12$

41. $\dfrac{1}{2 - i} \cdot \dfrac{2 + i}{2 + i} = \dfrac{2 + i}{5} = \dfrac{2}{5} + i\dfrac{1}{5}$

43. $\dfrac{-3i}{1 - i} \cdot \dfrac{1 + i}{1 + i} = \dfrac{-3i + 3}{2} = \dfrac{3}{2} - i\dfrac{3}{2}$

45. $\dfrac{-2 + 6i}{2} = \dfrac{-2}{2} + \dfrac{6i}{2} = -1 + 3i$

47.
$$\dfrac{-3 + 3i}{i} \cdot \dfrac{-i}{-i} = \dfrac{3i - 3i^2}{1} = 3i - 3(-1) = 3 + 3i$$

49.
$$\dfrac{1 - i}{3 + 2i} \cdot \dfrac{3 - 2i}{3 - 2i} = \dfrac{3 - 5i - 2}{13} = \dfrac{1}{13} - i\dfrac{5}{13}$$

51. $2i - 3i = -i$ **53.** $-4 + 2i$

55. $\left(i\sqrt{6}\right)^2 = -6$

57. $(i\sqrt{2})(i\sqrt{50}) = i^2\sqrt{2}\cdot 5\sqrt{2} = (-1)(2)(5) = -10$

59.
$$\dfrac{-2}{2} + \dfrac{i\sqrt{20}}{2} = -1 + i\dfrac{2\sqrt{5}}{2} = -1 + i\sqrt{5}$$

61. $-3 + \sqrt{9 - 20} = -3 + i\sqrt{11}$

63. $2i\sqrt{2}\left(i\sqrt{2} + 2\sqrt{2}\right) = 4i^2 + 8i = -4 + 8i$

65. $\dfrac{-2 + \sqrt{-16}}{2} = \dfrac{-2 + 4i}{2} = -1 + 2i$

67. $\dfrac{-4 + \sqrt{16 - 24}}{4} = \dfrac{-4 + 2\sqrt{2}i}{4} = \dfrac{-2 + i\sqrt{2}}{2}$

69. $\dfrac{-6 - \sqrt{-32}}{2} = \dfrac{-6 - 4i\sqrt{2}}{2} = -3 - 2i\sqrt{2}$

71. $\dfrac{-6 - \sqrt{36 + 48}}{-4} = \dfrac{-6 - 2\sqrt{21}}{-4} = \dfrac{3 + \sqrt{21}}{2}$

73. Since $x^2 = -1$, we obtain $x = \pm\sqrt{-1}$ by taking the square root of both sides. The solution set is $\{\pm i\}$.

75. Note, $x^2 = -8$. Then $x = \pm\sqrt{-8} = \pm i\sqrt{8}$. The solution set is $\{\pm 2i\sqrt{2}\}$.

77. Note, $x^2 = -\dfrac{1}{2}$. Since $x = \pm\sqrt{-\dfrac{1}{2}} = \pm i\sqrt{\dfrac{1}{2}}$, and $\sqrt{\dfrac{1}{2}} = \dfrac{1}{\sqrt{2}}$, the solution set is $\left\{\pm\dfrac{i\sqrt{2}}{2}\right\}$.

79. Note, $a = 1$, $b = -2$, and $c = 2$. By the quadratic formula, we obtain
$$\begin{aligned}\dfrac{-b \pm \sqrt{b^2 - 4ac}}{2a} &= \dfrac{2 \pm \sqrt{4 - 8}}{2}\\ &= \dfrac{2 \pm 2i}{2}.\end{aligned}$$
The solution set is $\{1 \pm i\}$.

81. Since $a = 1$, $b = -4$, and $c = 13$, we get
$$\begin{aligned}\dfrac{-b \pm \sqrt{b^2 - 4ac}}{2a} &= \dfrac{4 \pm \sqrt{16 - 52}}{2}\\ &= \dfrac{4 \pm 6i}{2}.\end{aligned}$$
The solution set is $\{2 \pm 3i\}$.

83. Note, $a = 1$, $b = -2$, and $c = 4$.
$$\begin{aligned}\dfrac{-b \pm \sqrt{b^2 - 4ac}}{2a} &= \dfrac{2 \pm \sqrt{4 - 16}}{2}\\ &= \dfrac{2 \pm 2i\sqrt{3}}{2}.\end{aligned}$$
The solution set is $\{1 \pm i\sqrt{3}\}$.

85. Rewrite the equation as $-2x^2 + 2x - 5 = 0$. Since $a = -2$, $b = 2$, and $c = -5$, we obtain
$$\begin{aligned}\dfrac{-b \pm \sqrt{b^2 - 4ac}}{2a} &= \dfrac{-2 \pm \sqrt{4 - 40}}{-4}\\ &= \dfrac{-2 \pm 6i}{-4}.\end{aligned}$$
The solution set is $\left\{\dfrac{1}{2} \pm \dfrac{3}{2}i\right\}$.

87. Note, $a = 4$, $b = -8$, and $c = 7$.

$$\frac{-b \pm \sqrt{b^2 - 4ac}}{2a} = \frac{8 \pm \sqrt{64 - 112}}{8}$$

$$= \frac{8 \pm 4i\sqrt{3}}{8}$$

The solution set is $\left\{ 1 \pm \frac{\sqrt{3}}{2}i \right\}$.

89. $x^2 - i^2 = x^2 + 1$

91. $(x - (1 + i))(x - (1 - i)) =$
$x^2 - (1 - i)x - (1 + i)x + (1 + i)(1 - i) =$
$x^2 - x + ix - x - ix + (1 - i^2) = x^2 - 2x + 2$

93. $(x - (2 + 3i))(x - (2 - 3i)) =$
$x^2 - (2 - 3i)x - (2 + 3i)x + (2 + 3i)(2 - 3i) =$
$x^2 - 2x + 3ix - 2x - 3ix + (4 - 9i^2) = x^2 - 4x + 13$

95. If r is the remainder when n is divided by 4, then $i^n = i^r$. The possible values of r are $0, 1, 2, 3$ and for i^r they are $1, i, -1, -i$, respectively.

97. Note, $w + \overline{w} = (a + bi) + (a - bi) = 2a$ is a real number and $w - \overline{w} = (a + bi) - (a - bi) = 2bi$ is an imaginary number.

When a complex number is added to its complex conjugate the sum is twice the real part of the complex number. When the complex conjugate of a complex number is subtracted from the complex number, the difference is an imaginary number.

99. Its reciprocal is $\dfrac{1}{a + bi} = \dfrac{a - bi}{(a + bi)(a - bi)}$

$= \dfrac{a - bi}{a^2 + b^2} = \dfrac{a}{a^2 + b^2} - \dfrac{b}{a^2 + b^2}i.$

For Thought

1. False   **2.**   True, by the Remainder Theorem.

3. True, by the Factor Theorem.

4. True, since $2^5 - 1 = 31$.

5. False, since $P(x) = 1$ has no zero.

6. False, rather $c^3 - c^2 + 4c - 5 = b$.

7. False, since $P(4) = -15$.   **8.**   True

9. True, since 1 is a root.

10. False, since -3 is not a root.

3.3 Exercises

1. Quotient $x - 3$, remainder 1

$$
\begin{array}{r}
x - 3 \\
x - 2 \overline{\smash{)}x^2 - 5x + 7} \\
\underline{x^2 - 2x} \\
-3x + 7 \\
\underline{-3x + 6} \\
1
\end{array}
$$

3. Quotient $-2x^2 + 6x - 14$, remainder 33

$$
\begin{array}{r}
-2x^2 + 6x - 14 \\
x + 3 \overline{\smash{)}-2x^3 + 0x^2 + 4x - 9} \\
\underline{-2x^3 - 6x^2} \\
6x^2 + 4x \\
\underline{6x^2 + 18x} \\
-14x - 9 \\
\underline{-14x - 42} \\
33
\end{array}
$$

5. Quotient $s^2 + 2$, remainder 16

$$
\begin{array}{r}
s^2 + 2 \\
s^2 - 5 \overline{\smash{)}s^4 - 3s^2 + 6} \\
\underline{s^4 - 5s^2} \\
2s^2 + 6 \\
\underline{2s^2 - 10} \\
16
\end{array}
$$

7. Quotient $x + 6$, remainder 13

$$
\begin{array}{r|rrr}
2 & 1 & 4 & 1 \\
  &   & 2 & 12 \\
\hline
  & 1 & 6 & 13
\end{array}
$$

9. Quotient $-x + 7$, remainder -12

$$
\begin{array}{r|rrr}
-3 & -1 & 4 & 9 \\
   &   & 3 & -21 \\
\hline
   & -1 & 7 & -12
\end{array}
$$

11. Quotient $4x^2 + 2x - 4$, remainder 0

$$
\begin{array}{r|rrrr}
1/2 & 4 & 0 & -5 & 2 \\
    &   & 2 & 1 & -2 \\
\hline
    & 4 & 2 & -4 & 0
\end{array}
$$

13. Quotient $2a^2 - 4a + 6$, remainder 0

$$
\begin{array}{r|rrrr}
-1/2 & 2 & -3 & 4 & 3 \\
      &   & -1 & 2 & -3 \\
\hline
      & 2 & -4 & 6 & 0
\end{array}
$$

15. Quotient $x^3 + x^2 + x + 1$, remainder -2

$$
\begin{array}{r|rrrrr}
1 & 1 & 0 & 0 & 0 & -3 \\
  &   & 1 & 1 & 1 & 1 \\
\hline
  & 1 & 1 & 1 & 1 & -2
\end{array}
$$

17. Quotient $x - 5/2$, remainder $-1/4$

$$
\begin{array}{r|rrr}
1/2 & 1 & -3 & 1 \\
    &   & 1/2 & -5/4 \\
\hline
    & 1 & -5/2 & -1/4
\end{array}
$$

19. $f(1) = 0$

$$
\begin{array}{r|rrrrrr}
1 & 1 & 0 & 0 & 0 & 0 & -1 \\
  &   & 1 & 1 & 1 & 1 & 1 \\
\hline
  & 1 & 1 & 1 & 1 & 1 & 0
\end{array}
$$

21. $f(-2) = -33$

$$
\begin{array}{r|rrrrrr}
-2 & 1 & 0 & 0 & 0 & 0 & -1 \\
   &   & -2 & 4 & -8 & 16 & -32 \\
\hline
   & 1 & -2 & 4 & -8 & 16 & -33
\end{array}
$$

23. $g(1) = 5$

$$
\begin{array}{r|rrrr}
1 & 1 & -4 & 0 & 8 \\
  &   & 1 & -3 & -3 \\
\hline
  & 1 & -3 & -3 & 5
\end{array}
$$

25. $g(-1/2) = 55/8$

$$
\begin{array}{r|rrrr}
-1/2 & 1 & -4 & 0 & 8 \\
     &   & -1/2 & 9/4 & -9/8 \\
\hline
     & 1 & -9/2 & 9/4 & 55/8
\end{array}
$$

27. $h(-1) = 0$

$$
\begin{array}{r|rrrrr}
-1 & 2 & 1 & -1 & 3 & 3 \\
   &   & -2 & 1 & 0 & -3 \\
\hline
   & 2 & -1 & 0 & 3 & 0
\end{array}
$$

29. $h(1) = 8$

$$
\begin{array}{r|rrrrr}
1 & 2 & 1 & -1 & 3 & 3 \\
  &   & 2 & 3 & 2 & 5 \\
\hline
  & 2 & 3 & 2 & 5 & 8
\end{array}
$$

31. Yes, $(x+3)(x^2+x-2) = (x+3)(x+2)(x-1)$

$$
\begin{array}{r|rrrr}
-3 & 1 & 4 & 1 & -6 \\
   &   & -3 & -3 & 6 \\
\hline
   & 1 & 1 & -2 & 0
\end{array}
$$

33. Yes, $(x-4)(x^2+8x+15) = (x-4)(x+5)(x+3)$

$$
\begin{array}{r|rrr}
4 & 1 & 4 & -17 & -60 \\
  &   & 4 & 32 & 60 \\
\hline
  & 1 & 8 & 15 & 0
\end{array}
$$

35. Yes, since the remainder below is zero

$$
\begin{array}{r|rrrr}
3 & 2 & -5 & -4 & 3 \\
  &   & 6 & 3 & -3 \\
\hline
  & 2 & 1 & -1 & 0
\end{array}
$$

37. No, since the remainder below is not zero

$$
\begin{array}{r|rrrr}
-2 & 1 & 2 & 3 & 1 \\
   &   & -2 & 0 & -6 \\
\hline
   & 1 & 0 & 3 & -5
\end{array}
$$

39. Yes, since the remainder below is zero

$$
\begin{array}{r|rrrrr}
-1 & 1 & 2 & 4 & 6 & 3 \\
   &   & -1 & -1 & -3 & -3 \\
\hline
   & 1 & 1 & 3 & 3 & 0
\end{array}
$$

41. No, since the remainder below is not zero

$$
\begin{array}{r|rrrr}
1/2 & 1 & 3 & -5 & 7 \\
 & & 1/2 & 7/4 & -13/8 \\
\hline
 & 1 & 7/2 & -13/4 & 43/8
\end{array}
$$

43. $\pm\{1, 2, 3, 4, 6, 8, 12, 24\}$ **45.** $\pm\{1, 3, 5, 15\}$

47. $\pm\left\{1, 3, 5, 15, \dfrac{1}{2}, \dfrac{1}{4}, \dfrac{1}{8}, \dfrac{3}{2}, \dfrac{3}{4}, \right.$

   $\left. \dfrac{3}{8}, \dfrac{5}{2}, \dfrac{5}{4}, \dfrac{5}{8}, \dfrac{15}{2}, \dfrac{15}{4}, \dfrac{15}{8}\right\}$

49. $\pm\left\{1, 2, \dfrac{1}{2}, \dfrac{1}{3}, \dfrac{1}{6}, \dfrac{1}{9}, \dfrac{1}{18}, \dfrac{2}{3}, \dfrac{2}{9}\right\}$

51. Zeros are $2, 3, 4$ since

$$
\begin{array}{r|rrrr}
2 & 1 & -9 & 26 & -24 \\
 & & 2 & -14 & 24 \\
\hline
 & 1 & -7 & 12 & 0
\end{array}
$$

 and $x^2 - 7x + 12 = (x - 4)(x - 3)$

53. Zeros are $-3, 2 \pm i$ since

$$
\begin{array}{r|rrrr}
-3 & 1 & -1 & -7 & 15 \\
 & & -3 & 12 & -15 \\
\hline
 & 1 & -4 & 5 & 0
\end{array}
$$

 $x^2 - 4x + 5 = (x - 2)^2 + 1 = 0$ or $x - 2 = \pm i$

55. Zeros are $1/2, 3/2, 5/2$ since

$$
\begin{array}{r|rrrr}
1/2 & 8 & -36 & 46 & -15 \\
 & & 4 & -16 & 15 \\
\hline
 & 8 & -32 & 30 & 0
\end{array}
$$

 and $8a^2 - 32a + 30 = 2(4a^2 - 16a + 15) = 2(2a - 3)(2a - 5)$

57. Zeros are $\dfrac{1}{2}, \dfrac{1 \pm i}{3}$ since

$$
\begin{array}{r|rrrr}
1/2 & 18 & -21 & 10 & -2 \\
 & & 9 & -6 & 2 \\
\hline
 & 18 & -12 & 4 & 0
\end{array}
$$

and the zeros of $18t^2 - 12t + 4$ are (by the quadratic formula) $\dfrac{1 \pm i}{3}$

59. Zeros are $1, -2, \pm i$ since

$$
\begin{array}{r|rrrrr}
1 & 1 & 1 & -1 & 1 & -2 \\
 & & 1 & 2 & 1 & 2 \\
\hline
 & 1 & 2 & 1 & 2 & 0
\end{array}
$$

$$
\begin{array}{r|rrrr}
-2 & 1 & 2 & 1 & 2 \\
 & & -2 & 0 & -2 \\
\hline
 & 1 & 0 & 1 & 0
\end{array}
$$

and the zeros of $w^2 + 1$ are $\pm i$

61. Zeros are $-1, \pm\sqrt{2}$ since

$$
\begin{array}{r|rrrrr}
-1 & 1 & 2 & -1 & -4 & -2 \\
 & & -1 & -1 & 2 & 2 \\
\hline
 & 1 & 1 & -2 & -2 & 0
\end{array}
$$

$$
\begin{array}{r|rrrr}
-1 & 1 & 1 & -2 & -2 \\
 & & -1 & 0 & 2 \\
\hline
 & 1 & 0 & -2 & 0
\end{array}
$$

and the zeros of $x^2 - 2$ are $\pm\sqrt{2}$

63. Zeros are $1/2, 1/3, 1/4$ since

$$
\begin{array}{r|rrrr}
1/2 & 24 & -26 & 9 & -1 \\
 & & 12 & -7 & 1 \\
\hline
 & 24 & -14 & 2 & 0
\end{array}
$$

 and $2(12x^2 - 7x + 1) = 2(4x - 1)(3x - 1)$

65. Rational zero is $1/16$ since

$$
\begin{array}{r|rrrr}
1/16 & 16 & -33 & 82 & -5 \\
 & & 1 & -2 & 5 \\
\hline
 & 16 & -32 & 80 & 0
\end{array}
$$

and by the quadratic formula $16x^2 - 32x + 80$ has imaginary zeros $1 \pm 2i$.

67. Rational zeros are $7/3, -6/7$ since

$$
\begin{array}{r|rrrrr}
7/3 & 21 & -31 & -21 & -31 & -42 \\
    &   & 49 & 42 & 49 & 42 \\
\hline
    & 21 & 18 & 21 & 18 & 0
\end{array}
$$

$$
\begin{array}{r|rrrr}
-6/7 & 21 & 18 & 21 & 18 \\
     &   & -18 & 0 & -18 \\
\hline
     & 21 & 0 & 21 & 0
\end{array}
$$

and $21x^2 + 21 = 0$ has imaginary zeros $\pm i$.

69.

$$
\frac{2x+1}{x-2} = 2 + \frac{5}{x-2} \text{ since}
$$

$$
\begin{array}{r|rr}
2 & 2 & 1 \\
  &   & 4 \\
\hline
  & 2 & 5
\end{array}
$$

71.

$$
\frac{a^2-3a+5}{a-3} = a + \frac{5}{a-3} \text{ since}
$$

$$
\begin{array}{r|rrr}
3 & 1 & -3 & 5 \\
  &   & 3 & 0 \\
\hline
  & 1 & 0 & 5
\end{array}
$$

73.

$$
\frac{c^2-3c-4}{c^2-4} = 1 + \frac{-3c}{c^2-4} \text{ since}
$$

$$
\begin{array}{r}
1 \\
c^2-4 \overline{\smash{\big)}\, c^2-3c-4} \\
\underline{c^2+0c-4} \\
-3c
\end{array}
$$

75.

$$
\frac{4t-5}{2t+1} = 2 + \frac{-7}{2t+1} \text{ since}
$$

$$
\begin{array}{r}
2 \\
2t+1 \overline{\smash{\big)}\, 4t-5} \\
\underline{4t+2} \\
-7
\end{array}
$$

77. a) Note, $\dfrac{P(t)}{t} = -t^3 + 12t^2 - 58t + 132.$

$$
\begin{array}{r|rrrr}
6 & -1 & 12 & -58 & 132 \\
  &   & -6 & 36 & -132 \\
\hline
  & -1 & 6 & -22 & 0
\end{array}
$$

The drug will be eliminated in $t = 6$ hrs.

b) About 120 ppm

c) About 3 hours

d) Between 1 and 5 hours approximately, the concentration is above 80 ppm. Thus, the concentration is above 80 ppm for about 4 hours.

79. If w is the width, then $w(w+4)(w+9) = 630$. This can be re-written as

$$
w^3 + 13w^2 + 36w - 630 = 0.
$$

Using synthetic division, we find

$$
\begin{array}{r|rrrr}
5 & 1 & 13 & 36 & -630 \\
  &   & 5 & 90 & 630 \\
\hline
  & 1 & 18 & 126 & 0
\end{array}
$$

Since $w^2 + 18w + 126 = 0$ has non-real roots, the width of the HP box is $w = 5$ in. The dimensions are 5 in. by 9 in. by 14 in.

81. Remainder is $c = 10$ since

$$
\begin{array}{r|rrr}
3 & 1 & -2 & 7 \\
  &   & 3 & 3 \\
\hline
  & 1 & 1 & 10
\end{array}
$$

83. If $x - c$ is a factor of $P(x)$ then $P(x) = Q(x) \cdot (x-c)$ for some polynomial $Q(x)$. Thus, $P(c) = Q(c) \cdot 0 = 0$ and c is a zero of $P(x)$.

For Thought

1. False, since 1 has multiplicity 1.   **2.**   True

3. True   **4.**   False, it factors as $(x-5)^4(x+2)$.

5. False, rather $4+5i$ is also a solution.   **6.**   True

7. False, since they are solutions to a polynomial with real coefficients of degree at least 4.

8. False, 2 is not a solution.

9. True, since $-x^3 - 5x^2 - 6x - 1 = 0$ has no sign changes.

10. True

3.4 Exercises

1. Degree 2, the root is 5 and its multiplicity is 2, since $(x-5)^2 = 0$

3. Degree 5, roots are 0 with multiplicity 3 and ± 3 since $x^3(x-3)(x+3) = 0$

5. Degree 4, roots are $x = 0, 1$ each of multiplicity 2 since $x^2(x^2 - 2x + 1) = x^2(x-1)^2$

7. Degree 4, roots $3/2, -4/3$ each of multiplicity 2

9. Degree 3, roots are $0, 2 \pm \sqrt{10}$ since $x(x^2 - 4x - 6) = x((x-2)^2 - 10) = 0$

11. $x^2 + 9$

13. $\left[(x-1) - \sqrt{2}\right]\left[(x-1) + \sqrt{2}\right] = (x-1)^2 - 2 = x^2 - 2x - 1$

15. $[(x-3) - 2i][(x-3) + 2i] = (x-3)^2 + 4 = x^2 - 6x + 13$

17. $(x-2)[(x-3) - 4i][(x-3) + 4i] = (x-2)[(x-3)^2 + 16] = x^3 - 8x^2 + 37x - 50$

19. $(x+3)(x-5) = 0$ or $x^2 - 2x - 15 = 0$

21. $(x+4i)(x-4i) = 0$ or $x^2 + 16 = 0$

23. $(x-(3-i))(x-(3+i)) = 0$ or $x^2 - 6x + 10 = 0$

25. $(x+2)(x-i)(x+i) = 0$ or $x^3 + 2x^2 + x + 2 = 0$

27. $x(x - i\sqrt{3})(x + i\sqrt{3}) = 0$ or $x^3 + 3x = 0$

29. $(x-3)[x-(1-i)][x-(1+i)] = 0$ or $x^3 - 5x^2 + 8x - 6 = 0$

31. $(x-1)(x-2)(x-3) = 0$ or $x^3 - 6x^2 + 11x - 6 = 0$

33. $(x-1)[x-(2-3i)][x-(2+3i)] = 0$ or $x^3 - 5x^2 + 17x - 13 = 0$

35. $(2x-1)(3x-1)(4x-1) = 0$ or $24x^3 - 26x^2 + 9x - 1 = 0$

37. $(x-i)(x+i)[x-(1+i)][x-(1-i)] = 0$ or $x^4 - 2x^3 + 3x^2 - 2x + 2 = 0$

39. $P(x) = x^3 + 5x^2 + 7x + 1$ has no sign change and $P(-x) = -x^3 + 5x^2 - 7x + 1$ has 3 sign changes. There are (a) 3 negative roots, or (b) 1 negative root & 2 imaginary roots.

41. $P(x) = -x^3 - x^2 + 7x + 6$ has 1 sign change and $P(-x) = x^3 - x^2 - 7x + 6$ has 2 sign changes. There are (a) 1 positive root & 2 negative roots, or (b) 1 positive root & 2 imginary roots.

43. $P(y) = y^4 + 5y^2 + 7 = P(-y)$ has no sign change. There are 4 imaginary roots.

45. $P(t) = t^4 - 3t^3 + 2t^2 - 5t + 7$ has 4 sign changes and $P(-t) = t^4 + 3t^3 + 2t^2 + 5t + 7$ has no sign change. There are (a) 4 positive roots, or (b) 2 positive roots & 2 imaginary roots, or (c) 4 imaginary roots.

47. $P(x) = x^5 + x^3 + 5x$ and $P(-x) = -x^5 - x^3 - 5x$ have no sign changes; 4 imaginary roots and 0.

49. Best integral bounds are $-1 < x < 3$. One checks that $1, 2$ are not upper bounds and that 3 is a bound.

3	2	-5	0	6
		6	3	9
	2	1	3	15

-1 is a lower bound since

-1	2	-5	0	6
		-2	7	-7
	2	-7	7	-1

51. Best integral bounds are $-3 < x < 2$. One checks that 1 is not an upper bound and that 2 is a bound.

2	4	8	-11	-15
		8	32	42
	4	16	21	27

$-1, -2$ are not lower bounds but -3 is a bound since

-3	4	8	-11	-15
		-12	12	-3
	4	-4	1	-18

53. Best integral bounds are $-1 < x < 5$. One checks that $1, 2, 3, 4$ are not upper bounds and that 5 is a bound

5	1	-5	3	2	-1
		5	0	15	85
	1	0	3	17	84

-1 is a lower bound since

-1	1	-5	3	2	-1
		-1	6	-9	7
	1	-6	9	-7	6

55. Best integral bounds are $-1 < x < 3$. Multiply equation by -1, this makes the leading coefficient positive. One checks that $1, 2$ are not upper bounds and that 3 is a bound.

3	2	-5	3	-9
		6	3	18
	2	1	6	9

-1 is a lower bound since

-1	2	-5	3	-9
		-2	7	-10
	2	-7	10	-19

57. Roots are $1, 5, -2$ since

1	1	-4	-7	10
		1	-3	-10
	1	-3	-10	0

and $x^2 - 3x - 10 = (x - 5)(x + 2)$

59. Roots are $-3, \dfrac{3 \pm \sqrt{13}}{2}$ since

-3	1	0	-10	-3
		-3	9	3
	1	-3	-1	0

and by the quadratic formula the roots of $x^2 - 3x - 1 = 0$ are $\dfrac{3 \pm \sqrt{13}}{2}$.

61. Roots are $2, -4, \pm i$ since

2	1	2	-7	2	-8
		2	8	2	8
	1	4	1	4	0

-4	1	4	1	4
		-4	0	-4
	1	0	1	0

and the root of $x^2 + 1 = 0$ are $\pm i$.

63. Roots are $1/3, 1/2, -5$ since

1/3	6	25	-24	5
		2	9	-5
	6	27	-15	0

and $6x^2 + 27x - 15 = 3(2x - 1)(x + 5)$.

65. Roots are $1, -2$ each with multiplicity 2 since

1	1	2	-3	-4	4
		1	3	0	-4
	1	3	0	-4	0

```
-2 | 1    3    0   -4
   |     -2   -2    4
   ------------------
     1    1   -2    0
```

and $x^2 + x - 2 = (x + 2)(x - 1)$.

67. Use synthetic division on the cubic factor in $x(x^3 - 6x^2 + 12x - 8) = 0$.

```
2 | 1   -6   12   -8
  |      2   -8    8
  ------------------
    1   -4    4    0
```

Since $x^2 - 4x + 4 = (x - 2)^2$, roots are 0, 2 (with multiplicity 3).

69. Use synthetic division on the 5th degree factor in $x(x^5 - x^4 - x^3 + x^2 - 12x + 12) = 0$.

```
1 | 1   -1   -1    1  -12   12
  |      1    0   -1    0  -12
  -----------------------------
    1    0   -1    0  -12    0
```

```
2 | 1    0   -1    0  -12
  |      2    4    6   12
  ----------------------
    1    2    3    6    0
```

```
-2 | 1    2    3    6
   |     -2    0   -6
   ------------------
     1    0    3    0
```

Since the roots of $x^2 + 3 = 0$ are $\pm i\sqrt{3}$, all the roots are $x = 0, 1, \pm 2, \pm i\sqrt{3}$.

71. By the Theorem of Bounds and the graph below

```
6 | 2   -3  -50   18
  |     12   54   24
  ------------------
    2    9    4   42
```

```
-5 | 2   -3  -50   18
   |    -10   65  -75
   ------------------
     2  -13   15  -57
```

the intervals are $-5 < x < 6, -5 < x < 6$

73. By the Theorem of Bounds and the graph below

```
6 | 1    0  -26    0  153
  |      6   36   60  360
  ----------------------
    1    6   10   60  513
```

```
-6 | 1    0  -26    0  153
   |     -6   36  -60  360
   ----------------------
     1   -6   10  -60  513
```

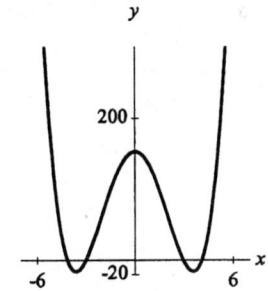

the intervals are $-6 < x < 6, -5 < x < 5$.

75. By the Theorem of Bounds and the graph below

```
23 | 4  -90   -2   45
   |     92   46 1012
   ------------------
     4    2   44 1057
```

```
-1 | 4  -90   -2   45
   |     -4   94  -92
   ------------------
     4  -94   92  -47
```

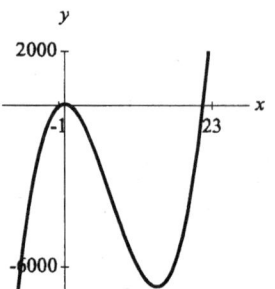

the intervals are $-1 < x < 23$, $-1 < x < 23$.

77. Multiplying by 100, $t^3 - 8t^2 + 11t + 20 = 0$.

$$
\begin{array}{r|rrrr}
4 & 1 & -8 & 11 & 20 \\
  &   & 4 & -16 & -20 \\
\hline
  & 1 & -4 & -5 & 0
\end{array}
$$

Since $t^2 - 4t - 5 = (t - 5)(t + 1) = 0$, we obtain $t = 4$ sec and $t = 5$ sec.

79. Let x be the radius of the cone.
The volume of the cone is $\dfrac{\pi}{3}x^2 \cdot 2$ and the volume of the cylinder with height $4x$ is $\pi x^2 \cdot (4x)$. So $114\pi = \dfrac{\pi}{3}2x^2 + \pi x^2(4x)$.

Divide by π, multiply by 3, and simplify to obtain $12x^3 + 2x^2 - 342 = 0$.

$$
\begin{array}{r|rrrr}
3 & 12 & 2 & 0 & -342 \\
  &   & 36 & 114 & 342 \\
\hline
  & 12 & 38 & 114 & 0
\end{array}
$$

The radius is $x = 3$ in. since $12x^2 + 38x + 114 = 0$ has no real roots.

81. From the following

$$\overline{(a + bi) + (c + di)} = \overline{(a + c) + i(b + d)}$$
$$= (a + c) - i(b + d)$$

and

$$\overline{a + bi} + \overline{c + di} = (a - bi) + (c - di)$$
$$= (a + c) - i(b + d)$$

we obtain that the conjugate of the sum of two complex numbers is equal to the sum of their conjugates.

83. If a is real then $\overline{a} = \overline{a + 0i} = a - 0i = a$.

85. Let $f(x) = a(x - 1)(x - 2)(x - 3)$.

Since $f(0) = -6a = 3$, we find $a = -\dfrac{1}{2}$.

Then substitute $a = -\dfrac{1}{2}$ and multiply out $f(x)$ to obtain $f(x) = -\dfrac{1}{2}x^3 + 3x^2 - \dfrac{11}{2}x + 3$.

For Thought

1. False, $(\sqrt{x-1} + \sqrt{x})^2 = (x-1) + 2\sqrt{x(x-1)} + x$.

2. False, since -1 is a solution of the first and not of the second equation.

3. False, since -27 is a solution of the first equation but not of the second.

4. False, rather let $u = x^{1/4}$ and $u^2 = x^{1/2}$.

5. True, since $x - 1 = \pm 4^{-3/2}$.

6. False, $\left(-\dfrac{1}{32}\right)^{-2/5} = (-32)^{2/5} = (-2)^2 = 4$.

7. False, $x = -2$ is not a solution.

8. True

9. True

10. False, $(x^3)^2 = x^6$.

3.5 Exercises

1. Factor: $x^2(x + 3) - 4(x + 3) = 0$
$(x^2 - 4)(x + 3) = (x - 2)(x + 2)(x + 3) = 0$
The solution set is $\{\pm 2, -3\}$.

3. Factor: $2x^2(x + 500) - (x + 500) = 0$
$(2x^2 - 1)(x + 500) = 0$
The solution set is $\left\{\pm\dfrac{\sqrt{2}}{2}, -500\right\}$.

5. Set the right-hand side to 0 and factor.

$$a(a^2 - 15a + 5) = 0$$

$$a = \frac{15 \pm \sqrt{15^2 - 4(1)(5)}}{2} \quad \text{or} \quad a = 0$$

$$a = \frac{15 \pm \sqrt{205}}{2} \quad \text{or} \quad a = 0$$

The solution set is $\left\{\dfrac{15 \pm \sqrt{205}}{2}, 0\right\}$.

7. Factor: $3y^2(y^2 - 4) = 3y^2(y - 2)(y + 2) = 0$
The solution set is $\{0, \pm 2\}$.

9. Factor: $(a^2 - 4)(a^2 + 4) =$
$(a - 2)(a + 2)(a - 2i)(a + 2i) = 0$.
The solution set is $\{\pm 2, \pm 2i\}$.

11. Squaring each side, we get

$$
\begin{aligned}
x + 1 &= x^2 - 10x + 25 \\
0 &= x^2 - 11x + 24 = (x - 8)(x - 3).
\end{aligned}
$$

Checking $x = 3$, we get $2 \neq -2$. Then $x = 3$ is an extraneous root. The solution set is $\{8\}$.

13. Isolate the radical and then square each side.

$$
\begin{aligned}
x &= x^2 - 40x + 400 \\
0 &= x^2 - 41x + 400 = (x - 25)(x - 16)
\end{aligned}
$$

Checking $x = 16$, we get $2 \neq -6$. Then $x = 16$ is an extraneous root. The solution set is $\{25\}$.

15. Isolate the radical and then square each side.

$$
\begin{aligned}
2w &= \sqrt{1 - 3w} \\
4w^2 &= 1 - 3w \\
4w^2 + 3w - 1 &= (4w - 1)(w + 1) = 0 \\
w &= \frac{1}{4}, -1
\end{aligned}
$$

Checking $w = -1$, we get $-1 \neq 1$.
Then $w = -1$ is an extraneous root.

The solution set is $\left\{\dfrac{1}{4}\right\}$.

17. Multiply both sides by $z\sqrt{4z + 1}$ and square each side.

$$
\begin{aligned}
\sqrt{4z + 1} &= 3z \\
4z + 1 &= 9z^2 \\
0 &= 9z^2 - 4z - 1
\end{aligned}
$$

By the quadratic formula,

$$
\begin{aligned}
z &= \frac{4 \pm \sqrt{16 - 4(9)(-1)}}{18} \\
z &= \frac{4 \pm \sqrt{52}}{18} = \frac{4 \pm 2\sqrt{13}}{18} = \frac{2 \pm \sqrt{13}}{9}
\end{aligned}
$$

Since $z = \dfrac{2 - \sqrt{13}}{9} < 0$ and the right-hand side of the original equation is nonnegative,

$z = \dfrac{2 - \sqrt{13}}{9}$ is an extraneous root.

The solution set is $\left\{\dfrac{2 + \sqrt{13}}{9}\right\}$.

19. Squaring each side, one obtains

$$
\begin{aligned}
x^2 - 2x - 15 &= 9 \\
x^2 - 2x - 24 &= (x - 6)(x + 4) = 0.
\end{aligned}
$$

The solution set is $\{-4, 6\}$.

21. Isolate a radical and square each side.

$$
\begin{aligned}
\sqrt{x + 40} &= \sqrt{x} + 4 \\
x + 40 &= x + 8\sqrt{x} + 16 \\
24 &= 8\sqrt{x} \\
3 &= \sqrt{x} \\
9 &= x
\end{aligned}
$$

The solution set is $\{9\}$.

23. Isolate a radical and square each side.

$$
\begin{aligned}
\sqrt{n + 4} &= 5 - \sqrt{n - 1} \\
n + 4 &= 25 - 10\sqrt{n - 1} + (n - 1) \\
-20 &= -10\sqrt{n - 1} \\
2 &= \sqrt{n - 1} \\
4 &= n - 1
\end{aligned}
$$

The solution set is $\{5\}$.

25. Isolate a radical and square each side.

$$
\begin{aligned}
\sqrt{2x + 5} &= 9 - \sqrt{x + 6} \\
2x + 5 &= 81 - 18\sqrt{x + 6} + (x + 6) \\
x - 82 &= -18\sqrt{x + 6} \\
x^2 - 164x + 6724 &= 324(x + 6) \\
x^2 - 488x + 4780 &= 0 \\
(x - 10)(x - 478) &= 0
\end{aligned}
$$

Checking $x = 478$ we get $53 \neq 9$ and $x = 478$ is an extraneous root. The solution set is $\{10\}$.

27. Raise each side to the power 3/2.
Then $x = \pm 2^{3/2} = \pm 8^{1/2} = \pm 2\sqrt{2}$.
The solution set is $\left\{\pm 2\sqrt{2}\right\}$.

29. Raise each side to the power $-3/4$.

Thus, $w = \pm (16)^{-3/4} = \pm (2)^{-3} = \pm \dfrac{1}{8}$.

The solution set is $\left\{ \pm \dfrac{1}{8} \right\}$.

31. Raise each side to the power -2.

So, $t = (7)^{-2} = \dfrac{1}{49}$. The solution set is $\left\{ \dfrac{1}{49} \right\}$.

33. Raise each side to the power -2.

Then $s - 1 = (2)^{-2} = \dfrac{1}{4}$ and $s = 1 + \dfrac{1}{4}$.

The solution set is $\left\{ \dfrac{5}{4} \right\}$.

35. Since $(x^2 - 9)(x^2 - 3) = 0$, the solution set is $\left\{ \pm 3, \pm \sqrt{3} \right\}$.

37. Let $u = \dfrac{2c - 3}{5}$ and $u^2 = \left(\dfrac{2c - 3}{5} \right)^2$. Then

$$
\begin{aligned}
u^2 + 2u - 8 &= 0 \\
(u + 4)(u - 2) &= 0 \\
u &= -4, 2
\end{aligned}
$$

$$
\begin{array}{ccc}
\dfrac{2c - 3}{5} = -4 & \text{or} & \dfrac{2c - 3}{5} = 2 \\
2c - 3 = -20 & \text{or} & 2c - 3 = 10 \\
c = -\dfrac{17}{2} & \text{or} & c = \dfrac{13}{2}.
\end{array}
$$

The solution set is $\left\{ -\dfrac{17}{2}, \dfrac{13}{2} \right\}$.

39. Let $u = \dfrac{1}{5x - 1}$ and $u^2 = \left(\dfrac{1}{5x - 1} \right)^2$. Then

$$
\begin{aligned}
u^2 + u - 12 &= (u + 4)(u - 3) = 0 \\
u &= -4, 3
\end{aligned}
$$

$$
\begin{array}{ccc}
\dfrac{1}{5x - 1} = -4 & \text{or} & \dfrac{1}{5x - 1} = 3 \\
1 = -20x + 4 & \text{or} & 1 = 15x - 3 \\
x = \dfrac{3}{20} & \text{or} & \dfrac{4}{15} = x.
\end{array}
$$

The solution set is $\left\{ \dfrac{3}{20}, \dfrac{4}{15} \right\}$.

41. Let $u = v^2 - 4v$ and $u^2 = (v^2 - 4v)^2$. Then

$$
u^2 - 17u + 60 = (u - 5)(u - 12) = 0
$$

$$
\begin{array}{ccc}
& u = & 5, 12 \\
v^2 - 4v = 5 & \text{or} & v^2 - 4v = 12 \\
v^2 - 4v - 5 = 0 & \text{or} & v^2 - 4v - 12 = 0 \\
(v - 5)(v + 1) = 0 & \text{or} & (v - 6)(v + 2) = 0.
\end{array}
$$

The solution set is $\{-1, -2, 5, 6\}$.

43. Factor the left-hand side.

$$
\begin{aligned}
(\sqrt{x} - 3)(\sqrt{x} - 1) &= 0 \\
\sqrt{x} &= 3, 1 \\
x &= 9, 1
\end{aligned}
$$

The solution set is $\{1, 9\}$.

45. Factor the left-hand side as $(\sqrt{q} - 4)(\sqrt{q} - 3) = 0$. Then $\sqrt{q} = 4, 3$ and the solution set is $\{16, 9\}$.

47. Set the right-hand side to 0 and factor.

$$
\begin{aligned}
x^{2/3} - 7x^{1/3} + 10 &= 0 \\
\left(x^{1/3} - 5 \right) \left(x^{1/3} - 2 \right) &= 0 \\
x^{1/3} &= 5, 2
\end{aligned}
$$

The solution set is $\{8, 125\}$.

49. An equivalent statement is

$$
\begin{array}{ccc}
w^2 - 4 = 3 & \text{or} & w^2 - 4 = -3 \\
w^2 = 7 & \text{or} & w^2 = 1.
\end{array}
$$

The solution set is $\left\{ \pm \sqrt{7}, \pm 1 \right\}$.

51. An equivalent statement assuming $5v \geq 0$ is

$$
\begin{array}{ccc}
v^2 - 3v = 5v & \text{or} & v^2 - 3v = -5v \\
v^2 - 8v = 0 & \text{or} & v^2 + 2v = 0 \\
v(v - 8) = 0 & \text{or} & v(v + 2) = 0 \\
v = & & 0, 8, -2.
\end{array}
$$

Since $5v \geq 0$, $v = -2$ is an extraneous root and the solution set is $\{0, 8\}$.

53. An equivalent statement is

$$
\begin{array}{ccc}
x^2 - x - 6 = 6 & \text{or} & x^2 - x - 6 = -6 \\
x^2 - x - 12 = 0 & \text{or} & x^2 - x = 0 \\
(x - 4)(x + 3) = 0 & \text{or} & x(x - 1) = 0.
\end{array}
$$

The solution set is $\{-3, 0, 1, 4\}$.

55. An equivalent statement is

$$x + 5 = 2x + 1 \quad \text{or} \quad x + 5 = -(2x + 1)$$
$$4 = x \quad \text{or} \quad x = -2.$$

The solution set is $\{-2, 4\}$.

57. Isolate a radical and square both sides.

$$\sqrt{16x + 1} = \sqrt{6x + 13} - 1$$
$$16x + 1 = (6x + 13) - 2\sqrt{6x + 13} + 1$$
$$10x - 13 = -2\sqrt{6x + 13}$$

$$100x^2 - 260x + 169 = 4(6x + 13)$$
$$100x^2 - 284x + 117 = 0$$

$$x = \frac{284 \pm \sqrt{284^2 - 4(100)(117)}}{200}$$
$$x = \frac{284 \pm 184}{200}$$
$$x = \frac{1}{2}, \frac{117}{50}$$

Checking $x = \dfrac{117}{50}$ we get $\sqrt{\dfrac{1922}{50}} - \sqrt{\dfrac{1352}{50}} > 0$

and so $x = \dfrac{117}{50}$ is an extraneous root.

The solution set is $\left\{\dfrac{1}{2}\right\}$.

59. Factor as a difference of two squares and then as a sum and difference of two cubes.

$$(v^3 - 8)(v^3 + 8) = 0$$
$$(v - 2)(v^2 + 2v + 4)(v + 2)(v^2 - 2v + 4) = 0$$

Then $v = \pm 2$ or

$$v = \frac{-2 \pm \sqrt{2^2 - 16}}{2} \quad \text{or} \quad v = \frac{2 \pm \sqrt{2^2 - 16}}{2}$$
$$v = \frac{-2 \pm 2i\sqrt{3}}{2} \quad \text{or} \quad v = \frac{2 \pm 2i\sqrt{3}}{2}$$

The solution set is $\left\{\pm 2, -1 \pm i\sqrt{3}, 1 \pm i\sqrt{3}\right\}$.

61. Raise both sides to the power 4. Then

$$7x^2 - 12 = x^4$$
$$0 = x^4 - 7x^2 + 12 = (x^2 - 4)(x^2 - 3)$$
$$x = \pm 2, \pm \sqrt{3}$$

Since the left-hand side of the given equation is nonnegative, $x = -2, -\sqrt{3}$ are extraneous roots. The solution set is $\left\{2, \sqrt{3}\right\}$.

63. Raise both sides to the power 3.

$$2 + x - 2x^2 = x^3$$
$$x^3 + 2x^2 - x - 2 = 0$$
$$x^2(x + 2) - (x + 2) = (x^2 - 1)(x + 2) = 0$$
$$x = \pm 1, -2$$

The solution set is $\{\pm 1, -2\}$.

65. Let $t = \dfrac{x - 2}{3}$ and $t^2 = \left(\dfrac{x - 2}{3}\right)^2$. Then

$$t^2 - 2t + 10 = 0$$
$$t^2 - 2t + 1 = -10 + 1$$
$$(t - 1)^2 = -9$$
$$t = 1 \pm 3i$$
$$\frac{x - 2}{3} = 1 \pm 3i$$
$$x - 2 = 3 \pm 9i$$
$$x = 5 \pm 9i.$$

The solution set is $\{5 \pm 9i\}$.

67. Raise both sides to the power 5/2. Then

$$3u - 1 = \pm 2^{5/2}$$
$$3u - 1 = \pm 32^{1/2}$$
$$3u = 1 \pm 4\sqrt{2}$$

The solution set is $\left\{\dfrac{1 \pm 4\sqrt{2}}{3}\right\}$.

69. Factor this quadratic type expression.

$$(x^2 + 1) - 11\sqrt{x^2 + 1} + 30 = 0$$
$$\left(\sqrt{x^2 + 1} - 5\right)\left(\sqrt{x^2 + 1} - 6\right) = 0$$
$$\sqrt{x^2 + 1} = 5 \quad \text{or} \quad \sqrt{x^2 + 1} = 6$$
$$x^2 = 24 \quad \text{or} \quad x^2 = 35$$
$$x = \pm 2\sqrt{6}, \pm\sqrt{35}$$

The solution set is $\left\{\pm\sqrt{35}, \pm 2\sqrt{6}\right\}$.

71. An equivalent statement is

$$x^2 - 2x = 3x - 6 \quad \text{or} \quad x^2 - 2x = -3x + 6$$
$$x^2 - 5x + 6 = 0 \quad \text{or} \quad x^2 + x - 6 = 0$$
$$(x - 3)(x - 2) = 0 \quad \text{or} \quad (x + 3)(x - 2) = 0$$
$$x = 2, \pm 3$$

The solution set is $\{2, \pm 3\}$.

73. Raise both sides to the power $-5/3$. Then

$$
\begin{aligned}
3m + 1 &= \left(-\frac{1}{8}\right)^{-5/3} \\
3m + 1 &= \left(-\frac{1}{2}\right)^{-5} \\
3m + 1 &= -32.
\end{aligned}
$$

The solution set is $\{-11\}$.

75. An equivalent statement assuming $x - 2 \geq 0$ is

$$
\begin{array}{llll}
x^2 - 4 = x - 2 & \text{or} & x^2 - 4 = -x + 2 \\
x^2 - x - 2 = 0 & \text{or} & x^2 + x - 6 = 0 \\
(x - 2)(x + 1) = 0 & \text{or} & (x + 3)(x - 2) = 0 \\
& x = & 2, -1, -3.
\end{array}
$$

Since $x - 2 \geq 0$, $x = -1, -3$ are extraneous roots and the solution set is $\{2\}$.

77. Solve for S.

$$
\begin{aligned}
21.24 + 1.25S^{1/2} - 9.8(18.34)^{1/3} &= 16.296 \\
1.25S^{1/2} - 25.84396 &\approx -4.944 \\
S^{1/2} &\approx 16.72 \\
S &\approx 279.56
\end{aligned}
$$

The maximum sailing area is 279.56 m^2.

79. Solve for x with $C = 83.50$.

$$
\begin{aligned}
0.5x + \sqrt{8x + 5000} &= 83.50 \\
\sqrt{8x + 5000} &= 83.50 - 0.5x \\
8x + 5000 &= 6972.25 - 83.50x + 0.25x^2 \\
0 &= 0.25x^2 - 91.50x + 1972.25
\end{aligned}
$$

$$
\begin{aligned}
x &= \frac{91.50 \pm \sqrt{(-91.50)^2 - 4(0.25)(1972.25)}}{0.5} \\
x &= \frac{91.50 \pm 80}{0.5} \\
x &= 23, 343
\end{aligned}
$$

Checking $x = 343$, the value of the left-hand side of the first equation exceeds 83.50 and so $x = 343$ is an extraneous root.
Thus, 23 loaves cost \$83.50.

81. Let x and $x + 6$ be two numbers. Then

$$
\begin{aligned}
\sqrt{x + 6} - \sqrt{x} &= 1 \\
\sqrt{x + 6} &= \sqrt{x} + 1 \\
x + 6 &= x + 2\sqrt{x} + 1 \\
5 &= 2\sqrt{x} \\
25 &= 4x \\
x &= \frac{25}{4}
\end{aligned}
$$

Since $\dfrac{25}{4} + 6 = \dfrac{49}{4}$, the numbers are $\dfrac{25}{4}$ and $\dfrac{49}{4}$.

83. Let x be the length of the short leg. Since $x + 7$ is the other leg, by the Pythagorean Theorem the hypotenuse is $\sqrt{x^2 + (x + 7)^2}$. Then

$$
x + (x + 7) + \sqrt{x^2 + (x + 7)^2} = 30
$$

$$
\begin{aligned}
2x - 23 &= -\sqrt{x^2 + (x + 7)^2} \\
4x^2 - 92x + 529 &= 2x^2 + 14x + 49 \\
2x^2 - 106x + 480 &= 0 \\
x^2 - 53x + 240 &= 0 \\
(x - 5)(x - 48) &= 0 \\
x &= 5, 48
\end{aligned}
$$

Since the perimeter is 30 in., $x = 48$ is an extraneous root. The short leg is $x = 5$ in.

85. Let x be the length of one side of the original square foundation. From the 2100 ft^2 we have

$$
\begin{aligned}
(x - 10)(x + 30) &= 2100 \\
x^2 + 20x - 2400 &= 0 \\
(x + 60)(x - 40) &= 0.
\end{aligned}
$$

Since x is nonnegative, $x = 40$ and the area of the square foundation is $x^2 = 1600$ ft^2.

87. Solving for d, we find

$$598.9 \left(\frac{d}{64}\right)^{-2/3} = 14.26$$

$$\left(\frac{d}{64}\right)^{-2/3} = \frac{14.26}{598.9}$$

$$\frac{d}{64} = \left(\frac{14.26}{598.9}\right)^{-3/2}$$

$$d = 64 \left(\frac{14.26}{598.9}\right)^{-3/2}$$

$$d \approx 17,419.3 \text{ lbs.}$$

89. Let $x^2(x+2)$ be the volume of shrimp to be shipped. Here, x is the length of one side of the square base and $x+2$ is the height. The volume of the box containing the styrofoam and shrimp is $(x+2)^2(x+4)$. Since the amount of shrimp is one-half of the volume of the box it is to be shipped in, we have

$$2x^2(x+2) = (x+2)^2(x+4)$$
$$2x^3 + 4x^2 = (x^2 + 4x + 4)(x+4)$$
$$x^3 - 4x^2 - 20x - 16 = 0.$$

By using the Rational Zero Theorem and synthetic division, we obtain

$$x^3 - 4x^2 - 20x - 16 = (x+2)(x^2 - 6x - 8).$$

Note, we have to exclude the zero of $x+2$ which is -2 since a dimension is a positive number. Then by using the method of completing the square, we get

$$x^2 - 6x = 8$$
$$x^2 - 6x + 9 = 17$$
$$(x-3)^2 = 17$$
$$x = 3 \pm \sqrt{17}$$
$$x \approx -1.123, 7.123$$

Again, we exclude $x = -1.123$.
So $x = 3 + \sqrt{17}$.
Thus, the volume of shrimp to be shipped is $(3+\sqrt{17})^2(5+\sqrt{17}) \approx 462.89$ in.^{3}

91. Let x be the number of hours after 10:00 a.m. so that the distance between Nancy & Edgar is 14 miles greater than the distance between Nancy & William.

By the Pythagorean Theorem, $\sqrt{(5x)^2 + (12x)^2}$ and $\sqrt{(5x)^2 + [4(x+2)]^2}$ are the distances between Nancy & Edgar and Nancy & William, respectively. Since these distances are equal, we have

$$\sqrt{(5x)^2 + (12x)^2} - 14 = \sqrt{(5x)^2 + [4(x+2)]^2}$$
$$\sqrt{169x^2} - 14 = \sqrt{41x^2 + 64x + 64}$$
$$13x - 14 = \sqrt{41x^2 + 64x + 64}$$
$$169x^2 - 364x + 196 = 41x^2 + 64x + 64$$
$$128x^2 - 428x + 132 = 0$$
$$32x^2 - 107x + 33 = 0$$

$$x = \frac{107 \pm \sqrt{(-107)^2 - 4(32)(33)}}{64} = 3, \frac{11}{32}.$$

Since the left-hand side of the first equation is negative when $x = \frac{11}{32}$, we find that $x = \frac{11}{32}$ is an extraneous root. Thus, $x = 3$ hours and the time is 1:00 p.m.

93. (a) $\sqrt[3]{(2)(1.4)(1.6)} \approx \$1.65$ billion

(b) Let p be the net income in the fourth quarter. Then

$$\sqrt[4]{(2)(1.4)(1.6)p} = 1.9$$
$$2(1.4)(1.6)p = 1.9^4$$
$$p = \frac{1.9^4}{2(1.4)(1.6)}$$
$$p \approx \$2.91 \text{ billion.}$$

95. The weight of a cylindrical tank is its volume times its density. Then

$$\pi \left(\frac{d}{2}\right)^2 \cdot d \cdot 1600 = 25,850,060$$
$$400\pi d^3 = 25,850,060$$
$$d = \sqrt[3]{\frac{25,850,060}{400\pi}} \approx 27.4.$$

The height of the tank is about 27.4 meters.

For Thought

1. False, to be symmetric about the origin one must have $P(-x) = -P(x)$ for *all* x in the domain.

2. True   3.   True

4. False, since $f(-x) = -f(x)$.

5. True 6.   True   7.   False

8. False, y-intercept is $(0, 38)$.   9.   True

10. False, only one x-intercept.

3.6 Exercises

1. Neither symmetry, crosses $(-2, 0)$, does not cross $(1, 0)$, $y \to \infty$ as $x \to \infty$, $y \to -\infty$ as $x \to -\infty$

3. Symmetric about y-axis, no x-intercepts, $y \to \infty$ as $x \to \infty$, $y \to \infty$ as $x \to -\infty$

5. y-axis, since $f(-x) = f(x)$

7. $x = 3/2$, since $\dfrac{3}{2}$ is the x-coordinate of the vertex of a parabola

9. None

11. Origin, since $f(-x) = -f(x)$

13. $x = 5$, since 5 is the x-coordinate of the vertex of a parabola

15. Origin, since $f(-x) = -f(x)$

17. It does not cross $(4, 0)$ since $x - 4$ is raised to an even power.

19. It crosses $(1/2, 0)$.

21. It crosses $(1/4, 0)$

23. No x-intercepts since $x^2 - 3x + 10 = 0$ has no real root.

25. It crosses $(3, 0)$ and not $(0, 0)$ since $x^3 - 3x^2 = x^2(x - 3)$.

27. It crosses $(1/2, 0)$ and not $(1, 0)$ since

1	2	-5	4	-1
		2	-3	1
	2	-3	1	0

and $2x^3 - 5x^2 + 4x - 1 = (x - 1)(2x^2 - 3x + 1) = (x - 1)^2(2x - 1)$.

29. It crosses $(2, 0)$ and not $(-3, 0)$ since

-3	-2	-8	6	36
		6	6	-36
	-2	-2	12	0

and $(x + 3)(-2x^2 - 2x + 12) = -2(x + 3)(x^2 + x - 6) = -2(x + 3)(x + 3)(x - 2) = -2(x + 3)^2(x - 2)$.

31. $y \to \infty$ 33. $y \to -\infty$

35. $y \to -\infty$ 37. $y \to \infty$

39. $y \to \infty$ 41.   e   43.   g

45. b   47.   c

49. $f(x) = x - 30$ has x-intercept $(30, 0)$ and y-intercept $(0, -30)$

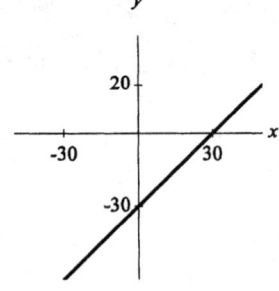

51. $f(x) = (x - 30)^2$ does not cross $(30, 0)$, y-intercept $(0, 900)$, $y \to \infty$ as $x \to \infty$ and as $x \to -\infty$

53. $f(x) = x^2(x - 40)$ crosses $(40, 0)$ but does not cross $(0, 0)$, $y \to \infty$ as $x \to \infty$, $y \to -\infty$ as $x \to -\infty$

55. $f(x) = (x - 20)^2(x + 20)^2$ does not cross $(20, 0)$, $(-20, 0)$, y-intercept $(0, 160000)$, $y \to \infty$ as $x \to \infty$ and as $x \to -\infty$

57. Since $f(x) = -x^3 - x^2 + 5x - 3$

1	-1	-1	5	-3
		-1	-2	3
	-1	-2	3	0

$f(x) = (x - 1)(-x^2 - 2x + 3) = -(x-1)(x+3)(x-1)$, the graph crosses $(-3, 0)$ but not $(1, 0)$, y-intercept $(0, -3)$, $y \to -\infty$ as $x \to \infty$, and $y \to \infty$ as $x \to -\infty$

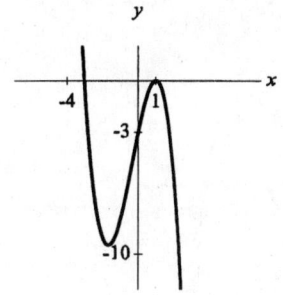

59. Since $x^3 - 10x^2 - 600x = x(x - 30)(x + 20)$, graph crosses $(0, 0), (30, 0), (-20, 0)$, $y \to \infty$ as $x \to \infty$, $y \to -\infty$ as $x \to -\infty$

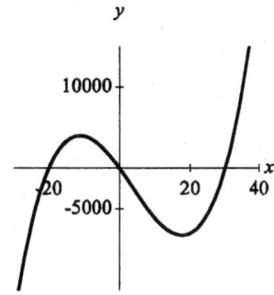

61. $f(x) = x^3 + 18x^2 - 37x + 60$ has only one x-intercept since

-20	1	18	-37	60
		-20	40	-60
	1	-2	3	0

and $x^2 - 2x + 3$ has no real root. Graph crosses $(-20, 0)$, y-intercept $(0, 60)$, $y \to \infty$ as $x \to \infty$, $y \to -\infty$ as $x \to -\infty$

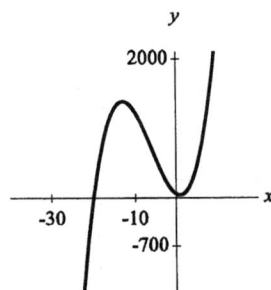

63. Since $-x^2(x^2 - 196) = -x^2(x - 14)(x + 14)$, graph crosses $(\pm 14, 0)$ and does not cross $(0,0)$, $y \to -\infty$ as $x \to \infty$ and as $x \to -\infty$

65. Use synthetic division on $f(x) = x^3 + 3x^2 + 3x + 1$ with $c = 1$.

$$
\begin{array}{r|rrrr}
-1 & 1 & 3 & 3 & 1 \\
   &   & -1 & -2 & -1 \\
\hline
   & 1 & 2 & 1 & 0
\end{array}
$$

Since $x^2 + 2x + 1 = (x + 1)^2$ then $f(x) = (x + 1)^3$. Graph crosses $(-1, 0)$, y-intercept $(0, 1)$, $y \to \infty$ as $x \to \infty$, and $y \to -\infty$ as $x \to -\infty$.

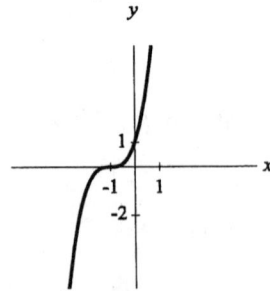

67. The roots of $x(x^2 - 3) = 0$ are $x = 0, \pm\sqrt{3}$.

If $x = 2$, then $(2)^2 - 3(2) > 0$.
If $x = 1$, then $(1)^2 - 3(1) < 0$.
If $x = -1$, then $(-1)^2 - 3(-1) > 0$.
If $x = -2$, then $(-2)^2 - 3(-2) < 0$.

The solution set is $(-\sqrt{3}, 0) \cup (\sqrt{3}, \infty)$.

69. The roots of $x^2(2 - x^2) = 0$ are $x = 0, \pm\sqrt{2}$.

If $x = 3$, then $2(3)^2 - (3)^4 < 0$.
If $x = 1$, then $2(1)^2 - (1)^4 > 0$.
If $x = -1$, then $2(-1)^2 - (-1)^4 > 0$.
If $x = -3$, then $2(-3)^2 - (-3)^4 < 0$.

The solution set is $(-\infty, -\sqrt{2}] \cup \{0\} \cup [\sqrt{2}, \infty)$.

71. Let $f(x) = x^3 + 4x^2 - x - 4$. Since

$$x^2(x + 4) - (x + 4) = (x + 4)(x^2 - 1) = 0,$$

the zeros of $f(x)$ are $x = -4, \pm 1$.

If $x = 2$, then $f(2) > 0$.
If $x = 0$, then $f(0) < 0$.
If $x = -2$, then $f(-2) > 0$.
If $x = -5$, then $f(-5) < 0$.

The solution set is $(-4, -1) \cup (1, \infty)$.

73. Let $f(x) = x^3 - 4x^2 - 20x + 48$. Since $f(x) = (x + 4)(x - 2)(x - 6)$, the zeros of $f(x)$ are $x = -4, 2, 6$.

If $x = 7$, then $f(7) > 0$.
If $x = 4$, then $f(4) < 0$.
If $x = 0$, then $f(0) > 0$.
If $x = -5$, then $f(-5) < 0$.

The solution set is $[-4, 2] \cup [6, \infty)$.

75. Note, $f(x) = x^3 - x^2 + x - 1 = x^2(x-1) + (x-1) = (x-1)(x^2+1) = 0$ has only one solution, namely, $x = 1$.

If $x = 2$, then $f(2) > 0$.
If $x = 0$, then $f(0) < 0$.

The solution set is $(-\infty, 1)$.

77. Let $f(x) = x^4 - 19x^2 + 90$.
Since $f(x) = (x^2 - 10)(x^2 - 9)$, the zeros of $f(x)$ are $x = \pm\sqrt{10}, \pm 3$.

If $x = 4$, then $f(4) > 0$.
If $x = 3.1$, then $f(3.1) < 0$.
If $x = 0$, then $f(0) > 0$.
If $x = -3.1$, then $f(-3.1) < 0$.
If $x = -4$, then $f(-4) > 0$.

The solution set is $(-\infty, -\sqrt{3}] \cup [2, \sqrt{3}] \cup [3, \infty)$.

79. Local max. ≈ 3.11, local min. ≈ 0.37

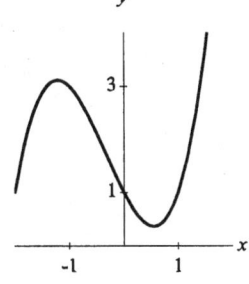

81. Local max. ≈ 23.74, local min. ≈ -163.74

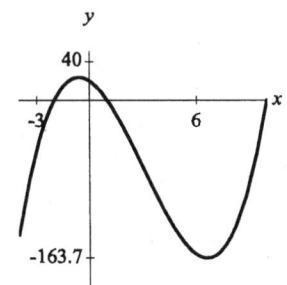

83. Local max. ≈ 21.01, local min. ≈ 13.99

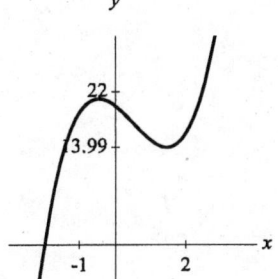

85. From the graph of the profit function, we find a local maximum of \$3400 and a local minimum of \$2600. To get a profit higher than \$3400, the company must spend more than \$2200 on advertising.

87. Note, $P = \dfrac{x}{10}(x^2 - 60x + 900) = \dfrac{x}{10}(x - 30)^2$.

The graph of P does not cross the x-intercept $(30, 0)$. Thus, the profit is \$0 when there are 30 stores, and then the profit increases as more stores (greater than 30) are opened.

89. Since $3x + 2y = 12$, we get $y = \dfrac{12 - 3x}{2}$.

The volume V of the block is given by

$$
\begin{aligned}
V &= xy\frac{4y}{3} \\
&= \frac{4x}{3}y^2 \\
&= \frac{4x}{3}\left(\frac{12-3x}{2}\right)^2 \\
&= 3x^3 - 24x^2 + 48x.
\end{aligned}
$$

One finds from the graph of

$$ V = \frac{4x}{3}\left(\frac{12-3x}{2}\right)^2 $$

that the optimal dimensions of the block are

$x = \dfrac{4}{3}$ in. by $y = \dfrac{12 - 4}{2} = 4$ in. by $\dfrac{16}{3}$ in..

91. If the surface area is 50 square feet, then no paint can be used (since all the paint is used to coat the inside of the can).

Since the amount of paint in a 1-pint can is $\dfrac{1}{8} \cdot \dfrac{1}{7.5}$ or $\dfrac{1}{60}$ feet^3, we get $\dfrac{1}{60} = \pi r^2 h$. Solving for h, one obtains $h = \dfrac{1}{60\pi r^2}$. Then

$$2\pi r^2 + 2\pi rh = 50$$
$$2\pi r^2 + 2\pi r\left(\dfrac{1}{60\pi r^2}\right) = 50$$
$$2\pi r^2 + \dfrac{1}{30r} - 50 = 0.$$

With a graphing calculator, one finds

$$r \approx \pm 2.82 \text{ or } r \approx 6.67 \times 10^{-4}.$$

If $r \approx 2.82$, then $h \approx 6.67 \times 10^{-4}$ feet since $h = \dfrac{1}{60\pi r^2}$. Similarly, if $r \approx 6.67 \times 10^{-4}$, then $h \approx 11,924.69$ feet. Thus, either the radius and height of the can are

$$r \approx 2.82 \text{ ft and } h \approx 6.67 \times 10^{-4} \text{ ft}$$

or

$$r \approx 6.67 \times 10^{-4} \text{ ft and } h \approx 11,924.69 \text{ ft.}$$

For Thought

1. False, $\sqrt{x} - 3$ is not a polynomial.

2. False, domain is $(-\infty, 2) \cup (2, \infty)$.

3. False

4. False, it has three vertical asymptotes.

5. True

6. False, $y = 5$ is the horizontal asymptote.

7. True   **8.**   False

9. True, it is an even function.

10. True, since $x = -3$ is not a vertical asymptote.

3.7 Exercises

1. $(-\infty, -2) \cup (-2, \infty)$

3. $(-\infty, -2) \cup (-2, 2) \cup (2, \infty)$

5. $(-\infty, 3) \cup (3, \infty)$

7. $(-\infty, 0) \cup (0, \infty)$

9. $(-\infty, -1) \cup (-1, 0) \cup (0, 1) \cup (1, \infty)$ since
$$f(x) = \dfrac{3x^2 - 1}{x(x^2 - 1)}$$

11. $(-\infty, -3) \cup (-3, -2) \cup (-2, \infty)$ since
$$f(x) = \dfrac{-x^2 + x}{(x + 3)(x + 2)}$$

13. Domain $(-\infty, 2) \cup (2, \infty)$, asymptotes $y = 0$ and $x = 2$

15. Domain $(-\infty, 0) \cup (0, \infty)$, asymptotes $y = x$ and $x = 0$

17. Asymptotes $x = 2$, $y = 0$

19. Asymptotes $x = \pm 3$, $y = 0$

21. Asymptotes $x = 1$, $y = 2$

23. Asymptotes $x = 0$, $y = x - 2$ since
$$f(x) = x - 2 + \dfrac{1}{x}$$

25. Asymptotes $x = -1$, $y = 3x - 3$ since

$$\begin{array}{r} 3x - 3 \\ x + 1 \overline{\smash{\big)}\ 3x^2 + 0x + 4} \\ \underline{3x^2 + 3x} \\ -3x + 4 \\ \underline{-3x - 3} \\ 7 \end{array}$$

and $f(x) = 3x - 3 + \dfrac{7}{x + 1}$

27. Asymptotes $x = -2$, $y = -x + 6$ since

$$
\begin{array}{r}
-x + 6 \\
x + 2 \overline{\smash{)}-x^2 + 4x + 0} \\
\underline{-x^2 - 2x} \\
6x + 0 \\
\underline{6x + 12} \\
-12
\end{array}
$$

and $f(x) = -x + 6 + \dfrac{-12}{x + 2}$

29. Asymptotes $x = 0$, $y = 0$, no x or y-intercept

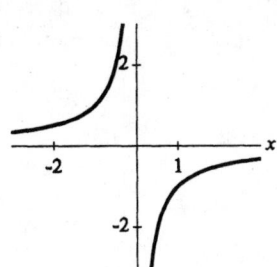

31. Asymptotes $x = 2$, $y = 0$, no x-intercept, y-intercept $(0, -1/2)$

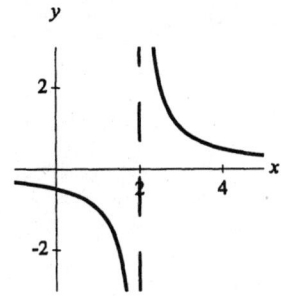

33. Asymptotes $x = \pm 2$, $y = 0$, no x-intercept, y-intercept $(0, -1/4)$

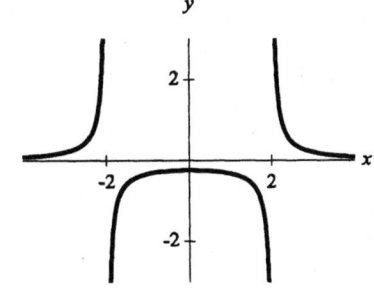

35. Asymptotes $x = -1$, $y = 0$, no x-intercept, y-intercept $(0, -1)$

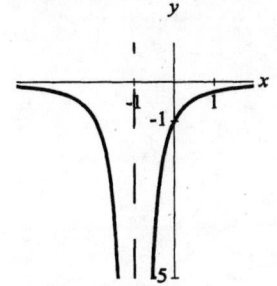

37. Asymptotes $x = 1$, $y = 2$, x-intercept $(-1/2, 0)$, y-intercept $(0, -1)$

39. Asymptotes $x = -2$, $y = 1$, x-intercept $(3, 0)$, y-intercept $(0, -3/2)$

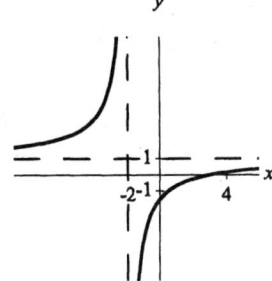

41. Asymptotes $x = \pm 1$, $y = 0$, x- and y-intercept is $(0, 0)$

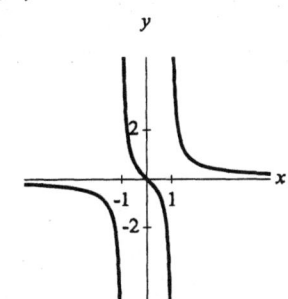

43. Since $f(x) = \dfrac{4x}{(x-1)^2}$, asymptotes are

$x = 1$, $y = 0$, x- and y-intercept is $(0,0)$

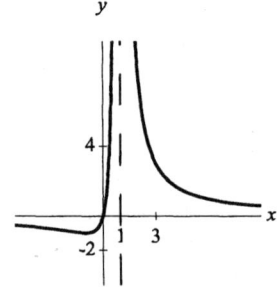

45. Asymptotes are $x = \pm 3$, $y = -1$,
x-intercept $(\pm 2\sqrt{2}, 0)$, y-intercept $(0, -8/9)$

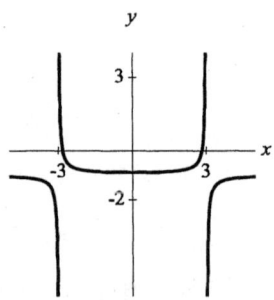

47. Since $f(x) = \dfrac{2x^2 + 8x + 2}{(x+1)^2}$,

asymptotes are $x = -1$, $y = 2$,
by solving $2x^2 + 8x + 2 = 0$ one gets the
x-intercepts $\left(-2 \pm \sqrt{3}, 0\right)$,
y-intercept $(0,2)$

49. Since $f(x) = x + \dfrac{1}{x}$, oblique asymptote is

$y = x$, asymptote $x = 0$, no x-intercept,
no y-intercept, graph goes through
$(1,2), (-1,-2)$

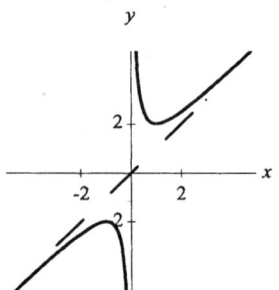

51. Since $f(x) = x - \dfrac{1}{x^2}$, oblique asymptote

is $y = x$, asymptote $x = 0$, x-intercept $(1,0)$,
no y-intercept, graph goes through $(-1,-2)$

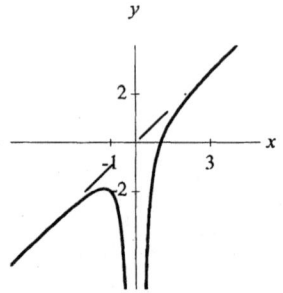

53. $f(x) = x - 1 + \dfrac{1}{x+1}$ since

$$
\begin{array}{r}
x - 1 \\
x+1 \overline{\smash{)}x^2 + 0x } \\
\underline{x^2 + x } \\
-x + 0 \\
\underline{-x - 1} \\
1
\end{array}
$$

Oblique asymptote $y = x - 1$,
asymptote $x = -1$, x-intercept $(0,0)$,
graph goes through $(-2,-4)$

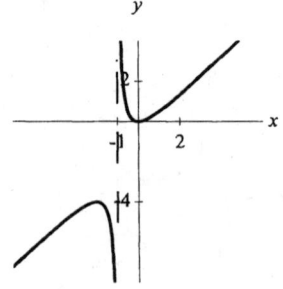

55. $f(x) = 2x + 1 + \dfrac{1}{x-1}$ since

$$
\begin{array}{r}
2x + 1 \\
x - 1 \overline{\smash{)}2x^2 - x + 0} \\
\underline{2x^2 - 2x} \\
x + 0 \\
\underline{x - 1} \\
1
\end{array}
$$

Oblique asymptote $y = 2x + 1$,
asymptote $x = 1$,
x-intercepts $(0,0), (1/2, 0)$,
graph goes through $(2, 6)$

57. e **59.** a

61. b **63.** c

65. Domain is $\{x : x \neq \pm 1\}$ since

$$f(x) = \frac{x+1}{(x+1)(x-1)} = \frac{1}{x-1} \text{ if } x \neq -1,$$

a 'hole' at $(-1, -1/2)$, asymptotes $x = 1$,
$y = 0$, no x-intercept, y-intercept $(0, -1)$

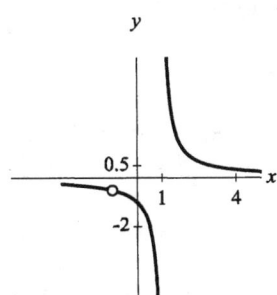

67. The domain is $\{x : x \neq 1\}$ since

$$f(x) = \frac{(x-1)(x+1)}{x-1} = x + 1 \text{ where } x \neq 1,$$

a line with a 'hole' at $(1, 2)$

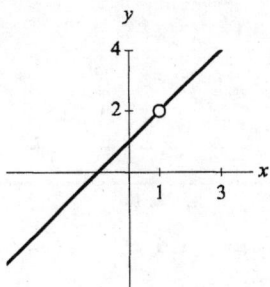

69. Symmetric about y-axis, asymptote $x = 0$,
goes through $(0, 2), (1, 1)$

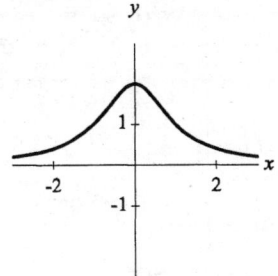

71. Since $f(x) = \dfrac{x-1}{x(x^2 - 9)}$, asymptotes are

$x = \pm 3, x = 0, y = 0$, crosses $(1, 0)$

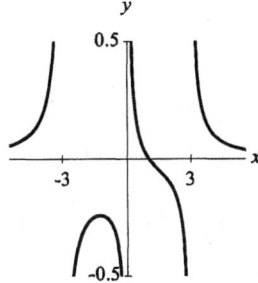

73. Asymptotes $x = 0, y = 0$, x-intercept $(-1, 0)$,
no y-intercept, graph goes through $(1, 2)$

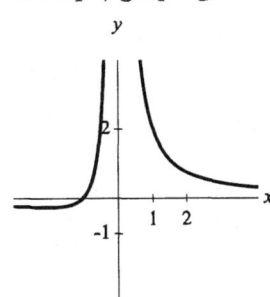

75. The sign graph of $\dfrac{x-4}{x+2} \le 0$ is

```
- - - - - - - - 0 + + + +
- - - - 0 + + + + + + + +
←————————————————————→
       -2        4
```

The solution set is $(-2, 4]$.

77. The sign graph of $\dfrac{q+8}{q+3} > 0$ is

```
- - - - - - - - 0 + + + +
- - - - 0 + + + + + + + +
←————————————————————→
     -8          -3
```

The solution set is $(-\infty, -8) \cup (-3, \infty)$.

79. The sign graph of $\dfrac{(w-3)(w+2)}{w-6} \ge 0$ is

```
- - - - - - - - - - - - - 0 + + +
- - - - - - - - - 0 + + + + + +
- - - - - 0 + + + + + + + + +
←————————————————————————→
      -2        3        6
```

The solution set is $(6, \infty) \cup [-2, 3]$.

81. The sign graph of $\dfrac{-5}{(x+2)(x-3)} > 0$ is

```
- - - - - - - - - - - - - - - - - - - - -
- - - - - - - - - 0 + + + + + +
- - - - - - - 0 + + + + + + + + + +
←————————————————————————→
      -2        3
```

The solution set is $(-2, 3)$.

83. The sign graph of $\dfrac{(x-4)(x+2)}{5-x} > 0$ is

```
+ + + + + + + + + + + + + + 0 - - -
- - - - - - - - - 0 + + + + + +
- - - - - 0 + + + + + + + + +
←————————————————————————→
      -2        4        5
```

The solution set is $(-\infty, -2) \cup (4, 5)$.

85. Let $R(x) = \dfrac{(x-3)(x+1)}{x-5} \ge 0$.

Since $R(-2) < 0, R(0) > 0, R(4) < 0$,
and $R(7) > 0$, we obtain

```
- 0 + 0 - U +
←————————————————————→
-2  -1  0  3  4  5  7
```

The solution set is $[-1, 3] \cup (5, \infty)$.

87. Let $f(x) = \dfrac{(x-\sqrt{7})(x+\sqrt{7})}{(\sqrt{2}-x)(\sqrt{2}+x)}$.

If $x = 3$, then $f(3) < 0$.
If $x = 2$, then $f(2) > 0$.
If $x = 0$, then $f(0) < 0$.
If $x = -2$, then $f(-2) > 0$.
If $x = -3$, then $f(-3) < 0$.

```
- 0 + U - U + 0 -
←————————————————————→
 -√7   -√2  √2   √7
```

The solution set is

$$(-\infty, -\sqrt{7}] \cup (-\sqrt{2}, \sqrt{2}) \cup [\sqrt{7}, \infty).$$

89. Let $R(x) = \dfrac{(x+1)^2}{(x-5)(x+3)} \ge 0$.

Since $R(-4) > 0, R(-2) < 0, R(0) < 0$,
and $R(6) > 0$, we get

```
+ U - 0 - U +
←————————————————————→
-4  -3  -2  -1  0  5  6
```

The solution set is $(-\infty, -3) \cup \{-1\} \cup (5, \infty)$.

91. Since the sign graph of $\dfrac{w-1}{w^2} > 0$ is

```
- - - - - - - - 0 + + + +
+ + + + 0 + + + + + + + +
←————————————————————→
        0        1
```

then the solution set is the interval $(1, \infty)$.

93. Since the sign graph of $\dfrac{w^2 - 4w + 5}{w-3} > 0$ is

```
- - - - - 0 + + + + + + + +
←————————————————————→
          3
```

the solution is the interval $(3, \infty)$.

95. Average cost is $C = \dfrac{100 + x}{x}$. If she visits the zoo 100 times, then her average cost per visit is $C = \dfrac{200}{100} = \$2$. Since $C \to 1$ as $x \to \infty$, over a long period her average cost per visit is $\$1$.

97. Since the second half is 100 miles long and it must be completed in $4 - x$ hours, the average speed for the second half is $S(x) = \dfrac{100}{4 - x}$. The asymptote $x = 4$ implies the average speed in the second half goes to ∞ as the completion time in the first half shortens to four hours.

99. a) About 15 minutes

   b) About 200 ppm

   c) The horizontal asymptote is PPM $= 0$ and the vertical asymptote is $t = 0$. In simple terms, a long exposure to low level carbon monoxide still leads to permanent brain damage, and a very high level carbon monoxide can lead to permanent brain damage in a few minutes.

101.

   (a) $h = \dfrac{500}{\pi r^2}$ since $500 = \pi r^2 h$

   (b) $S = 2\pi r^2 + 2\pi r h = 2\pi r^2 + 2\pi r \dfrac{500}{\pi r^2}$

or equivalently $S = 2\pi r^2 + \dfrac{1000}{r}$

   (c) As can be seen from the graph of $S = 2\pi r^2 + \dfrac{1000}{r}$, S is minimized when $r \approx 4.3$ feet.

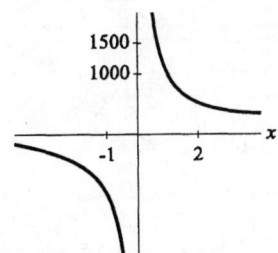

   (d) If it costs \$8 per ft^2 to construct the tank in part (c), then the cost of the tank is given by

$$8\left(2\pi(4.3)^2 + \frac{1000}{4.3}\right) \approx \$2789.87.$$

Chapter 3 Review Exercises

1. By using the method of completing the square, we obtain

$$\begin{aligned} f(x) &= 3\left(x^2 - \frac{2}{3}x + \frac{1}{9}\right) - \frac{1}{3} + 1 \\ &= 3\left(x - \frac{1}{3}\right)^2 + \frac{2}{3} \end{aligned}$$

3. Since $y = 2(x^2 - 2x + 1) - 2 - 1 = 2(x - 1)^2 - 3$, the vertex is $(1, -3)$ and axis of symmetry is $x = 1$. Setting $y = 0$, we get $x - 1 = \pm\dfrac{\sqrt{6}}{2}$ and the x-intercepts are $\left(\dfrac{2 \pm \sqrt{6}}{2}, 0\right)$. The y-intercept is $(0, -1)$.

5. From the x-intercepts, $y = a(x + 1)(x - 3)$. Substitute $(0, 6)$, so $6 = a(-3)$ or $a = -2$. The equation of the parabola is $y = -2(x^2 - 2x - 3)$ or $y = -2x^2 + 4x + 6$.

7. $-1 - i$

9. $16 - 40i + 25i^2 = 16 - 25 - 40i = -9 - 40i$

11. $2 + 6i - 6i - 18i^2 = 20$

13.
$$\frac{2}{i} - \frac{3i}{i} = \frac{2(-i)}{i(-i)} - 3 = \frac{-2i}{1} - 3 = -3 - 2i$$

15.
$$\frac{1-i}{2+i} \cdot \frac{2-i}{2-i} = \frac{1-3i}{5} = \frac{1}{5} - i\frac{3}{5}$$

17.
$$\frac{1+i}{2-3i} \cdot \frac{2+3i}{2+3i} = \frac{-1+5i}{4+9} = -\frac{1}{13} + i\frac{5}{13}$$

19.
$$\frac{6+2i\sqrt{2}}{2} = 3 + i\sqrt{2}$$

21.
$$\frac{-6+\sqrt{-20}}{-8} = \frac{-6+2i\sqrt{5}}{-8} = \frac{3}{4} - i\frac{\sqrt{5}}{4}$$

23. $(i^4)^8 \cdot i^2 + (i^4)^4 \cdot i^3 = 1 \cdot (-1) + 1 \cdot (-i) = -1 - i$

25. $1/3$ **27.** $\pm 2\sqrt{2}$

29. Factoring, we obtain
$$m(x) = (2x-1)(4x^2 + 2x + 1).$$

By using the quadratic formula, we obtain the zeros to the second factor. Namely,
$$x = \frac{-2 \pm \sqrt{-12}}{8} = \frac{-2 \pm 2i\sqrt{3}}{8}.$$

The zeros are $\dfrac{1}{2}, \dfrac{-1 \pm i\sqrt{3}}{4}$.

31. Since $P(t) = (t^2 - 10)(t^2 + 10)$, the zeros are $\pm\sqrt{10}, \pm i\sqrt{10}$

33. Factoring, we obtain
$$R(s) = 4s^2(2s-1) - (2s-1) = (4s^2-1)(2s-1).$$
The zeros are $\dfrac{1}{2}$ (with multiplicity 2) and $-\dfrac{1}{2}$.

35. Find the roots of the second factor of $f(x) = x(x^2 + 2x - 6)$. Since $x^2 + 2x - 6 = (x^2 + 2x + 1) - 6 - 1 = (x+1)^2 - 7$, the zeros are $0, -1 \pm \sqrt{7}$.

37. $P(3) = 108 - 27 + 3 - 1 = 83$.
By synthetic division, we get

```
3 |  4   -3    1    -1
  |      12   27    84
  ----------------------
     4    9   28    83
```

Remainder is $P(3) = 83$.

39. $P(-1/2) = \dfrac{1}{4} - \dfrac{1}{4} + 3 + 2 = 5$.

By synthetic division, we obtain

```
-1/2 | -8    0    2    0   -6    2
     |       4   -2    0    0    3
     ------------------------------
       -8    4    0    0   -6    5
```

So remainder is $P(-1/2) = 5$.

41. $\pm\left\{1, \dfrac{1}{3}, 2, \dfrac{2}{3}\right\}$

43. $\pm\left\{1, \dfrac{1}{2}, \dfrac{1}{3}, \dfrac{1}{6}, 3, \dfrac{3}{2}\right\}$

45.
$$\begin{aligned}
2\left(x + \frac{1}{2}\right)(x-3) &= (2x+1)(x-3) \\
&= 2x^2 - 5x - 3
\end{aligned}$$

An equation is $2x^2 - 5x - 3 = 0$.

47. $(x - (3-2i))(x - (3+2i)) =$
$$\begin{aligned}
&= ((x-3) + 2i)((x-3) - 2i) \\
&= (x-3)^2 + 4 \\
&= x^2 - 6x + 13
\end{aligned}$$

An equation is $x^2 - 6x + 13 = 0$.

49. $(x-2)(x - (1-2i))(x - (1+2i)) =$
$$\begin{aligned}
&= (x-2)((x-1) + 2i)((x-1) - 2i) \\
&= (x-2)\left((x-1)^2 + 4\right) \\
&= (x-2)(x^2 - 2x + 5) \\
&= x^3 - 4x^2 + 9x - 10
\end{aligned}$$

Thus, an equation is $x^3 - 4x^2 + 9x - 10 = 0$.

51. $\left(x - (2-\sqrt{3})\right)\left(x - (2+\sqrt{3})\right) =$
$$\begin{aligned}
&= \left((x-2) + \sqrt{3}\right)\left((x-2) - \sqrt{3}\right) \\
&= \left((x-2)^2 - 3\right) \\
&= x^2 - 4x + 1
\end{aligned}$$

Thus, an equation is $x^2 - 4x + 1 = 0$.

53. $P(x) = P(-x) = x^8 + x^6 + 2x^2$ has no sign variation. There are 6 imaginary roots and 0 has multiplicity 2.

55. $P(x) = 4x^3 - 3x^2 + 2x - 9$ has 3 sign variations and $P(-x) = -4x^3 - 3x^2 - 2x - 9$ has no sign variation. There are
(a) 3 positive roots, or
(b) 1 positive & 2 imaginary roots.

57. $P(x) = x^3 + 2x^2 + 2x + 1$ has no sign variation and $P(-x) = -x^3 + 2x^2 - 2x + 1$ has 3 sign variations. There are
(a) 3 negative roots, or
(b) 1 negative root & 2 imaginary roots.

59. Best integral bounds: $-4 < x < 3$. One checks that $1, 2$ are not upper bounds and that 3 is a bound.

3	6	5	-50
		18	69
	6	23	19

One checks that $-1, -2, -3$ are not lower bounds and that -4 is a bound since

-4	6	5	-50
		-24	76
	6	-19	26

61. Best integral bounds: $-1 < x < 8$. One checks that $1, 2, 3, 4, 5, 6, 7$ are not upper bounds and that 8 is a bound.

8	2	-15	31	-12
		16	8	312
	2	1	39	300

-1 is a lower bound since

-1	2	-15	31	-12
		-2	17	-48
	2	-17	48	-60

63. Best integral bounds: $-1 < x < 1$

1	12	-4	-3	1
		12	8	5
	12	8	5	6

-1	12	-4	-3	1
		-12	16	-13
	12	-16	13	-12

65. Roots are $1, 2, 3$ since

1	1	-6	11	-6
		1	-5	6
	1	-5	6	0

and $x^2 - 5x + 6 = (x - 3)(x - 2)$.

67. Roots are $\pm i, 1/3, 1/2$ since

1/2	6	-5	7	-5	1
		3	-1	3	-1
	6	-2	6	-2	0

1/3	6	-2	6	-2
		2	0	2
	6	0	6	0

and the zeros of $6x^2 + 6 = 6(x^2 + 1) = 0$ are $\pm i$.

69. Roots are $3, 3 \pm i$ since

3	1	-9	28	-30
		3	-18	30
	1	-6	10	0

and the zeros of $x^2 - 6x + 10 = (x - 3)^2 + 1 = 0$ are $3 \pm i$.

71. Roots are $2, 1 \pm i\sqrt{2}$ since

2	1	-4	7	-6
		2	-4	6
	1	-2	3	0

and the zeros of $x^2 - 2x + 3 = (x - 1)^2 + 2 = 0$ are $1 \pm i\sqrt{2}$.

73. Apply synthetic division to the second factor in $x(2x^3 - 5x^2 - 2x + 2) = 0$.

```
1/2 | 2    -5    -2    2
    |       1    -2   -2
    -----------------------
      2    -4    -4    0
```

By completing the square, the zeros of
$2x^2 - 4x - 4 = 2(x^2 - 2x) - 4 = 2(x - 1)^2 - 6 = 0$
are $1 \pm \sqrt{3}$. All the roots are $x = 0, 1/2, 1 \pm \sqrt{3}$.

75. Solve an equivalent statement assuming $3v \geq 0$.

$$2v - 1 = 3v \quad \text{or} \quad 2v - 1 = -3v$$
$$-1 = v \quad \text{or} \quad 5v = 1$$

Since $3v \geq 0$, $v = -1$ is an extraneous root.
The solution set is $\{1/5\}$.

77. Let $w = x^2$ and $w^2 = x^4$.

$$w^2 + 7w = 18$$
$$(w + 9)(w - 2) = 0$$
$$w = -9, 2$$
$$x^2 = -9 \quad \text{or} \quad x^2 = 2.$$

Since $x^2 = -9$ has no real solution, the solution set is $\{\pm\sqrt{2}\}$.

79. Isolate a radical and square both sides.

$$\sqrt{x + 6} = \sqrt{x - 5} + 1$$
$$x + 6 = x - 5 + 2\sqrt{x - 5} + 1$$
$$5 = \sqrt{x - 5}$$
$$25 = x - 5$$

The solution set is $\{30\}$.

81. Let $w = \sqrt[4]{y}$ and $w^2 = \sqrt{y}$.

$$w^2 + w - 6 = 0$$
$$(w + 3)(w - 2) = 0$$
$$w = -3 \quad \text{or} \quad w = 2$$
$$y^{1/4} = -3 \quad \text{or} \quad y^{1/4} = 2$$

Since $y^{1/4} = -3$ has no real solution, the solution set is $\{16\}$.

83. Let $w = x^2$ and $w^2 = x^4$.

$$w^2 - 3w - 4 = 0$$
$$(w + 1)(w - 4) = 0$$
$$x^2 = -1 \quad \text{or} \quad x^2 = 4$$

Since $x^2 = -1$ has no real solution, the solution set is $\{\pm 2\}$.

85. Raise to the power $3/2$ and get
$x - 1 = \pm(4^{1/2})^3$. So, $x = 1 \pm 8$.
The solution set is $\{-7, 9\}$.

87. No solution since $(x + 3)^{-3/4}$ is nonnegative.

89. Since $3x - 7 = 4 - x$, we obtain $4x = 11$.
The solution set is $\{11/4\}$.

91. Symmetric about $x = 3/4$ since $\dfrac{-b}{2a} = \dfrac{3}{4}$.

93. Symmetric about y-axis since $f(-x) = f(x)$.

95. Symmetric about the origin for
$f(-x) = -f(x)$.

97. $(-\infty, -5/2) \cup (-5/2, \infty)$

99. $(-\infty, \infty)$

101. Since $f(x) = (x - 2)(x + 1)$, the x-intercepts
are $(2, 0), (-1, 0)$, y-intercept is $(0, -2)$.
Since $\dfrac{-b}{2a} = \dfrac{1}{2}$, the vertex is $(1/2, -9/4)$.

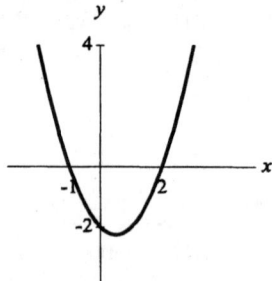

103. Use synthetic division on $f(x) = x^3 - 3x - 2$.

```
-1 | 1    0    -3    -2
   |     -1     1     2
   -----------------------
     1   -1    -2     0
```

Since $x^2 - x - 2 = (x - 2)(x + 1)$,
$f(x) = (x + 1)^2(x - 2)$. Graph crosses $(2, 0)$

but does not cross $(-1,0)$. y-intercept is $(0,-2)$.

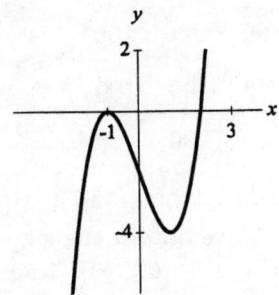

105. Use synthetic division on
$$f(x) = \frac{1}{2}x^3 - \frac{1}{2}x^2 - 2x + 2.$$

$$
\begin{array}{r|rrrr}
1 & 1/2 & -1/2 & -2 & 2 \\
  &   & 1/2 & 0 & -2 \\
\hline
  & 1/2 & 0   & -2 & 0
\end{array}
$$

Since the roots of $\frac{1}{2}x^2 - 2 = 0$ are ± 2,
the graph crosses x-intercepts $(\pm 2, 0), (1, 0)$,
y-intercept is $(0, 2)$.

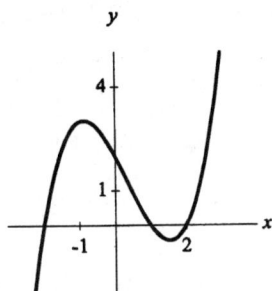

107. Factoring, we get $f(x) = \frac{1}{4}(x^4 - 8x^2 + 16) = \frac{1}{4}(x^2 - 4)^2$. The graph does not cross through x-intercepts $(\pm 2, 0)$, y-intercept is $(0, 4)$.

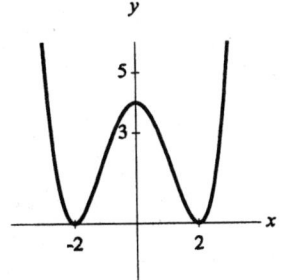

109. $f(x) = \dfrac{2}{x+3}$ has no x-intercept, y-intercept is $(0, 2/3)$, asymptotes are $x = -3$, $y = 0$

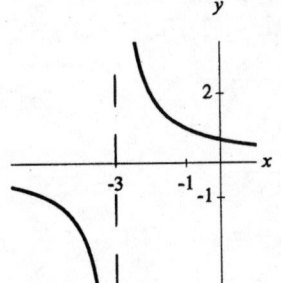

111. $f(x) = \dfrac{2x}{x^2 - 4}$ has x-intercept $(0,0)$,
asymptotes are $x = \pm 2$, $y = 0$.
Symmetric about the origin.
Graph goes through $(1, -2/3)$, $(3, 6/5)$, and $(-3, -6/5)$.

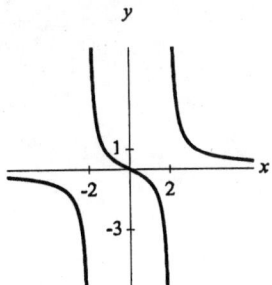

113. $f(x) = \dfrac{(x-1)^2}{x-2}$ has x-intercept $(1,0)$,
y-intercept is $(0, -1/2)$, asymptote $x = 2$,
and oblique asymptote $y = x$ since

$$
\begin{array}{r|rrr}
2 & 1 & -2 & 1 \\
  &   & 2 & 0 \\
\hline
  & 1 & 0 & 1
\end{array}
$$

and $f(x) = x + \dfrac{1}{x-2}$.

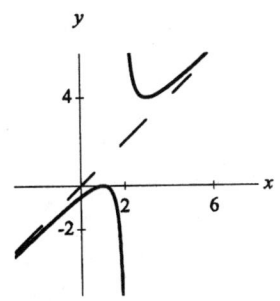

115. $f(x) = \dfrac{2x-1}{2-x}$ has x-intercept $(1/2, 0)$,

y-intercept is $(0, -1/2)$,
asymptotes are $x = 2$ and $y = -2$,
graph goes through $(1,1), (3,-5)$.

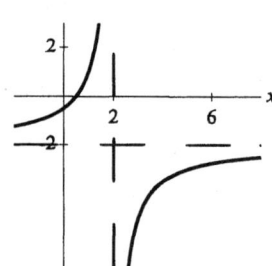

117. Since $f(x) = \dfrac{x^2-4}{x-2} = x + 2$ if $x \neq 2$,

x-intercept is $(-2, 0)$, y-intercept is $(0, 2)$,
no asymptotes

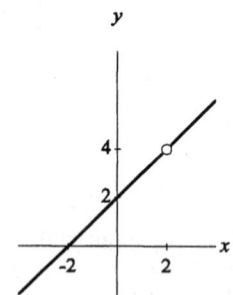

119. Set the right side to zero to obtain

$$8x^2 - 6x + 1 < 0.$$

Factoring, we get $(4x-1)(2x-1) < 0$.
The sign graph of which is given below.

```
- - - - - - - - - 0 + + + +
- - - - 0 + + + + + + + + +
←————————————————————————→
       1/4        1/2
```

The solution set is the interval $(1/4, 1/2)$.

121. The sign graph of $(3-x)(x+5) \geq 0$ is shown below.

```
+ + + + + + + + 0 - - - -
- - - - 0 + + + + + + + +
←————————————————————————→
        -5        3
```

The solution set is the interval $[-5, 3]$.

123. Using the Rational Zero Theorem and synthetic division, we obtain

```
1/2 | 4    -400    -1    100
    |        2   -199  -100
    ────────────────────────
      4    -398   -200     0
```

Note, $4x^2 - 398x - 200 = (4x+2)(x-100)$.
Using the zeros, we obtain the sign graph of $4x^3 - 400x^2 - x + 100 \geq 0$. Namely,

```
- - - - - - - - - - - - - - 0 + + +
- - - - - - - - - - 0 + + + + + +
- - - - - 0 + + + + + + + + + +
←————————————————————————————————→
      -1/2       1/2      100
```

The solution set is $\left[-\dfrac{1}{2}, \dfrac{1}{2}\right] \cup [100, \infty)$.

125. Let $R(x) = \dfrac{x+10}{x+2} - 5 = \dfrac{-4x}{x+2} < 0$.

Using the test-point method, we find $R(-3) < 0, R(-1) > 0$, and $R(1) < 0$.

```
    -   U   +   0   -
←——————————————————————→
   -3  -2  -1   0   1
```

The solution is $(-\infty, -2) \cup (0, \infty)$.

127. $R(x) = \dfrac{12-7x}{x^2} + 1 = \dfrac{(x-3)(x-4)}{x^2} > 0$.

Using the test-point method, we get $R(-1) > 0, R(1) > 0, R(3.5) < 0$, and $R(5) > 0$.

```
    +   U   +   0   -   0   +
←——————————————————————————————→
   -1   0   1   3  3.5  4   5
```

the solution is $(-\infty, 0) \cup (0, 3) \cup (4, \infty)$.

129. Let $R(x) = \dfrac{(x-1)(x-2)}{(x-3)(x-4)}$. We will use

the test-point method. Note, $R(0) > 0$, $R(1.5) < 0, R(2.5) > 0, R(3.5) < 0$, and $R(5) > 0$. Then we get the following.

$$
\begin{array}{ccccccccc}
+ & 0 & - & 0 & + & U & - & U & + \\
\end{array}
$$

$$
\begin{array}{ccccccccc}
0 & 1 & 1.5 & 2 & 2.5 & 3 & 3.5 & 4 & 5
\end{array}
$$

The solution is $(-\infty, 1] \cup [2, 3) \cup (4, \infty)$.

131. Quotient $x^2 - 3x$, remainder -15

$$
\begin{array}{r|rrrr}
3 & 1 & -6 & 9 & -15 \\
  &   & 3 & -9 & 0 \\
\hline
  & 1 & -3 & 0 & -15
\end{array}
$$

133. Since $\dfrac{-b}{2a} = \dfrac{-156}{-32} = 4.875$, the maximum

height is $-16(4.875)^2 + 156(4.875) = 380.25$ ft.

135. Let w be the height and let l be the length of the given cross-section of the room. Since the ratios of corresponding sides of similar triangles are equal, we obtain

$$
\begin{aligned}
\frac{7}{12} &= \frac{w}{(48 - l)/2} \\
w &= \frac{7}{24}(48 - l)
\end{aligned}
$$

and the area, A, of the cross-section of the room is $A = wl = \dfrac{7}{24}(48 - l)l$. Since $l = 24$ maximizes A, we obtain $w = \dfrac{7}{24}(48 - 24) = 7$ feet. The dimensions that will maximize the area are 24 ft by 7 ft.

137.

(a) $V(10) = \dfrac{10,000}{58} \approx 172.4$ feet per second

(b) $V = 200$

(c) 200 feet per second

Chapter 3 Test

1. Use the method of completing the square.

$$
y = 3(x^2 - 4x + 4) + 1 - 12 = 3(x - 2)^2 - 11
$$

2. $y = 3(x - 2)^2 - 11$ has vertex $(2, -11)$, axis of symmetry $x = 2$, y-intercept $(0, 1)$,

the x-intercepts are $\left(\dfrac{6 \pm \sqrt{33}}{3}, 0\right)$,

range $[-11, \infty)$

3. Minimum value is -11 since vertex is $(2, -11)$

4. Quotient $2x^2 - 6x + 14$, remainder -37

$$
\begin{array}{r|rrrr}
-3 & 2 & 0 & -4 & 5 \\
   &   & -6 & 18 & -42 \\
\hline
   & 2 & -6 & 14 & -37
\end{array}
$$

5. By the Remainder Theorem, remainder is -14.

6. $\pm\left\{1, \dfrac{1}{3}, 2, \dfrac{2}{3}, 3, 6\right\}$

7.

$$
\begin{aligned}
(x + 3)(x - 4i)(x + 4i) &= (x + 3)(x^2 + 16) \\
&= x^3 + 3x^2 + 16x + 48
\end{aligned}
$$

An equation is $x^3 + 3x^2 + 16x + 48 = 0$.

8. $P(x) = x^3 - 3x^2 + 5x + 7$ has 2 sign variations and $P(-x) = -x^3 - 3x^2 - 5x + 7$ has 1 sign variation. There are
(a) 2 positive roots & 1 negative root, or
(b) 1 negative root & 2 imaginary roots.

9. Since $\dfrac{-b}{2a} = \dfrac{-128}{-32} = 4$,

the maximum height is $S(4) = 256$ feet.

10. $16 - 24i + 9i^2 = 7 - 24i$

11.

$$
\frac{(2 - i)(3 - i)}{(3 + i)(3 - i)} = \frac{6 - 5i + i^2}{10} = \frac{5 - 5i}{10} = \frac{1}{2} - \frac{1}{2}i
$$

12. $i^4 i^2 - (i^4)^8 i^3 = 1(-1) - (1)(-i) = -1 + i$

13. $2i\sqrt{2}(i\sqrt{2} + \sqrt{6}) = -4 + 2i\sqrt{12} = -4 + 4i\sqrt{3}$

14. ± 3 **15.** $\pm 2, \pm 2i$

16.

$$
\begin{array}{r|rrrr}
2 & 1 & -4 & -1 & 10 \\
  &   & 2 & -4 & -10 \\
\hline
  & 1 & -2 & -5 & 0
\end{array}
$$

Since $x^2 - 2x - 5 = (x-1)^2 - 6$, the zeros are $2, 1 \pm \sqrt{6}$.

17. Zeros are $x = \pm i$, each with multiplicity 2 since $f(x) = (x^2+1)^2 = (x-i)^2(x+i)^2$

18. The zeros are 2, 0 with multiplicity 2, and $-3/2$ with mulitiplicity 3 since the equation can be written as $x^2(x-2)(2x+3)^3 = 0$.

19.

$$
\begin{array}{r|rrrr}
1/2 & 2 & -9 & 14 & -5 \\
    &   & 1 & -4 & 5 \\
\hline
    & 2 & -8 & 10 & 0
\end{array}
$$

By completing the square, we find that the roots of $2(x^2 - 4x) + 10 = 2(x-2)^2 + 2 = 0$ are $2 \pm i$. The zeros are $x = 2 \pm i, 1/2$.

20. Parabola $y = 2(x-3)^2 + 1$ with vertex $(3,1)$

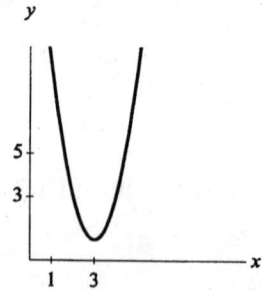

21. $y = (x-2)^2(x+1)$ crosses $(-1, 0)$ but does not cross $(2, 0)$, y-intercept is $(0, 4)$

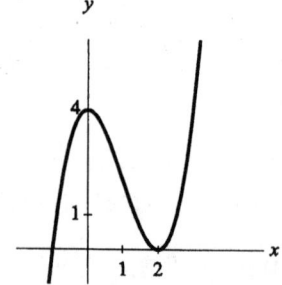

22. $y = x(x-2)(x+2)$ crosses $(0,0), (\pm 2, 0)$, and goes through $(1, -3)$

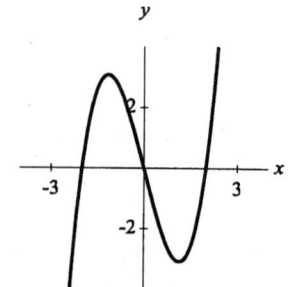

23. $y = \dfrac{1}{x-2}$ has asymptotes $x = 2, y = 0$, and goes through $(1, -1), (3, 1), (4, 1/2)$

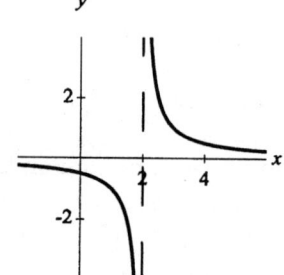

24. $y = \dfrac{2x - 3}{x - 2}$ has asymptotes $x = 2, y = 2$, x-intercept $(3/2, 0)$, y-intercept $(0, 3/2)$

25. $f(x) = x + \dfrac{1}{x}$ has oblique asymptote $y = x$, asymptote $x = 0$, goes through $(1, 2), (-1, -2)$

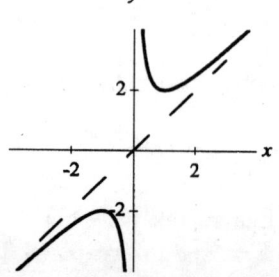

26. $y = \dfrac{4}{x^2 - 4}$ has asymptotes $x = \pm 2, y = 0$,

y-intercept $(0, -1)$

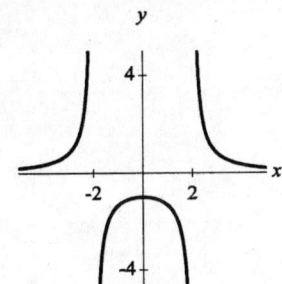

27. Note, $\dfrac{x^2 - 2x + 1}{x - 1} = x - 1$ provided $x \neq 1$;

no x-intercept, y-intercept $(0, -1)$

28. The sign graph of $(x-4)(x+2) < 0$ is

```
- - - - - - - - - 0 + + + +
- - - - 0 + + + + + + + +
←—————————————————→
     -2        4
```

The solution set is the interval $(-2, 4)$.

29. Set the right-hand side to zero to

obtain $\dfrac{2x-1}{x-3} > 0$. The sign chart is

```
- - - - - - - - - U + + + +
- - - - 0 + + + + + + + +
←—————————————————→
    1/2        3
```

The solution set is $(-\infty, 1/2) \cup (3, \infty)$.

30. The sign chart of $\dfrac{x+3}{(4-x)(x+1)} \geq 0$ is

```
+++++++++++++++++U— — —
——————————————U+++++++
—————0 + + + + + + + +
←—————————————————→
  -3      -1      4
```

The solution set is $(-\infty, -3] \cup (-1, 4)$.

31. The sign chart of $x(x^2 - 7) > 0$ is

```
- - - - - - - - - - - - 0+++
- - - - - - - - - - 0 ++++++
- - - - - 0 +++++++++++
←—————————————————→
   -√7     0     √7
```

The solution set is $(-\sqrt{7}, 0) \cup (\sqrt{7}, \infty)$.

32. If we raise each side of $(x-3)^{-2/3} = \dfrac{1}{3}$ to

the power -3, then we obtain $(x-3)^2 = 27$.

Thus, $x = 3 \pm \sqrt{27}$ and the solution set is $\left\{ 3 \pm 3\sqrt{3} \right\}$.

33. Isolate a radical and square each side.

$$\begin{aligned} \sqrt{x} - 1 &= \sqrt{x-7} \\ x - 2\sqrt{x} + 1 &= x - 7 \\ -2\sqrt{x} &= -8 \\ 4x &= 64 \end{aligned}$$

The solution set is $\{16\}$.

Tying It All Together

1. Since $2x - 3 = 0$ is equivalent to $2x = 3$, the solution is $x = \dfrac{3}{2}$.

2. Factoring, we obtain $2x^2 - 3x = x(2x - 3) = 0$.

Thus, the solutions are $x = 0, \dfrac{3}{2}$.

3. Using synthetic division, we obtain

```
-1 | 1    3    3    1
   |     -1   -2   -1
   —————————————————
     1    2    1    0
```

Since $x^2 + 2x + 1 = (x+1)^2$, we find
$x^3 + 3x^2 + 3x + 1 = (x+1)^3$.
Thus, the only solution is $x = -1$.

4. Multiply both sides of the equation of $\dfrac{x-1}{x-2} = 0$
   by $x - 2$. Then we obtain $x - 1 = 0$.
   The solution is $x = 1$.

5. If we raise each side of $x^{2/3} = 9$ to
   the power 3, then we obtain $x^2 = 729$.
   Thus, $x = \pm\sqrt{729}$ and the solution set is
   $\{\pm 27\}$.

6. The graph is a straight line with domain
   $(-\infty, \infty)$, range $(-\infty, \infty)$, x-intercept $\left(\dfrac{3}{2}, 0\right)$
   (see answer to number 1), and from the graph
   we obtain that the function is increasing
   in $(-\infty, \infty)$

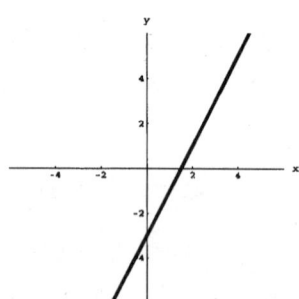

7. The graph is a parabola with domain $(-\infty, \infty)$,
   range $\left[-\dfrac{9}{8}, \infty\right)$, x-intercepts are $(0,0)$ and
   $\left(\dfrac{3}{2}, 0\right)$ (see answer to number 2), and the
   function is increasing in $\left(\dfrac{3}{4}, \infty\right)$,
   and decreasing in $\left(-\infty, \dfrac{3}{4}\right)$

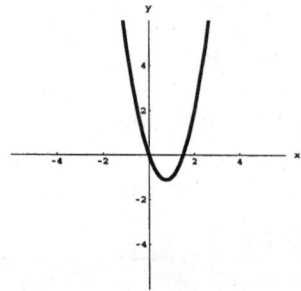

8. domain $(-\infty, \infty)$, range $(-\infty, \infty)$, the
   x-intercept is $(-1, 0)$ (see answer to
   number 3), and the function is increasing
   in $(-\infty, \infty)$

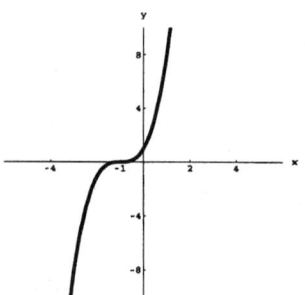

9. domain $(-\infty, 2) \cup (2, \infty)$,
   range $(-\infty, 1) \cup (1, \infty)$, the x-intercept is $(1, 0)$
   (see answer to number 4), and the function is
   decreasing in $(-\infty, 2) \cup (2, \infty)$

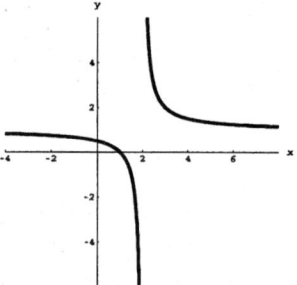

10. domain $(-\infty, \infty)$, range $[-9, \infty)$,
    the x-intercepts are $(27, 0)$ and $(-27, 0)$
    (see answer to number 5),
    decreasing in $(-\infty, 0)$, increasing in $(0, \infty)$

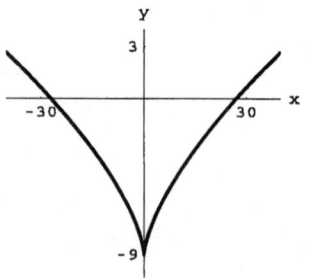

11. Based on the portion of the graph of $y = 2x - 3$ below the x-axis, the solution set of $2x - 3 < 0$ is the interval $\left(-\infty, \dfrac{3}{2}\right)$.

12. By taking the part of the graph of $y = 2x^2 - 3x$ below the x-axis, the solution set of $2x^2 - 3x < 0$ is the interval $\left(0, \dfrac{3}{2}\right)$.

13. By taking the part of the graph of $y = x^3 + 3x^2 + 3x + 1$ above the x-axis, the solution set of $x^3 + 3x^2 + 3x + 1 > 0$ is the interval $(-1, \infty)$.

14. By taking the part of the graph of $y = \dfrac{x-1}{x-2}$ above the x-axis and by considering the x-intercepts, the solution set of $\dfrac{x-1}{x-2} > 0$ is the $(-\infty, 1) \cup (2, \infty)$.

15. By taking the part of the graph of $y = x^{2/3} - 9$ below the x-axis and by conidering the x-intercepts, the solution set of $x^{2/3} - 9 < 0$ is the interval $(-27, 27)$.

16. Let $y = f(x)$, and interchange x and y. Then $x = 2y - 3$. Solving for y, we obtain $y = \dfrac{x+3}{2}$. The inverse is $f^{-1}(x) = \dfrac{x+3}{2}$.

17. Since the graph of $f(x) = 2x^2 - 3x$ does not satisfy the Horizontal Line Test, the function is not one-to-one. Thus, the inverse does not exist.

18. From the solution of number 3, we see that $f(x) = (x+1)^3$. Let $y = f(x)$, and interchange x and y. Then $x = (y+1)^3$. Taking the cube root of both sides, we get $\sqrt[3]{x} = y + 1$. Thus, the inverse is $f^{-1}(x) = \sqrt[3]{x} - 1$.

19. Let $y = f(x)$, and interchange x and y. Then $x = \dfrac{y-1}{y-2}$. Solving for y, we obtain

$$
\begin{aligned}
x(y-2) &= y - 1 \\
xy - 2x &= y - 1 \\
xy - y &= 2x - 1 \\
y(x-1) &= 2x - 1
\end{aligned}
$$

$$ y = \frac{2x-1}{x-1}. $$

The inverse is $f^{-1}(x) = \dfrac{2x-1}{x-1}$.

20. Since the graph of $f(x) = x^{2/3} - 9$ does not satisfy the Horizontal Line Test, the function is not one-to-one. Thus, the inverse does not exist.

For Thought

1. False, the base of an exponential function is positive.   **2.**   True

3. False, since there is no solution.

4. True   **5.**   True   **6.**   True   **7.**   True

8. False, since it is decreasing.

9. True, since $0.25 = 4^{-1}$.

10. True, since $\sqrt[100]{2^{173}} = \left(2^{173}\right)^{1/100}$.

4.1 Exercises

1. 27   **3.** $-(2^0) = -1$

5. $\dfrac{1}{2^3} = \dfrac{1}{8}$ **7.** $\left(\dfrac{2}{1}\right)^4 = 2^4 = 16$

9. $\left(8^{1/3}\right)^2 = 2^2 = 4$

11. $-\left(9^{1/2}\right)^{-3} = -(3)^{-3} = -\dfrac{1}{3^3} = -\dfrac{1}{27}$

13. $3^2 = 9$ **15.** $3^{-2} = 1/9$

17. $2^{-1} = 1/2$ **19.** $2^3 = 8$

21. $(1/4)^{-1} = 4$

23. $4^{1/2} = 2$

25. $f(x) = 5^x$ goes through $(-1, 1/5), (0, 1), (1, 5)$, domain is $(-\infty, \infty)$, range is $(0, \infty)$, increasing

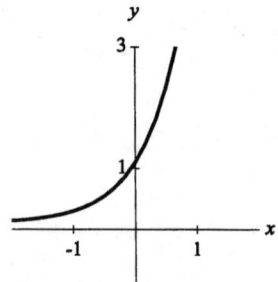

27. $f(x) = 10^{-x}$ goes through $(-1, 10), (0, 1)$, $(1, 1/10)$, domain is $(-\infty, \infty)$, range is $(0, \infty)$, decreasing

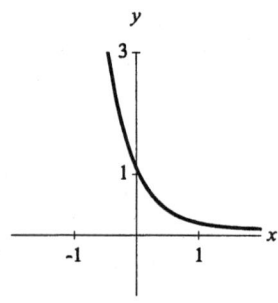

29. $f(x) = (1/4)^x$ goes through $(-1, 4), (0, 1)$, $(1, 1/4)$, domain is $(-\infty, \infty)$, range is $(0, \infty)$, decreasing

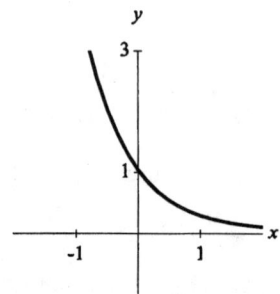

31. Shift $y = 2^x$ down by 3 units; $f(x) = 2^x - 3$ goes through $(-1, -2.5), (0, -2), (2, 1)$, domain $(-\infty, \infty)$, range $(-3, \infty)$, asymptote $y = -3$, increasing

33. Shift $y = 2^x$ to left by 3 units and down by 5 units; $f(x) = 2^{x+3} - 5$ goes through $(-4, -4.5), (-3, -4), (0, 3)$, domain $(-\infty, \infty)$, range $(-5, \infty)$, asymptote $y = -5$, increasing

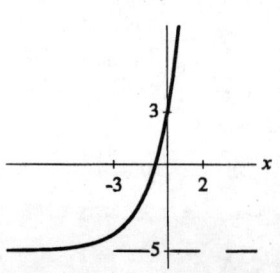

35. Reflect $y = 2^{-x}$ about x-axis, $f(x) = -2^{-x}$
goes through $(-1, -2), (0, -1), (1, -1/2)$,
domain $(-\infty, \infty)$, range $(-\infty, 0)$,
asymptote $y = 0$, increasing

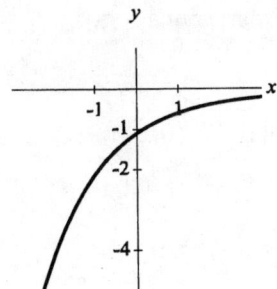

37. Reflect $y = 2^x$ about x-axis and shift
up by 1 unit, $f(x) = 1 - 2^x$ goes through
$(-1, 0.5), (0, 0), (1, -1)$, domain $(-\infty, \infty)$,
range $(-\infty, 1)$, asymptote $y = 1$, decreasing

39. Shift $y = 3^x$ to right by 2 and shrink by a
factor of 0.5, $f(x) = 0.5 \cdot 3^{x-2}$ goes through
$(0, 1/18), (2, 0.5), (3, 1.5)$, domain $(-\infty, \infty)$,
range $(0, \infty)$, asymptote $y = 0$, increasing

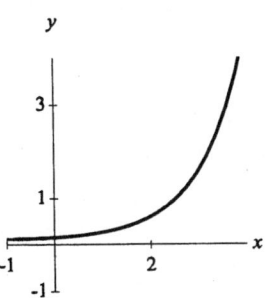

41. Stretch $y = (0.5)^x$ by a factor of 500,
$f(x) = 500 \cdot (0.5)^x$ goes through $(0, 500)$,
$(1, 250)$, domain $(-\infty, \infty)$, range $(0, \infty)$,
asymptote $y = 0$, decreasing

43. $y = 2^{x-5} - 2$

45. $y = -\left(\dfrac{1}{4}\right)^{x-1} - 2$

47. Since $2^x = 2^6$, solution set is $\{6\}$.

49. $\{-1\}$

51. Multiplying the equation by -1, $3^x = 3^3$
and the solution set is $\{3\}$.

53. $\{-2\}$

55. Since $(2^3)^x = 2^{3x} = 2$, $3x = 1$.
The solution set is $\left\{\dfrac{1}{3}\right\}$.

57. $\{-2\}$

59. Since $(2^{-1})^x = 2^{-x} = 2^3$, $-x = 3$.
The solution set is $\{-3\}$.

61. Since $10^{x-1} = 10^{-2}$, $x - 1 = -2$.
The solution set is $\{-1\}$.

63. Since $2^x = 4$, we find $x = 2$.

65. Since $2^x = \dfrac{1}{2}$, we obtain $x = -1$.

67. Since $\left(\dfrac{1}{3}\right)^x = 1$, we get $x = 0$.

69. Since $\left(\dfrac{1}{3}\right)^x = 3^3$, we find $x = -3$.

71. Since $10^x = 1000$, we obtain $x = 3$.

73. Since $10^x = 0.1 = 10^{-1}$, we find $x = -1$.

75. 1   **77.** -1

79. $(2, 9), (1, 3), (-1, 1/3), (-2, 1/9)$

81. $(0, 1), (-2, 25), (-1, 5), (1, 1/5)$

83. $(4, -16), (-2, -1/4), (-1, -1/2), (5, -32)$

85. The amount at the end of 6 years is

$$4500 \left(1 + \frac{0.08}{4}\right)^{24} = \$7,237.97.$$

The interest earned in 6 years is

$$\$7,237.97 - \$4,500 = \$2,737.97.$$

87. Assume there are 365 days in a year and 30 days in a month. The present value is

$$3000 \left(1 + \frac{0.065}{365}\right)^{-(365)(5+120/365)} = \$2121.82.$$

89. At 5% compounded continuously, it will amount to

$$100,000 e^{0.05 \cdot (5+45/365)} = \$129,196.51.$$

Compounded daily it will amount to

$$100,000 \left(1 + \frac{0.05}{365}\right)^{(365)(5+45/365)} = \$129,194.24.$$

91. At time $t = 0$, $A = 200e^0 = 200$ g are present. If $t = 500$, then $A = 200e^{-0.001(500)} \approx 121$ g.

93. **a)** Using a graphing calculator, we find that the equation is $y \approx 330.1(1.0366)^x$ where x is the number of years since 1980.

b) The data and the exponential curve are shown below.

c) In 2005, when $t = 25$, the predicted number of passengers is $330.1(1.0366)^{25} \approx 811$ million.

95. $P = 10 \left(\frac{1}{2}\right)^n$

97. When $t = 31$, the number of damaged O-rings is $n = 644e^{-0.15(31)} \approx 6$.

For Thought

1. True   **2.** False, since $\log_{100}(10) = 1/2$.

3. True   **4.** True

5. False, the domain is $(0, \infty)$.   **6.** True

7. True

8. False, since $\log_a(a) = 1$.

9. True   **10.** True

4.2 Exercises

1. 6   **3.** -4

5. $\dfrac{1}{4}$ **7.** -3

9. Since $2^6 = 64$, $\log_2(64) = 6$.

11. Since $3^{-4} = \dfrac{1}{81}$, $\log_3\left(\dfrac{1}{81}\right) = -4$.

13. Since $16^{1/4} = 2$, $\log_{16}(2) = \dfrac{1}{4}$.

15. Since $\left(\dfrac{1}{5}\right)^{-3} = 125$, $\log_{1/5}(125) = -3$.

17. Since $10^{-1} = 0.1$, $\log(0.1) = -1$

19. Since $10^0 = 1$, $\log(1) = 0$.

21. Since $e^1 = e$, $\ln(e) = 1$.

23. -5

25. $y = \log_3(x)$ goes through $(1/3, -1)$, $(1, 0)$, $(3, 1)$, domain $(0, \infty)$, range $(-\infty, \infty)$

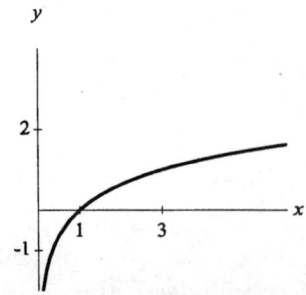

27. $f(x) = \log_5(x)$ goes through $(1/5, -1), (1, 0)$, and $(5, 1)$, domain $(0, \infty)$, range $(-\infty, \infty)$

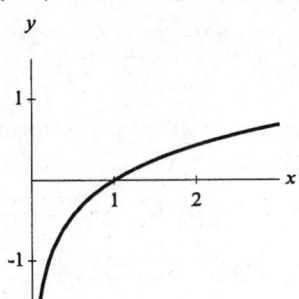

29. $y = \log_{1/2}(x)$ goes through $(2, -1), (1, 0)$, $(1/2, 1)$, domain $(0, \infty)$, range $(-\infty, \infty)$

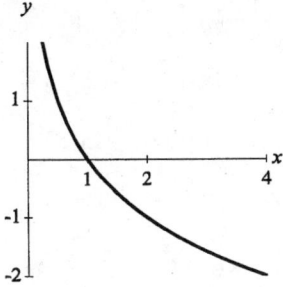

31. $h(x) = \log_{1/5}(x)$ goes through $(5, -1)$, $(1, 0), (1/5, 1)$, domain $(0, \infty)$, range $(-\infty, \infty)$

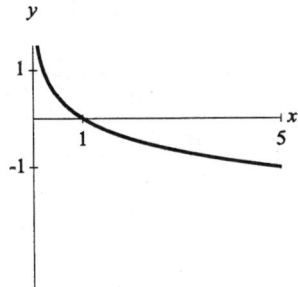

33. $f(x) = \ln(x - 1)$ goes through $\left(1 + \dfrac{1}{e}, -1\right)$, $(2, 0)$, $(1 + e, 1)$, domain $(1, \infty)$, range $(-\infty, \infty)$

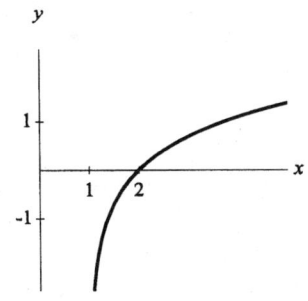

35. $f(x) = -3 + \log(x + 2)$ goes through $(-1.9, -4)$, $(-1, -3)$, $(8, -2)$, domain $(-2, \infty)$, range $(-\infty, \infty)$

37. $f(x) = -\dfrac{1}{2}\log(x - 1)$ goes through $(1.1, 0.5)$, $(2, 0)$, $(11, -0.5)$, domain $(1, \infty)$, range $(-\infty, \infty)$

39. $y = \ln(x - 3) - 4$

41. $y = -\log_2(x - 5) - 1$

43. $2^5 = 32$ **45.** $5^y = x$

47. $10^z = 1000$ **49.** $\log_5(125) = 3$

51. $\ln(y) = 3$ **53.** $\log_a(n) = x - 1$

55. $f^{-1}(x) = \log_2(x)$ **57.** $f^{-1}(x) = 7^x$

59. Replace $f(x)$ by y, interchange x and y, solve for y, and replace y by $f^{-1}(x)$.

$$
\begin{aligned}
y &= \ln(x - 1) \\
x &= \ln(y - 1) \\
e^x &= y - 1 \\
y = f^{-1}(x) &= e^x + 1
\end{aligned}
$$

61. Replace $f(x)$ by y, interchange x and y, solve for y, and replace y by $f^{-1}(x)$.

$$
\begin{aligned}
y &= 3^{x+2} \\
x &= 3^{y+2} \\
y + 2 &= \log_3(x) \\
y = f^{-1}(x) &= \log_3(x) - 2
\end{aligned}
$$

63. Replace $f(x)$ by y, interchange x and y, solve for y, and replace y by $f^{-1}(x)$.

$$
\begin{aligned}
x &= \frac{1}{2}10^{y-1} + 5 \\
x - 5 &= \frac{1}{2}10^{y-1} \\
2x - 10 &= 10^{y-1} \\
\log(2x - 10) &= y - 1 \\
\log(2x - 10) + 1 &= y \\
f(x)^{-1} &= \log(2x - 10) + 1
\end{aligned}
$$

65. Since $2^8 = x$, the solution is $x = 256$.

67. Since $3^{1/2} = x$, the solution is $x = \sqrt{3}$.

69. Since $x^2 = 16$ and the base of a logarithm is positive, $x = 4$.

71. By using the definition of the logarithm, we get $x = \log_3(77)$.

73. Since $y = \ln(x)$ is one-to-one, $x - 3 = 2x - 9$. The solution is $x = 6$.

75. Since $x^2 = 18$, we obtain $x = \sqrt{18} = 3\sqrt{2}$. The base of a logarithm is positive.

77. Since $x + 1 = \log_3(7)$, $x = \log_3(7) - 1$.

79. Since $y = \log(x)$ is one-to-one, we get

$$
\begin{aligned}
x &= 6 - x^2 \\
x^2 + x - 6 &= 0 \\
(x + 3)(x - 2) &= 0 \\
x &= -3, 2.
\end{aligned}
$$

But $\log(-3)$ is undefined, so the solution is $x = 2$.

81. Since $x^{-2/3} = \dfrac{1}{9}$, we find

$$
x = \pm \left(\frac{1}{9}\right)^{-3/2} = \pm \left(\frac{1}{3}\right)^{-3} = \pm 27.
$$

The solution is $x = 27$ since a negative number cannot be a base.

83. Since $(2^2)^{2x-1} = 2^{4x-2} = 2^{-1}$, we get $4x - 2 = -1$. The solution is $x = \dfrac{1}{4}$.

85. $x = \log(25) \approx 1.3979$

87. Solving for x, we obtain

$$
\begin{aligned}
e^{2x} &= 3 \\
2x &= \ln(3) \\
x &= \frac{1}{2}\ln(3) \\
x &\approx 0.5493.
\end{aligned}
$$

89. Solving for x, we get

$$
\begin{aligned}
e^x &= \frac{4}{5} \\
x &= \ln\left(\frac{4}{5}\right) \\
x &\approx -0.2231.
\end{aligned}
$$

91. Solving for x, we find

$$
\begin{aligned}
\frac{1}{10^x} &= 2 \\
\frac{1}{2} &= 10^x \\
x &= \log\left(\frac{1}{2}\right) \\
x &\approx -0.3010.
\end{aligned}
$$

93. Let t be the number of years.

$$
\begin{aligned}
1000 \cdot e^{0.14t} &= 10^6 \\
e^{0.14t} &= 1000 \\
0.14t &= \ln(1000) \\
t &\approx 49.341 \text{ years}
\end{aligned}
$$

Note, $0.341(365) \approx 125$.
It will take 49 years and 125 days.

95. Since $e^{rt} = A/P$, $rt = \ln(A/P)$ and

$$r = \frac{\ln(A/P)}{t}.$$

Thus, $1000 will double in 3 years if the rate

is $r = \dfrac{\ln(2000/1000)}{3} \approx 0.231$ or 23.1%.

97.

(a) Let t be the number of years.

$$
\begin{aligned}
P \cdot e^{0.1t} &= 2P \\
e^{0.1t} &= 2 \\
0.1t &= \ln(2) \\
t = \frac{\ln(2)}{0.1} &\approx 6.9
\end{aligned}
$$

An investment at 10% doubles every 6.9 years.

(b) If t is the number of years it takes before an investment doubles, then

$$
\begin{aligned}
P \cdot e^{rt} &= 2P \\
e^{rt} &= 2 \\
rt &= \ln(2) \\
t &= \frac{\ln(2)}{r} \\
t &\approx \frac{0.70}{r} \\
t &\approx \frac{70}{100r}
\end{aligned}
$$

Thus, the number of years it takes an investment to double is approximately 70 divided by the interest rate expressed in percentage.

99. Let r be the interest rate.

$$
\begin{aligned}
4,000 \cdot e^{200r} &= 4,500,000 \\
e^{200r} &= 1,125 \\
200r &= \ln(1,125) \\
r = \frac{\ln(1,125)}{200} &\approx 0.035
\end{aligned}
$$

The rate is 3.5%

101. Let t be the number of years.

$$
\begin{aligned}
F_o \cdot e^{-0.052t} &= 0.6F_o \\
e^{-0.052t} &= 0.6 \\
-0.052t &= \ln(0.6) \\
t = \frac{\ln(0.6)}{-0.052} &\approx 9.8
\end{aligned}
$$

Only 60% of the present forest will remain after 9.8 years.

103. Let r be the annual rate from 1950 to 1987.

$$
\begin{aligned}
2.5 \cdot e^{37r} &= 5 \\
e^{37r} &= 2 \\
37r &= \ln(2) \\
r = \frac{\ln(2)}{37} &\approx 0.0187
\end{aligned}
$$

The annual rate is 1.87%.

If the annual rate is 1.63% and the initial population is 5 billion in 1987, the world population in year 2010 will be $5 \cdot e^{0.0163(23)} \approx 7.3$ billion.

105.

(a) $0.1e^{0.46(15)} \approx 99.2$ billion gigabits per year.

(b) Solving for t, we find

$$
\begin{aligned}
.1e^{.46t} &= 14 \\
e^{.46t} &= 140 \\
.46t &= \ln(140) \\
t &= \frac{\ln(140)}{.46} \\
t &\approx 11.
\end{aligned}
$$

In 2005, the long distance data transmission is expected to be 14 billion gigabits/yr.

107.

a) The function is given by

$$
\begin{aligned}
p - 100 &= \frac{100 - 10}{2 - 4}(x - 2) \\
p - 100 &= -45(x - 2) \\
p &= -45x + 90 + 100 \\
p &= -45x + 190 \\
p &= -45\log(I) + 190.
\end{aligned}
$$

b) $p = -45\log(100,000) + 190 = -35\%$ or 0% of the population is expected to be without safe drinking water, i.e., everyone is expected to have safe water.

109. $pH = -\log\left(10^{-4.1}\right) = 4.1$

111. $pH = -\log\left(10^{-3.7}\right) = 3.7$

113. By substituting $x = 1$, we find a formula for c.

$$
\begin{aligned}
a \cdot b^x &= a \cdot e^{cx} \\
b^x &= e^{cx} \\
b &= e^c \\
c &= \ln(b).
\end{aligned}
$$

By using this formula, we find

$$y = 50(1.036)^x = 500e^{\ln(1.036)x}$$

and the continuous growth rate is

$$\ln(1.036) \cdot 100 \approx 3.54\%.$$

For Thought

1. False, since $\log(8) - \log(3) = \log(8/3) \neq \dfrac{\log(8)}{\log(3)}$.

2. True, since $\ln(3^{1/2}) = \dfrac{1}{2} \cdot \ln(3) = \dfrac{\ln(3)}{2}$.

3. True, since $\dfrac{\log_{19}(8)}{\log_{19}(2)} = \log_2(8) = 3 = \log_3(27)$.

4. True, because of the base-change formula.

5. False

6. False, since $\log(x) - \log(2) = \log(x/2)$.

7. False, since the solution of the first equation is $x = -2$ and the second equation is not defined when $x = -2$.

8. True   **9.**   False, since x can be negative.

10. False, since a can be negative and so $\ln(a)$ will not be a real number.

4.3 Exercises

1. $\log(15)$

3. $\log_2((x-1)x) = \log_2(x^2 - x)$

5. $\log_4(6)$ **7.** $\ln\left(\dfrac{x^8}{x^3}\right) = \ln(x^5)$

9. $\log_2(3x) = \log_2(3) + \log_2(x)$

11. $\log\left(\dfrac{x}{2}\right) = \log(x) - \log(2)$

13. $\log((x-1)(x+1)) = \log(x-1) + \log(x+1)$

15. $\ln\left(\dfrac{x-1}{x}\right) = \ln(x-1) - \ln(x)$

17. $\log(5^3) = 3\log(5)$

19. $\log(5^{1/2}) = \dfrac{1}{2} \cdot \log(5)$

21. $\log(5^{-1}) = -\log(5)$

23. $\sqrt{y}$ **25.** $y+1$

27. 999

29. $\log(2) + \log(5)$

31. $\log(5/2) = \log(5) - \log(2)$

33. $\log(\sqrt{2^2 \cdot 5}) = \dfrac{1}{2}(2\log(2) + \log(5)) = \log(2) + \dfrac{1}{2}\log(5)$

35. $\log(4) - \log(25) = \log(2^2) - \log(5^2) = 2\log(2) - 2\log(5)$

37. $\log_3(5) + \log_3(x)$

39. $\log_2(5) - \log_2(2y) = \log_2(5) - \log_2(2) - \log_2(y)$

41. $\log(3) + \dfrac{1}{2}\log(x)$

43. $\log(3) + (x-1)\log(2)$

45. $\dfrac{1}{3} \cdot \ln(xy) - \dfrac{4}{3}\ln(t) = \dfrac{1}{3} \cdot \ln(x) + \dfrac{1}{3} \cdot \ln(y) - \dfrac{4}{3}\ln(t)$

47. $\ln(6\sqrt{x-1}) - \ln(5x^3) = \ln(6) + \dfrac{1}{2} \cdot \ln(x-1) - \ln(5) - 3 \cdot \ln(x)$

49. $\log_2(5x^3)$

51. $\log_7(x^5) - \log_7(x^8) = \log(x^5/x^8) = \log_7(x^{-3})$

53. $\log(2xy/z)$

55. $\log\left(\dfrac{\sqrt{x}}{y}\right) + \log\left(\dfrac{z}{\sqrt[3]{w}}\right) = \log\left(\dfrac{z\sqrt{x}}{y\sqrt[3]{w}}\right)$

57. $\log_4(x^6) + \log_4(x^{12}) + \log_4(x^2) = \log_4(x^{20})$

59. Since $2^x = 9$, we get $x = \dfrac{\log 9}{\log 2} \approx 3.1699$.

61. Since $0.56^x = 8$, we get
$$x = \frac{\log 8}{\log 0.56} \approx -3.5864.$$

63. Since $1.06^x = 2$, we get
$$x = \frac{\log 2}{\log 1.06} \approx 11.8957.$$

65. Since $0.73^x = 0.5$, we get
$$x = \frac{\log 0.5}{\log 0.73} \approx 2.2025.$$

67. $\dfrac{\ln(9)}{\ln(4)} \approx \dfrac{2.1972246}{1.3862944} \approx 1.5850$

69. $\dfrac{\ln(2.3)}{\ln(9.1)} \approx \dfrac{0.8329091}{2.2082744} \approx 0.3772$

71. $\dfrac{\ln(12)}{\ln(1/2)} \approx -3.5850$

73. Since $4t = \log_{1.02}(3) = \dfrac{\ln(3)}{\ln(1.02)}$,

we find $t = \dfrac{\ln(3)}{4 \cdot \ln(1.02)} \approx 13.8695$.

75. Since $365t = \log_{1.0001}(3.5) = \dfrac{\ln(3.5)}{\ln(1.0001)}$,

we get $t = \dfrac{\ln(3.5)}{365 \cdot \ln(1.0001)} \approx 34.3240$.

77. $1 + r = \sqrt[3]{2.3}$, so $r = \sqrt[3]{2.3} - 1 \approx 0.3200$

79.
$$\left(1 + \frac{r}{12}\right)^{360} = 4.2$$
$$1 + \frac{r}{12} = \sqrt[360]{4.2}$$
$$r = 12\left(\sqrt[360]{4.2} - 1\right)$$
$$r \approx 0.0479$$

81. Since $x^5 = 33.4$, we get $x = \sqrt[5]{33.4} \approx 2.0172$.

83. Since $x^{-1.3} = 0.546$, we have $x = 0.546^{1/(-1.3)}$ or $x \approx 1.5928$.

85. Let t be the number of years.

$$800\left(1 + \frac{0.08}{365}\right)^{365t} = 2000$$
$$\left(1 + \frac{0.08}{365}\right)^{365t} = 2.5$$
$$(1.0002192)^{365t} \approx 2.5$$
$$365t \approx \log_{1.0002192}(2.5)$$
$$t \approx \frac{1}{365} \cdot \frac{\ln(2.5)}{\ln(1.0002192)}$$
$$t \approx 11.454889$$
$$t \approx 11 \text{ years}, 166 \text{ days}$$

87. Let t be the number of years.

$$W\left(1 + \frac{0.1}{4}\right)^{4t} = 3W$$
$$(1.025)^{4t} = 3$$
$$4t \approx \log_{1.025}(3)$$
$$t \approx \frac{1}{4} \cdot \frac{\ln(3)}{\ln(1.025)}$$
$$t \approx 11.123 \text{ years}$$
$$t \approx 11.123(4) \approx 44 \text{ quarters}$$

89. Let t be the number of years.

$$500(1 + r)^{25} = 2000$$
$$(1 + r)^{25} = 4$$
$$1 + r \approx \sqrt[25]{4}$$
$$r = \sqrt[25]{4} - 1$$
$$r \approx 0.057 \text{ or } 5.7\%$$

91. Let t be the number of years.

$$4000(1 + r)^{200} = 4.5 \times 10^6$$
$$(1 + r)^{200} = 1125$$
$$r = \sqrt[200]{1125} - 1$$
$$r \approx 0.035752 \text{ or } 3.58\%$$

93. The Richter scale rating is

$$\log(I) - \log(I_o) = \log\left(\frac{I}{I_o}\right).$$

When $I = 1000 \cdot I_o$, the Richter scale rating is

$$\log\left(\frac{1000 \cdot I_o}{I_o}\right) = \log(1000) = 3.$$

95. $t = \frac{1}{r}\ln(P/P_o) = \frac{1}{r}\ln(P) - \frac{1}{r}\ln(P_o)$

97.

(a) p decreases as n increases

(b) Solving for n, one can take the logarithm of both sides and to note that $y = \log(x)$ is an increasing function.

$$\left(\frac{7,059,051}{7,059,052}\right)^n > \frac{1}{2}$$

$$\log\left(\left(\frac{7,059,051}{7,059,052}\right)^n\right) > \log\left(\frac{1}{2}\right)$$

$$n\log\left(\frac{7,059,051}{7,059,052}\right) > \log\left(\frac{1}{2}\right)$$

$$n < \frac{\log(1/2)}{\log(7,059,051/7,059,052)}$$

$$n < 4,892,962$$

If at most $4,892,961$ tickets are purchsed, then the probability of a rollover is greater than 50%.

99. We note that $MR(x) = R(x+1) - R(x) =$
$500 \cdot \log(x+2) - 500 \cdot \log(x+1) =$

$$500 \cdot \log\left(\frac{x+2}{x+1}\right) = \log\left(\left(\frac{x+2}{x+1}\right)^{500}\right).$$

Hence, as $x \to \infty$, then $\dfrac{x+2}{x+1} \to 1$

and $MR(x) \to 0$.

101.

a) Let x be the number of years since 1985 and let y be the number of computers per 1000 people. With the aid of a calculator, we find that an exponential model is

$$y = 30.316815061562(1.2350959471194)^x$$

or approximately

$$y = 30.32(1.235)^x.$$

b) Since

$$\ln(1.2350959471194) \approx 0.211148657037,$$

we obtain

$$y = 30.316815061562e^{0.211148657037x}$$

or approximately

$$y = 30.32e^{0.21115x}.$$

c) Since $r \approx 0.211148657037$, the annual rate is 21.1% componded continuously

d) Let $y = 2000$ and solve for x.

$$2000 = 30.316815061562(1.2350959471194)^x$$

$$\ln\left(\frac{2000}{30.316815061562}\right) = \ln(1.2350959471194^x)$$

$$\ln\left(\frac{2000}{30.316815061562}\right) = x\ln(1.2350959471194)$$

$$\ln\left(\frac{2000}{30.316815061562}\right)\div\ln(1.2350959471194) = x$$

Thus, $x \approx 20$. In the year 2005, it is predicted there will be 2000 computers per 1000 people. It maybe reasonable to have two computers per person.

e) No, the data looks more linear.

103. Let $a > 0$, $a \neq 1$. Using the definition of a logarithm and properties of exponents, we have

$$\frac{M}{N} = \frac{a^{\log_a(M)}}{a^{\log_a(N)}} = a^{\log_a(M)-\log_a(N)}$$

and

$$\frac{M}{N} = a^{\log_a(M/N)}$$

Thus, $a^{\log_a(M/N)} = a^{\log_a(M) - \log_a(N)}$
and $\log_a(M) - \log_a(N) = \log_a(M/N)$.

105.

a) The graphs of $y_1 = \log_3(\sqrt{3}x)$ and
$y_2 = 0.5 + \log_3(x)$ are identical.

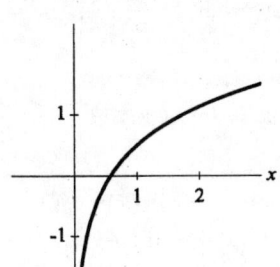

b) The graphs of $y_1 = \log_2(1/x)$ and
$y_2 = -\log_2(x)$ are identical.

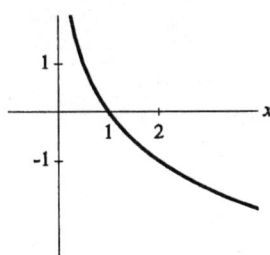

c) $y_1 = 3^{x-1}$ and $y_2 = \log_3(x) + 1$ are inverses
of each other

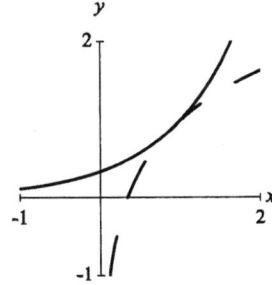

d) $y_1 = 3 + 2^{x-4}$ and $y_2 = \log_2(x-3) + 4$ are
inverses of each other

For Thought

1. True, since $(1.02)^x = 7$ is equivalent
to $x = \log_{1.02}(7)$.

2. True, for $x(1 - \ln(3)) = 8$ implies $x = \dfrac{8}{1 - \ln(3)}$.

3. False, $\ln(1 - \sqrt{6})$ is undefined.

4. True, by the definition of a logarithm.

5. False, the exact solution is $x = \log_3(17)$
which is not the same as 2.5789.

6. False, since $x = -2$ is not a solution of the first
equation but is a solution of the second one.

7. True, since $4^x = 2^{2x} = 2^{x-1}$.   **8.**   True

9. True, since $\dfrac{\ln(2)}{\ln(7)} = \dfrac{\log(2)}{\log(7)}$.

10. True, since $\log(e) \cdot \ln(10) = \ln\left(10^{\log(e)}\right) = \ln(e) = 1$.

4.4 Exercises

1. Since $x = 2^3$, we get $x = 8$.

3. Since $10^2 = x + 20$, $x = 80$.

5. Since $10^1 = x^2 - 15$, we get $x^2 = 25$.
The solutions are $x = \pm 5$.

7. Note, $x^2 = 9$ and $x = \pm 3$. Since the base of
a logarithm is positive, the solution is $x = 3$.

9. Note, $x^{-2} = 4$ or $\dfrac{1}{4} = x^2$. Since the base of
a logarithm is positive, the solution is $x = \dfrac{1}{2}$.

11.

$$\log_2(x^2 - 4) = 5$$
$$x^2 - 4 = 2^5$$
$$x^2 = 36$$
$$x = \pm 6$$

Checking $x = -6$, one gets $\log_2(-6 + 2)$ which is undefined. The solution is $x = 6$.

13.

$$\log(5) + \log(x) = 2$$
$$\log(5x) = 2$$
$$5x = 10^2$$
$$x = 20$$

15. Since $\ln(x(x + 2)) = \ln(8)$ and $y = \ln(x)$ is one-to-one, we get

$$x^2 + 2x = 8$$
$$x^2 + 2x - 8 = 0$$
$$(x + 4)(x - 2) = 0$$
$$x = -4, 2$$

Since $\ln(-4)$ is undefined, the solution is $x = 2$.

17. Since $\log(4x) = \log\left(\dfrac{5}{x}\right)$ and $y = \log(x)$ is one-to-one, we obtain

$$4x = \frac{5}{x}$$
$$4x^2 = 5$$
$$x^2 = \frac{5}{4}$$
$$x = \pm\frac{\sqrt{5}}{2}.$$

But $\log\left(-\dfrac{\sqrt{5}}{2}\right)$ is undefined, so $x = \dfrac{\sqrt{5}}{2}$.

19. Since $\log_2\left(\dfrac{x}{3x - 1}\right) = 0$, we get

$$\frac{x}{3x - 1} = 1$$
$$x = 3x - 1$$
$$1 = 2x$$
$$x = \frac{1}{2}.$$

21.

$$x \cdot \ln(3) + x \cdot \ln(2) = 2$$
$$x(\ln(3) + \ln(2)) = 2$$
$$x = \frac{2}{\ln(3) + \ln(2)}$$

23. Since $x - 1 = \log_2(7)$, $x = \dfrac{\ln(7)}{\ln(2)} + 1 \approx 3.8074$.

25. Since $4x = \log_{1.09}(3.4)$, $x = \dfrac{1}{4} \cdot \dfrac{\ln(3.4)}{\ln(1.09)} \approx 3.5502$.

27. Since $-x = \log_3(30)$, $x = -\dfrac{\ln(30)}{\ln(3)} \approx -3.0959$

29. Note, $-3x^2 = \ln(9)$. There is no solution since the left-hand side is non-negative and the right-hand side is positive.

31.

$$\ln(6^x) = \ln(3^{x+1})$$
$$x \cdot \ln(6) = (x + 1) \cdot \ln(3)$$
$$x \cdot \ln(6) = x \cdot \ln(3) + \ln(3)$$
$$x(\ln(6) - \ln(3)) = \ln(3)$$
$$x = \frac{\ln(3)}{\ln(6) - \ln(3)}$$
$$x \approx 1.5850$$

33.

$$\ln(e^{x+1}) = \ln(10^x)$$
$$(x + 1) \cdot \ln(e) = x \cdot \ln(10)$$
$$x + 1 = x \cdot \ln(10)$$
$$1 = x(\ln(10) - 1)$$
$$x = \frac{1}{\ln(10) - 1}$$
$$x \approx 0.7677$$

35.

$$2^{x-1} = (2^2)^{3x}$$
$$2^{x-1} = 2^{6x}$$
$$x - 1 = 6x$$
$$-1 = 5x$$
$$x = -0.2$$

37.

$$
\begin{aligned}
\ln(6^{x+1}) &= \ln(12^x) \\
(x+1)\cdot\ln(6) &= x\cdot\ln(12) \\
x\cdot\ln(6)+\ln(6) &= x\cdot\ln(12) \\
\ln(6) &= x(\ln(12)-\ln(6)) \\
x &= \frac{\ln(6)}{\ln(12)-\ln(6)} \\
x &\approx 2.5850
\end{aligned}
$$

39. Since $3 = e^{-\ln(w)} = e^{\ln(1/w)} = 1/w$, we have $\dfrac{1}{w} = 3$. Then $w = \dfrac{1}{3}$.

41.

$$
\begin{aligned}
(\log(z))^2 &= 2\cdot\log(z) \\
(\log(z))^2 - 2\cdot\log(z) &= 0 \\
\log(z)\cdot(\log(z)-2) &= 0 \\
\log(z)=0 \quad &\text{or} \quad \log(z)=2 \\
z=10^0 \quad &\text{or} \quad z=10^2 \\
z=1 \quad &\text{or} \quad z=100
\end{aligned}
$$

The solutions are $z = 1, 100$.

43. Divide the equation by $4(1.03)^x$.

$$
\begin{aligned}
\left(\frac{1.02}{1.03}\right)^x &= \frac{3}{4} \\
\ln\left(\left(\frac{1.02}{1.03}\right)^x\right) &= \ln\left(\frac{3}{4}\right) \\
x\cdot\ln\left(\frac{1.02}{1.03}\right) &= \ln\left(\frac{3}{4}\right) \\
x &= \frac{\ln\left(\dfrac{3}{4}\right)}{\ln\left(\dfrac{1.02}{1.03}\right)} \\
x &\approx 29.4872
\end{aligned}
$$

45. Note that $e^{\ln((x^2)^3)-\ln(x^2)} = e^{\ln(x^6)-\ln(x^2)} = e^{\ln(x^6/x^2)} = e^{\ln(x^4)} = x^4$.
Thus, $x^4 = 16$ and $x = \pm 2$.
But $\ln(-2)$ is undefined, so $x = 2$.

47. Since $\left(\dfrac{1}{2}\right)^2 = \dfrac{1}{4}$, we find

$$
\left(\frac{1}{2}\right)^{2x-1} = \left(\frac{1}{2}\right)^{6x+4}
$$

$$
\begin{aligned}
2x-1 &= 6x+4 \\
-5 &= 4x \\
x &= -\frac{5}{4}.
\end{aligned}
$$

49. By approximating the x-intercepts of the graph $y = 2^x - 3^{x-1} - 5^{-x}$, we find that the solutions are $x \approx 0.194, 2.70$.

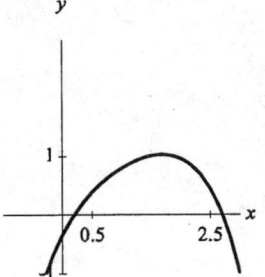

51. By approximating the x-intercept of the graph $y = \ln(x+51) - \log(-48-x)$, we obtain that the solution is $x \approx -49.73$.

53. By approximating the x-intercepts of the graph $y = x^2 - 2^x$, we find that the solutions are $x \approx -0.767, 2, 4$.

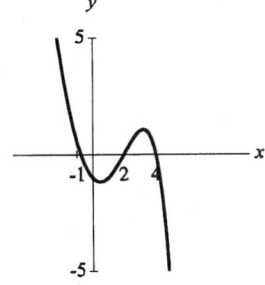

55. Using $A = A_o e^{rt}$ with $A_o = 1$ and the half-life, we obtain $\dfrac{1}{2} = e^{5730r}$. Thus, $5730r = \ln\left(\dfrac{1}{2}\right)$ and $r \approx -0.000120968$. When $A = 0.1$, we get

$$
0.1 = e^{-0.000120968t}
$$

$$\ln(0.1) = -0.000120968t$$
$$t = \frac{\ln(0.1)}{-0.000120968} \approx 19{,}035 \text{ years}$$

57. From Number 55, $r \approx -0.000120968$.
When $A = 10$ and $A_o = 12$, we obtain

$$10 = 12 \cdot e^{-0.000120968t}$$
$$\ln(10/12) = -0.000120968t$$
$$t = \frac{\ln(5/6)}{-0.000120968} \approx 1507 \text{ years.}$$

59. Let $A = A_o e^{rt}$ where $A_o = 25$, $A = 20$, and $t = 8000$. Then

$$20 = 25 \cdot e^{8000r}$$
$$\ln(0.8) = 8000r$$
$$r \approx -0.000027893.$$

To find the half-life, let $A = 12.5$. Thus, we have

$$12.5 = 25 \cdot e^{-0.000027893t}$$
$$\ln(0.5) = -0.000027893t$$
$$t \approx 24{,}850 \text{ years.}$$

61. $\dfrac{2.5(0.5)^{24/14}}{2.5} \times 100 \approx 30.5\%$, the percentage of the last dosage that remains before the next dosage is taken

63. Since the half-life is $t = 5730$ years and $A_o = 1$ and $A = 0.5$, we get

$$0.5 = e^{(5730)\cdot r}$$
$$\ln(0.5) = 5730 \cdot r$$
$$r \approx -0.000121.$$

If 79.3% of the carbon is still present, then

$$0.793 = e^{(-0.000121)\cdot t}$$
$$\ln(0.793) = -0.000121 \cdot t$$
$$t \approx 1917.$$

The scrolls were made in the year $1951 - 1917 = 34$ AD.

65. The initial difference in temperature is $325 - 35 = 290$, and after $t = 3$ hours the difference is $325 - 140 = 185$. Then

$$185 = 290 \cdot e^{3k}$$
$$\ln\left(\frac{185}{290}\right) = 3k$$
$$k \approx -0.1498417.$$

The difference in temperature when the roast is well-done is $325 - 170 = 155$. Thus,

$$155 = 290 \cdot e^{(-0.1498417)\cdot t}$$
$$\ln\left(\frac{155}{290}\right) = (-0.1498417) \cdot t$$
$$t \approx 4.18 \text{ hr}$$
$$t \approx 4 \text{ hr and } 11 \text{ min.}$$

James must wait 1 hour and 11 minutes longer. If the oven temperature is set at 170°, then the initial and final differences are 135 and 0, respectively. Since $0 = 135 \cdot e^{(-0.1498417)\cdot t}$ has no solution, James has to wait forever.

67. At 7:00 a.m., the difference in temperature is $80 - 40 = 40$, and $t = 1$ hour later the difference in temperature is $72 - 40 = 32$. Then

$$32 = 40 \cdot e^{1 \cdot k}$$
$$\ln\left(\frac{32}{40}\right) = k$$
$$k \approx -0.2231436.$$

Let n be the number of hours before 7:00 a.m. when death occured. At the time of death, the difference in temperature is $98 - 40 = 58$. Then

$$40 = 58 \cdot e^{-0.2231436 \cdot n}$$
$$\ln\left(\frac{40}{58}\right) = -0.2231436 \cdot n$$
$$n \approx 1.665 \text{ hr}$$
$$n \approx 1 \text{ hr and } 40 \text{ min.}$$

Death occured at 5:20 a.m.

69. Since $R = P\dfrac{i}{1 - (1+i)^{-nt}}$, we obtain

$$1 - (1+i)^{-nt} = Pi/R$$
$$1 - Pi/R = (1+i)^{-nt}$$
$$\ln(1 - Pi/R) = -nt\ln(1+i)$$
$$\frac{-\ln(1 - Pi/R)}{n\ln(1+i)} = t.$$

Let $i = 0.09/12 = 0.0075$, $P = 100,000$, and $R = 1250$. Then

$$t = \frac{-\ln(1 - Pi/R)}{n\ln(1+i)} \approx 10.219.$$

It will take 10 yr, 3 mo to pay off the loan.

71. The future values of the $1000 and $1100 investments are equal. Then

$$1,000 \cdot e^{0.06t} = 1100\left(1 + \frac{0.06}{365}\right)^{365t}$$

$$e^{0.06t} = 1.1\left(1 + \frac{0.06}{365}\right)^{365t}$$

$$.06t = \ln(1.1) + 365t\ln\left(1 + \frac{0.06}{365}\right)$$

$$.06t - 365t\ln\left(1 + \frac{0.06}{365}\right) = \ln(1.1)$$

$$t = \frac{\ln(1.1)}{.06 - 365\ln\left(1 + \frac{0.06}{365}\right)}$$

$$t \approx 19,328.84173 \text{ years}$$

They will be equal after $19,328$ yr, 307 days.

73. a) Let $t = 0$. The present number of rabbits is

$$P = 12,300 + 1000 \cdot \ln(1) = 12,300 + 0 = 12,300.$$

b) In about 15 years.

c) The number of years before there will be $15,000$ rabbits is given by

$$12,300 + 1000 \cdot \ln(t+1) = 15,000$$
$$1000 \cdot \ln(t+1) = 2,700$$
$$\ln(t+1) = 2.7$$
$$t + 1 = e^{2.7}$$
$$t \approx 13.9 \text{ yr.}$$

75. a) Let $n = 2500$ and $A = 400$.
Since $n = k\log(A)$, $2500 = k\log(400)$.
Then $k = \dfrac{2500}{\log(400)}$. When $A = 200$,
the number of species left is

$$n = \frac{2500}{\log(400)}\log(200) \approx 2211 \text{ species.}$$

b) Let $n = 3500$ and $A = 1200$.
Since $n = k\log(A)$, $3500 = k\log(1200)$.
Then $k = \dfrac{3500}{\log(1200)}$. When $n = 1000$,
the remaining forest area is given by

$$1000 = \frac{3500}{\log(1200)}\log(A)$$

$$\frac{1000}{3500}\log(1200) = \log(A)$$

$$\frac{2}{7}\log(1200) = \log(A)$$

$$\log\left(1200^{2/7}\right) = \log(A)$$

$$1200^{2/7} = A$$

Thus, the percentage of forest that has been destroyed is $100 - \dfrac{A}{1200}(100) \approx 99\%$.

77. Since $m = 0$ and $M_v = 4.39$, the distance to α Centauri is given by

$$4.39 - 5 + 5 \cdot \log(d) = 0$$
$$5 \cdot \log(d) = 0.61$$
$$\log(d) = 0.122$$
$$d = 10^{0.122} \approx 1.32 \text{ parsecs.}$$

79.

a) Let $y = \log(P)$. A formula for P is

$$y - 4.5 = \frac{4.5 - 1.5}{0 - 18}(x - 0)$$

$$y = -\frac{1}{6}x + 4.5$$

$$\log(P) = -\frac{1}{6}x + 4.5$$

$$P = 10^{(-\frac{1}{6}x+4.5)}$$

b) In 1991, $x = 9$ and $P = 10^{(-\frac{1}{6} \cdot 9 + 4.5)} = 10^3 = \$1000$ per gigabit.

c) Solving for x, we derive

$$10^1 = 10^{(-\frac{1}{6}x+4.5)}$$
$$1 = -\frac{1}{6}x + 4.5$$
$$\frac{1}{6}x = 3.5$$
$$x = 21$$

In the year 2003, according to this model a hard drive with one gigabit will cost $10.

81. If the sound level is 90 db, then the intensity of the sound is given by

$$
\begin{aligned}
10 \cdot \log(I \times 10^{12}) &= 90 \\
\log(I) + \log(10^{12}) &= 9 \\
\log(I) + 12 &= 9 \\
\log(I) &= -3 \\
I &= 10^{-3} \text{ watts per/m}^2.
\end{aligned}
$$

83. Since $P \cdot e^{(0.06)18} = 20{,}000$, the investment will grow to $P = \dfrac{20{,}000}{e^{(0.06)18}} \approx \$6791.91$.

85.

a) The logarithmic regression line is

$$y = 114.7865206 - 12.75457939 \ln(x)$$

or approximately is

$$y = 114.8 - 12.8 \ln(x)$$

where the year is $1960 + x$.

b) Let $x = 50$. Since

$$114.7865206 - 12.75457939 \ln(50) \approx 65,$$

the percentage of households in 2010 with two parents is 65%.

c) Let $y = 50$. Then

$$50 = 114.7865206 - 12.75457939 \ln(x)$$

$$
\begin{aligned}
\ln(x) &= \frac{114.7865206 - 50}{12.75457939} \\
x &\approx 161.
\end{aligned}
$$

Since $1960 + 161 = 2121$, the percentage of two-parent families will reach 50% in the year 2121.

d) The data looks like a quadratic or exponential model.

87. By using the first five terms of the formula,

$$
\begin{aligned}
e^{0.1} &\approx 1 + 0.1 + \frac{(0.1)^2}{2} + \frac{(0.1)^3}{6} + \frac{(0.1)^4}{24} \\
&\approx 1.105170833
\end{aligned}
$$

From a calculator, $e^{0.1} \approx 1.105170918$.

Chapter 4 Review Exercises

1. 64   **3.**   6   **5.**   0

7. 17   **9.** $2^5 = 32$ **11.** $\log(10^3) = 3$

13. $\log_2(2^9) = 9$ **15.** $\log(1000) = 3$

17. $\log_2(1) - \log_2(8) = 0 - 3 = -3$

19. $\log_2(8) = 3$

21. $\log((x-3)x) = \log(x^2 - 3x)$

23. $\ln(x^2) + \ln(3y) = \ln(3x^2 y)$

25. $\log(3) + \log(x^4) = \log(3) + 4 \cdot \log(x)$

27. $\log_3(5) + \log_3(x^{1/2}) - \log_3(y^4) =$
  $\log_3(5) + \dfrac{1}{2} \cdot \log_3(x) - 4 \cdot \log_3(y)$

29. $\ln(2 \cdot 5) = \ln(2) + \ln(5)$

31. $\ln(5^2 \cdot 2) = \ln(5^2) + \ln(2) = 2 \cdot \ln(5) + \ln(2)$

33. Since $\log_{10}(x) = 10$, we get $x = 10^{10}$.

35. Since $x^4 = 81$ and $x > 0$, $x = 3$.

37. Since $\log_{1/3}(27) = -3 = x + 2$, we get $x = -5$.

39. Since $3^{x+2} = 3^{-2}$, $x + 2 = -2$. So $x = -4$.

41. Since $x - 2 = \ln(9)$, we obtain $x = 2 + \ln(9)$.

43. Since $(2^2)^{x+3} = 2^{2x+6} = 2^{-x}$, we get
$2x + 6 = -x$. Then $6 = -3x$ and so $x = -2$.

45.

$$
\begin{aligned}
\log(2x^2) &= 5 \\
2x^2 &= 10^5 \\
x^2 &= 50,000 \\
x &= \pm 100\sqrt{5}
\end{aligned}
$$

Since $\log(-100\sqrt{5})$ is undefined, $x = 100\sqrt{5}$.

47.

$$
\begin{aligned}
\log_2\left(x^2 - 4x\right) &= \log_2(x + 24) \\
x^2 - 4x &= x + 24 \\
x^2 - 5x - 24 &= 0 \\
(x - 8)(x + 3) &= 0 \\
x &= 8, -3
\end{aligned}
$$

Since $\log(-3)$ is undefined, $x = 8$.

49. Since $\ln((x + 2)^2) = \ln(4^3)$ and $y = \ln(x)$ is a one-to-one function, we obtain

$$
\begin{aligned}
(x + 2)^2 &= 64 \\
x + 2 &= \pm 8 \\
x &= -2 \pm 8 \\
x &= 6, -10.
\end{aligned}
$$

Checking $x = -10$ one gets $2\ln(-8)$ which is undefined. So $x = 6$.

51.

$$
\begin{aligned}
x \cdot \log(4) + x \cdot \log(25) &= 6 \\
x(\log(4) + \log(25)) &= 6 \\
x \cdot \log(100) &= 6 \\
x \cdot 2 &= 6 \\
x &= 3
\end{aligned}
$$

53. The missing coordinates are

(i) 3 since $\left(\dfrac{1}{3}\right)^{-1} = 3,$

(ii) -3 since $\left(\dfrac{1}{3}\right)^{-3} = 27,$

(iii) $\sqrt{3}$ since $\left(\dfrac{1}{3}\right)^{-1/2} = \sqrt{3}$, and

(iv) 0 since $\left(\dfrac{1}{3}\right)^{0} = 1.$

55. c **57.** b **59.** d **61.** e

63. Domain $(-\infty, \infty)$, range $(0, \infty)$, increasing, asymptote $y = 0$

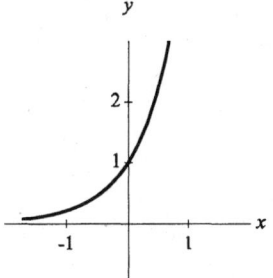

65. Domain $(-\infty, \infty)$, range $(0, \infty)$, decreasing, asymptote $y = 0$

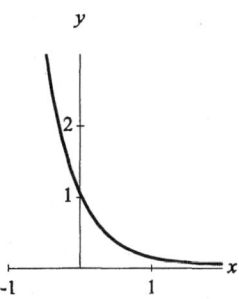

67. Domain $(0, \infty)$, range $(-\infty, \infty)$, increasing, asymptote $x = 0$

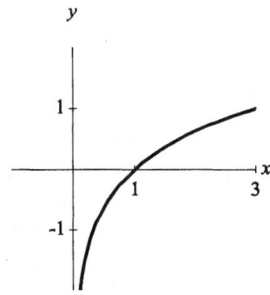

69. Domain $(-3, \infty)$, range $(-\infty, \infty)$, increasing, asymptote $x = -3$

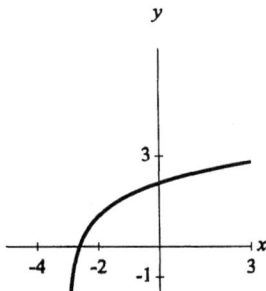

71. Domain $(-\infty, \infty)$, range $(1, \infty)$, increasing, asymptote $y = 1$

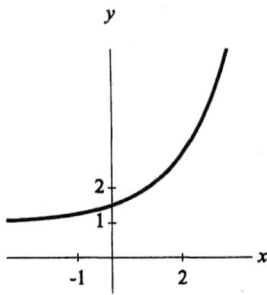

73. Domain $(-\infty, 2)$, range $(-\infty, \infty)$, decreasing, asymptote $x = 2$

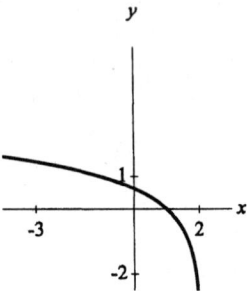

75. $f^{-1}(x) = \log_7(x)$ **77.** $f^{-1}(x) = 5^x$

79. Replace $f(x)$ by y, interchange x and y, solve for y, and replace y by $f^{-1}(x)$.

$$
\begin{aligned}
y &= 3\log(x-1) \\
x &= 3\log(y-1) \\
\frac{x}{3} &= \log(y-1) \\
10^{x/3} &= y-1 \\
10^{x/3} + 1 &= y \\
f^{-1}(x) &= 10^{x/3} + 1
\end{aligned}
$$

81. Replace $f(x)$ by y, interchange x and y, solve for y, and replace y by $f^{-1}(x)$.

$$
\begin{aligned}
y &= e^{x+2} - 3 \\
x &= e^{y+2} - 3 \\
x+3 &= e^{y+2} \\
\ln(x+3) &= y+2 \\
\ln(x+3) - 2 &= y \\
f^{-1}(x) &= \ln(x+3) - 2
\end{aligned}
$$

83. Since $3^x = 10$, $x = \log_3(10) = \dfrac{\ln(10)}{\ln(3)} \approx 2.0959$

85. Since $\log_3(x) = 1.876$, $x = 3^{1.876} \approx 7.8538$.

87. After taking the natural logarithm of both sides,
we have

$$
\begin{aligned}
\ln(5^x) &= \ln(8^{x+1}) \\
x \cdot \ln(5) &= (x+1) \cdot \ln(8) \\
x \cdot \ln(5) &= x \cdot \ln(8) + \ln(8) \\
x \cdot (\ln(5) - \ln(8)) &= \ln(8) \\
x &= \frac{\ln(8)}{\ln(5) - \ln(8)} \\
x &\approx -4.4243.
\end{aligned}
$$

89. True, since $\log_3(81) = 4$ and $2 = \log_3(9)$.

91. False, since $\ln(3^2) = 2 \cdot \ln(3) \neq (\ln(3))^2$.

93. True, since $4 \cdot \log_2(8) = 4 \cdot 3 = 12$.

95. False, since $3 + \log(6) = \log(10^3) + \log(6) = \log(6000)$.

97. False, since $\log_2(16) = 4$, $\log_2(8) = 3$, and $\dfrac{3}{4} \neq 3 - 4$.

99. False, since $\log_2(25) = 2\log_2(5) = 2 \cdot \dfrac{\log(5)}{\log(2)} \neq 2 \cdot \log(5)$.

101. True, because of the base-changing formula.

103. If H_B^+ is the hydrogen ion concentration of liquid B then $10 \cdot H_B^+$ is the hydrogen ion concentration of liquid A. The pH of A is

$$
-\log(10 \cdot H_B^+) = -1 - \log(H_B^+)
$$

i.e. the pH of A is one less than the pH of B.

105. The value at the end of 18 years is

$$50,000\left(1+\frac{0.05}{4}\right)^{18\cdot4} \approx \$122,296.01.$$

107. Let t be the number of years.

$$50,000\left(1+\frac{0.05}{4}\right)^{4t} = 100,000$$

$$(1.0125)^{4t} = 2$$

$$4t = \log_{1.0125}(2)$$

$$t = \frac{1}{4}\cdot\frac{\ln(2)}{\ln(1.0125)}$$

$$t \approx 13.9 \ \text{years}$$

It doubles in $4\cdot13.9 \approx 56$ quarters.

109. The present amount is $A = 25\cdot e^0 = 25$ g.
After $t = 1000$ years, the amount left
is $A = 25\cdot e^{-0.32} \approx 18.15$ g.

To find the half-life, let $A = 12.5$.

$$25\cdot e^{-0.00032t} = 12.5$$

$$e^{-0.00032t} = 0.5$$

$$-0.00032t = \ln(0.5)$$

$$t \approx 2166$$

The half-life is 2166 years.

111. Let $f(t) = 10,000$.

$$40,000\cdot(1 - e^{-0.0001t}) = 10,000$$

$$1 - e^{-0.0001t} = 0.25$$

$$0.75 = e^{-0.0001t}$$

$$\ln(0.75) = -0.0001t$$

$$t \approx 2877 \ \text{hr}$$

It takes 2877 hours. to learn $10,000$ words.

113. In the following equations, we solve for x.

$$1026\left(\frac{25,005}{64}\right)^x = 19.2$$

$$\left(\frac{25,005}{64}\right)^x = \frac{19.2}{1026}$$

$$x\ln\left(\frac{25,005}{64}\right) = \ln\left(\frac{19.2}{1026}\right)$$

$$x = \frac{\ln\left(\frac{19.2}{1026}\right)}{\ln\left(\frac{25,005}{64}\right)}$$

$$x \approx -0.667 \quad\text{or}\quad x \approx -\frac{2}{3}$$

In the following equations, we obtain y.

$$\frac{25005}{2240}\left(\frac{35+\frac{1}{12}}{100}\right)^y = 258.51$$

$$\left(\frac{35+\frac{1}{12}}{100}\right)^y = \frac{258.51(2240)}{25005}$$

$$y = \frac{\ln\left(\frac{258.51(2240)}{25005}\right)}{\ln\left(\frac{35+\frac{1}{12}}{100}\right)}$$

$$y \approx -3$$

Next, we solve for z.

$$13.5\left(\frac{25005}{64}\right)^z = 1.85$$

$$\left(\frac{25005}{64}\right)^z = \frac{1.85}{13.5}$$

$$z\ln\left(\frac{25005}{64}\right) = \ln\left(\frac{1.85}{13.5}\right)$$

$$z = \frac{\ln\left(\frac{1.85}{13.5}\right)}{\ln\left(\frac{25005}{64}\right)}$$

$$z \approx -0.333 \quad\text{or}\quad z \approx -\frac{1}{3}$$

Chapter 4 Test

1. 3 **2.** -2 **3.** 6.47 **4.** $\sqrt{2}$

5. $f^{-1}(x) = e^x$

6. Replace $f(x)$ by y, interchange x and y,
solve for y, and replace y by $f^{-1}(x)$.

$$y = 8^{x+1} - 3$$

$$x = 8^{y+1} - 3$$

$$x + 3 = 8^{y+1}$$

$$\log_8(x+3) = y + 1$$

$$\log_8(x+3) - 1 = y$$

$$f^{-1}(x) = \log_8(x+3) - 1$$

7. $\log(x) + \log(y^3) = \log(xy^3)$

8. $\ln(\sqrt{x-1}) - \ln(33) = \ln\left(\dfrac{\sqrt{x-1}}{33}\right)$

9. $\log_a(2^2 \cdot 7) = \log_a(2^2) + \log_a(7) = 2 \cdot \log_a(2) + \log_a(7)$

10. $\log_a \left(\dfrac{7}{2} \right) = \log_a(7) - \log_a(2)$

11.

$$
\begin{aligned}
\log_2(x^2 - 2x) &= 3 \\
x^2 - 2x &= 2^3 \\
x^2 - 2x - 8 &= 0 \\
(x - 4)(x + 2) &= 0 \\
x &= 4, -2
\end{aligned}
$$

But $\log_2(-2)$ is undefined, so $x = 4$.

12.

$$
\begin{aligned}
\log \left(\dfrac{10x}{x + 2} \right) &= \log(3^2) \\
\dfrac{10x}{x + 2} &= 9 \\
10x &= 9x + 18 \\
x &= 18
\end{aligned}
$$

13.

$$
\begin{aligned}
\ln(3^x) &= \ln(5^{x-1}) \\
x \cdot \ln(3) &= (x - 1) \cdot \ln(5) \\
x \cdot \ln(3) &= x \cdot \ln(5) - \ln(5) \\
x(\ln(3) - \ln(5)) &= -\ln(5) \\
x(\ln(5) - \ln(3)) &= \ln(5) \\
x &= \dfrac{\ln(5)}{\ln(5) - \ln(3)} \\
x &\approx 3.1507
\end{aligned}
$$

14. By the definition of a logarithm, we obtain $x - 1 = 3^{5.46}$. Then $x = 1 + 3^{5.46} \approx 403.7931$.

15. Domain $(-\infty, \infty)$, range $(1, \infty)$, increasing, asymptote $y = 1$

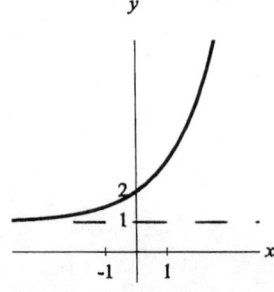

16. Domain $(1, \infty)$, range $(-\infty, \infty)$, decreasing, asymptote $x = 1$

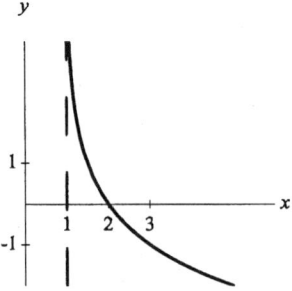

17. $(1, 0)$

18. Compounded quarterly, the investment is worth $2000 \left(1 + \dfrac{0.08}{4} \right)^{80} \approx \$9750.88$.

Compounded continuously, the investment is worth $2000 \cdot e^{0.08(20)} \approx \$9906.06$

19. The amount of power at the end of $t = 200$ days is $P = 50 \cdot e^{-200/250} \approx 22.5$ watts.

To find the half-life, let $P = 25$.

$$
\begin{aligned}
50 \cdot e^{-t/250} &= 25 \\
e^{-t/250} &= 0.5 \\
-\dfrac{t}{250} &= \ln(0.5) \\
t &\approx 173.3
\end{aligned}
$$

The half-life is 173.3 days.

The operational life for a power of $P = 9$ watts is given by

$$
\begin{aligned}
50 \cdot e^{-t/250} &= 9 \\
e^{-t/250} &= \dfrac{9}{50} \\
-\dfrac{t}{250} &= \ln \left(\dfrac{9}{50} \right) \\
t &\approx 428.7 \text{ days.}
\end{aligned}
$$

20.

$$
\begin{aligned}
4{,}000 \cdot \left(1 + \dfrac{0.06}{4} \right)^{4t} &= 10{,}000 \\
(1.015)^{4t} &= 2.5 \\
4t &= \log_{1.015}(2.5) \\
t &\approx 15.38576 \text{ years} \\
t &\approx 61.5 \text{ quarters}
\end{aligned}
$$

21. Substituting $t = 100$, we obtain

$$-50 \cdot \ln(1 - p) = 100$$
$$\ln(1 - p) = -2$$
$$1 - p = e^{-2}$$
$$p = 1 - e^{-2}$$
$$p \approx 0.86.$$

The level reached after 100 hr is $p = 0.86$. When $p = 1$, then $t = -50 \ln(0)$ which is undefined, so it is impossible to master MGM.

Tying It All Together

1. Since $x - 3 = \pm 2$, we get $x = 3 \pm 2 = 1, 5$.

2. Since $\log((x - 3)^2) = \log(4)$, we obtain

$$(x - 3)^2 = 4$$
$$x - 3 = \pm 2$$
$$x = 3 \pm 2$$
$$x = 5, 1.$$

Checking $x = 1$, one gets $2\log(-2)$ which is undefined, so $x = 5$.

3. Using the definition of a logarithm, we obtain $x - 3 = 2^4$. The solution is $x = 16 + 3 = 19$.

4. Since $2^{x-3} = 2^2$, we have $x - 3 = 2$. So $x = 5$.

5. Square both sides of $\sqrt{x - 3} = 4$. Then $x - 3 = 16$. The solution is $x = 19$.

6. An equivalent equation is $x - 3 = \pm 4$. Thus, $x = 3 \pm 4 = -1, 7$.

7. Completing the square, we obtain

$$x^2 - 4x + 4 = -2 + 4$$
$$(x - 2)^2 = 2$$
$$x - 2 = \pm\sqrt{2}$$
$$x = 2 \pm \sqrt{2}.$$

8. Since $2^{x-3} = 2^{2x}$, $x - 3 = 2x$. So $x = -3$.

9. Raise both sides to the third power.

$$(\sqrt[3]{x - 5})^3 = 5^3$$
$$x - 5 = 125$$
$$x = 130$$

10. By using the definition of a logarithm, we have $x = \log_2(3)$.

11.

$$\log(4x - 12) = \log(x)$$
$$4x - 12 = x$$
$$3x = 12$$
$$x = 4$$

12. Use synthetic division with $c = -1$.

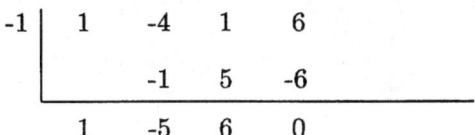

The quotient factors as $x^2 - 5x + 6 = (x - 3)(x - 2)$. The solutions are $x = -1, 2, 3$.

13. Parabola $y = x^2$ goes through $(0, 0), (\pm 1, 1)$

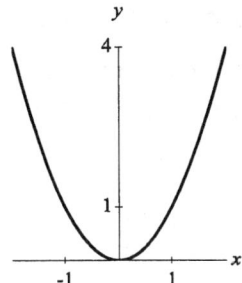

14. Parabola $y = (x - 2)^2$ goes through $(0, 4)$, $(1, 1)$, $(3, 1)$

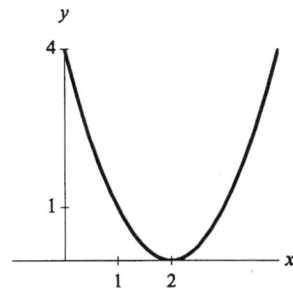

15. $y = 2^x$ goes through $\left(-1, \dfrac{1}{2}\right)$, $(0,1)$, $(1,2)$

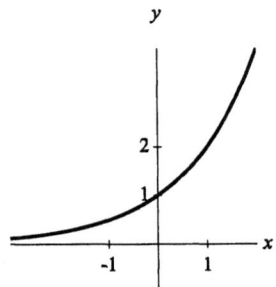

16. $y = x^{-2}$ goes through $(\pm 1, 1)$, $\left(\pm 2, \dfrac{1}{4}\right)$,

and $\left(\pm \dfrac{1}{2}, 4\right)$

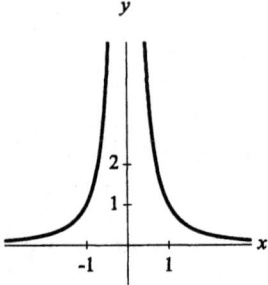

17. $y = \log_2(x - 2)$ goes through $(3,0)$, $(4,1)$, and $(2.5, -1)$

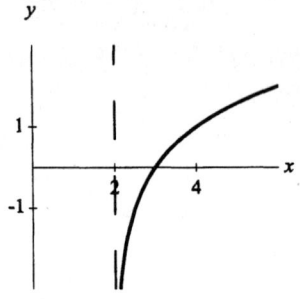

18. Line $y = x - 2$ goes through $(0,-2)$, $(2,0)$

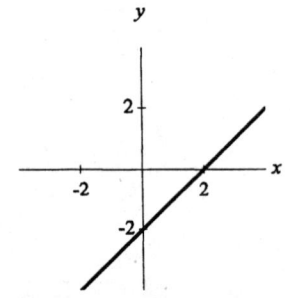

19. Line $y = 2x$ goes through $(0,0)$, $(1,2)$

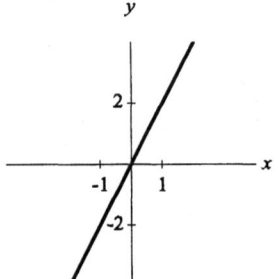

20. Line $y = x \cdot \log(2)$ goes through $(0,0)$, $(1, \log(2))$

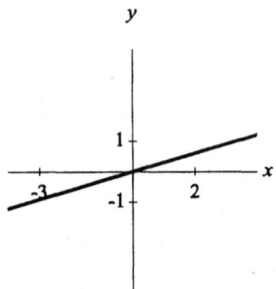

21. Horizontal line $y = e^2$ goes through $(0, e^2)$

22. Parabola $y = 2 - x^2$ goes through $(0,2)$, $(\pm 1, 1)$

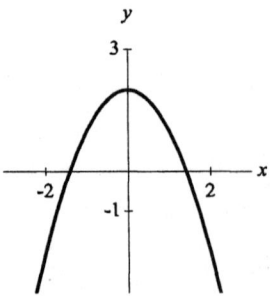

23. $y = \dfrac{2}{x}$ is symmetric about the origin and

goes through $(1,2)$, $(2,1)$, $\left(\dfrac{1}{2}, 4\right)$, $\left(4, \dfrac{1}{2}\right)$

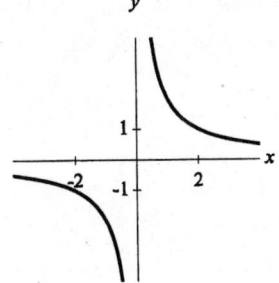

24. Graph of $y = \dfrac{1}{x-2}$ is obtained by shifting

$y = \dfrac{1}{x}$ to the right by 2 units.

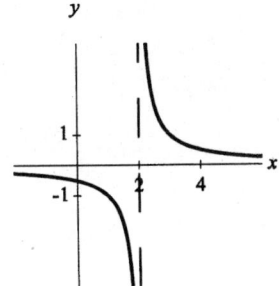

25. $f^{-1}(x) = 3x$

26. $f^{-1}(x) = -\log_3(x)$ since $f(x) = 3^{-x}$

27. $f^{-1}(x) = x^2 + 2$ for $x \geq 0$

28. $f^{-1}(x) = \sqrt[3]{x-2} + 5$

29. $f^{-1}(x) = (10^x + 3)^2$ **30.** $\{(1,3), (4,5)\}$

31. $f^{-1}(x) = 5 + \dfrac{1}{x-3}$

32. $f^{-1}(x) = (\ln(3-x))^2$ for $x \leq 2$

33. $(p \circ m)(x) = e^{x+5}$, domain $(-\infty, \infty)$,
range $(0, \infty)$

34. $(p \circ q)(x) = e^{\sqrt{x}}$, domain $[0, \infty)$, range $[1, \infty)$

35. $(q \circ p \circ m)(x) = q\left(e^{x+5}\right) = \sqrt{e^{x+5}}$,
domain $(-\infty, \infty)$, range $(0, \infty)$

36. $(m \circ r \circ q)(x) = m\left(\ln(\sqrt{x})\right) = \ln(\sqrt{x}) + 5$,
domain $(0, \infty)$, range $(-\infty, \infty)$

37. $(p \circ r \circ m)(x) = p\left(\ln(x+5)\right) = e^{\ln(x+5)}$,
domain $(-5, \infty)$, range $(0, \infty)$

38. $(r \circ q \circ p)(x) = r\left(\sqrt{e^x}\right) = \ln\left(\sqrt{e^x}\right) = \dfrac{1}{2}x$,
domain $(-\infty, \infty)$, range $(-\infty, \infty)$

39. $F(x) = (f \circ g \circ h)(x)$

40. $H(x) = (g \circ h \circ f)(x)$

41. $G(x) = (h \circ g \circ f)(x)$

42. $M(x) = (h \circ f \circ g)(x)$

For Thought

1. True, since if $x = 2$ and $y = 5$ we get $2 + 3 = 5$.

2. False, since $(2, 3)$ does not satisfy $x - y = 1$.

3. False, it is independent since the two lines are perpendicular and intersect at only one point.

4. True. Multiply $x - 2y = 4$ by -3 and add to the second equation.

$$
\begin{array}{rcr}
-3x + 6y &=& -12 \\
3x - 6y &=& 8 \\
\hline
0 &=& -4
\end{array}
$$

Since $0 = -4$ is false, there is no solution.

5. True, adding gives $2x = 6$.   **6.**   True

7. False, it is dependent because substituting $x = 5 + 3y$ results in an identity.

$$
\begin{array}{rcl}
9y - 3(5 + 3y) &=& -15 \\
9y - 15 - 9y &=& -15 \\
-15 &=& -15
\end{array}
$$

8. False, it is dependent and the solution set is $\{(5 + 3y, y) \mid y \text{ is any real number}\}$.

9. False, it is dependent, there are an infinite number of solutions.

10. True, since both lines have slope $1/2$.

5.1 Exercises

1. Note, $x = 1$ and $y = 3$ satisfies both $x + y = 4$ and $x - y = -2$. Yes, $(1, 2)$ is a solution.

3. Note, $x = -1$ and $y = 5$ does not satisfy $x - 2y = -9$. Thus, $(-1, 5)$ is not a solution.

5. $\{(1, 2)\}$

7. No solution since the lines are parallel. The solution set is $\emptyset$.

9. $\{(3, 2)\}$

11. $\{(3, 1)\}$

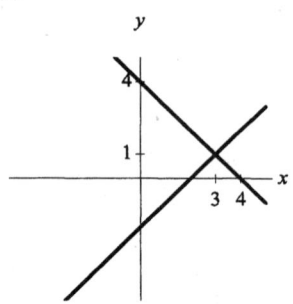

13. No solution since the lines are parallel. The solution set is $\emptyset$.

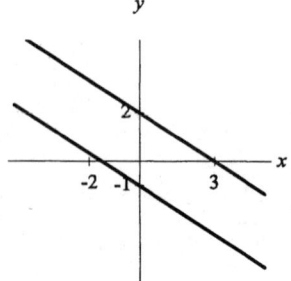

15. Since the lines are identical, the solution set is $\{(x, y) \mid x - 2y = 6\}$.

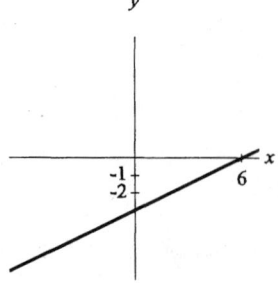

17. Substitute $y = 2x + 1$ into $3x - 4y = 1$.

$$
\begin{aligned}
3x - 4(2x + 1) &= 1 \\
3x - 8x - 4 &= 1 \\
-5x &= 5 \\
x &= -1
\end{aligned}
$$

From $y = 2x + 1$, $y = 2(-1) + 1 = -1$.
Independent and solution set is $\{(-1, -1)\}$.

19. Substitute $y = 1 - x$ into $2x - 3y = 8$.

$$
\begin{aligned}
2x - 3(1 - x) &= 8 \\
2x - 3 + 3x &= 8 \\
5x &= 11 \\
x &= 11/5
\end{aligned}
$$

From $y = 1 - x$, $y = 1 - 11/5 = -6/5$.
Independent and solution set is
$\{(11/5, -6/5)\}$.

21. Substitute $y = 3x + 5$ into $3(x + 1) = y - 2$.

$$
\begin{aligned}
3x + 3 &= (3x + 5) - 2 \\
3x + 3 &= 3x + 3 \\
3 &= 3
\end{aligned}
$$

Dependent and solution set is
$\{(x, y) \mid y = 3x + 5\}$.

23. Multiplying $\frac{1}{2}x + \frac{1}{3}y = 3$ by 6, we get

$3x + 2y = 18$. Then substitute $2y = 6 - 3x$

into $3x + 2y = 18$. So

$$
\begin{aligned}
3x + (6 - 3x) &= 18 \\
6 &= 18
\end{aligned}
$$

Inconsistent and the solution set is $\emptyset$.

25. Multiplying $0.05x + 0.06y = 10.50$ by 100, we obtain $5x + 6y = 1050$. Then substitute $y = 200 - x$ into $5x + 6y = 1050$.

$$
\begin{aligned}
5x + 6(200 - x) &= 1050 \\
5x + 1200 - 6x &= 1050 \\
-x &= -150 \\
x &= 150
\end{aligned}
$$

From $y = 200 - x$, $y = 200 - 150 = 50$.
Independent and solution set is $\{(150, 50)\}$.

27. Since $3x + 1 = 3x - 7$ leads to $1 = -7$, the system is inconsistent and the solution set is $\emptyset$.

29. Multiplying the first and second equations by 6 and 4, respectively, we have
$3x - 2y = 72$ and $x - 2y = 4$.
Substitute $2y = x - 4$ into $3x - 2y = 72$.

$$
\begin{aligned}
3x - (x - 4) &= 72 \\
2x + 4 &= 72 \\
2x &= 68 \\
x &= 34
\end{aligned}
$$

From $2y = x - 4$, $y = \dfrac{34 - 4}{2} = 15$.
Independent and solution set is $\{(34, 15)\}$.

31. Adding the two equations, we get $2x = 26$.
So $x = 13$ and from $x + y = 20$,
we obtain $13 + y = 20$ or $y = 7$.
Independent and solution set is $\{(13, 7)\}$.

33. Multiplying $x - y = 5$ by 2 and by adding to the second equation, we obtain

$$
\begin{aligned}
2x - 2y &= 10 \\
3x + 2y &= 10 \\
\hline
5x &= 20 \\
x &= 4
\end{aligned}
$$

From $x - y = 5$, $4 - y = 5$ or $y = -1$.
Independent and solution set is $\{(4, -1)\}$.

35. Adding the two equations leads to $0 = 12$.
Inconsistent and the solution set is $\emptyset$.

37. Multiply $2x + 3y = 1$ by -3 and $3x - 5y = -8$ by 2. Then add the equations.

$$
\begin{aligned}
-6x - 9y &= -3 \\
6x - 10y &= -16 \\
\hline
-19y &= -19 \\
y &= 1
\end{aligned}
$$

Since $2x + 3y = 1$, $2x + 3 = 1$ or $x = -1$.
Independent and solution set is $\{(-1, 1)\}$.

39. Multiply $0.05x + 0.1y = 0.6$ by -100 and $x + 2y = 12$ by 5. Then add the equations.

$$
\begin{aligned}
-5x - 10y &= -60 \\
5x + 10y &= 60 \\
\hline
0 &= 0
\end{aligned}
$$

Dependent and the solution set is
$\{(x, y) | x + 2y = 12\}$.

41. Multiplying $\frac{x}{2} + \frac{y}{2} = 5$ by 2 and

$\frac{3x}{2} - \frac{2y}{3} = 2$ by 6, we have $x + y = 10$

and $9x - 4y = 12$, respectively. Then multiply
$x + y = 10$ by 4 and add to the second equation.

$$
\begin{array}{rcl}
4x + 4y & = & 40 \\
9x - 4y & = & 12 \\
\hline
13x & = & 52 \\
x & = & 4
\end{array}
$$

Since $x + y = 10$, $4 + y = 10$ and $y = 6$.
Independent and the solution set is $\{(4, 6)\}$.

43. Multiply $3x - 2.5y = -4.2$ by -4 and
$0.12x + 0.09y = 0.4932$ by 100.
Then add the equations.

$$
\begin{array}{rcl}
-12x + 10y & = & 16.8 \\
12x + 9y & = & 49.32 \\
\hline
19y & = & 66.12 \\
y & = & 3.48
\end{array}
$$

From $3x - 2.5y = -4.2$, we find

$$
\begin{array}{rcl}
3x - 2.5(3.48) & = & -4.2 \\
3x - 8.7 & = & -4.2 \\
3x & = & 4.5 \\
x & = & 1.5.
\end{array}
$$

Independent and the solution set is
$\{(1.5, 3.48)\}$.

45. Independent   47. Dependent

49. The point of intersection is $(-1000, -497)$.

51. The point of intersection is approximately
$(6.18, -0.54)$

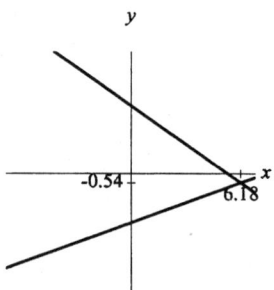

53. Let x and y be Althea's and Vaughn's incomes,
respectively. Then $x + y = 82,000$ and $x - y =
16,000$. By adding these two equations, we get
$2x = 98,000$ or $x = 49,000$. Thus, Althea's
income is \$49,000 and Vaughn's income
is \$33,000.

55. Let x and y be the amounts invested at 10%
and 8%, respectively.

$$
\begin{array}{rcl}
x + y & = & 25,000 \\
0.1x + 0.08y & = & 2200
\end{array}
$$

Multiply the second equation by -10 and
add to the first.

$$
\begin{array}{rcl}
x + y & = & 25,000 \\
-x - 0.8y & = & -22,000 \\
\hline
0.2y & = & 3000 \\
y & = & 15,000
\end{array}
$$

Carmen invested \$15,000 at 8% and \$10,000
at 10%.

57. Let x and y be the prices of an adult ticket
and child ticket, respectively.

$$
\begin{array}{rcl}
2x + 5y & = & 33 \\
x + 3y & = & 18.50
\end{array}
$$

Multiply the second equation by -2 and add
to the first equation.

$$
\begin{array}{rcl}
2x + 5y & = & 33 \\
-2x - 6y & = & -37 \\
\hline
-y & = & -4 \\
y & = & 4
\end{array}
$$

A child's ticket costs \$4. Since $x + 3y = 18.50$,
$x + 12 = 18.50$ and so an adult's ticket
costs $x = \$6.50$.

59. If m and f are the number of male and female memberships, respectively, then a system of equations is

$$m + f = 12$$
$$500m + 500f = 6,000.$$

Dividing the second equation by 500, one finds $m + f = 12$; the system of equations is dependent. Therefore, one cannot conclude the number of female memberships and male memberships.

All we know is that m and $12 - m$ are the number of male and female memberships where $0 \le m \le 12$.

61. If c and m are the prices of a coffee and a muffin, respectively, then a system of equations is

$$3c + 7m = 7.77$$
$$6c + 14m = 14.80.$$

But if one multiplies the second equation by two, then one obtains $6c + 14m = 15.54$; this last equation contradicts the second equation in the system. Therefore, the system of equations is inconsistent and has no solution.

63. Let x and y be the number of male and female students, respectively.

$$0.5x + 0.3y = 230$$
$$0.2x + 0.6y = 260$$

Multiply the first equation by -2 and the second by 5. Then add the resulting equations.

$$\begin{array}{rcl} -x - 0.6y &=& -460 \\ x + 3y &=& 1300 \\ \hline 2.4y &=& 840 \\ y &=& 350 \end{array}$$

From $0.2x + 0.6y = 260$, $0.2x + 210 = 260$ and so $x = 250$. There are $250 + 350 = 600$ students at CHS.

65. Let x and y be the number of nickels and pennies, respectively. We obtain

$$x + y = 87$$
$$0.05x + 0.01y = 1.75.$$

Multiply the second equation by -100 and then add it to the first equation.

$$\begin{array}{rcl} x + y &=& 87 \\ -5x - y &=& -175 \\ \hline -4x &=& -88 \\ x &=& 22 \end{array}$$

From $x + y = 87$, $22 + y = 87$ and $y = 65$. Isabelle has 22 nickels and 65 pennies.

67. The weights x and y must satisfy

$$5x = 3y$$
$$3(4 + x + y) = 6y.$$

The second equation can be written as $12 + 3x = 3y$. Substituting into the first equation one finds

$$\begin{array}{rcl} 12 + 3x &=& 5x \\ 12 &=& 2x \\ 6 &=& x. \end{array}$$

Then $x = 6$ oz and $y = 10$ oz since

$$y = \frac{5x}{3} = \frac{5(6)}{3}.$$

69. Let x be the number of months. Plan A costs $150x + 800$ and Plan B costs $200x + 200$. Thus, Plan A is cheaper in the long run. The number of months for which the costs are the same is given by

$$\begin{array}{rcl} 150x + 800 &=& 200x + 200 \\ 600 &=& 50x \\ x &=& 12 \text{ months.} \end{array}$$

71. If we set the formulas equal to each other, then we find

$$\begin{array}{rcl} 0.08aD &=& \dfrac{D(a+1)}{24} \\[2mm] 0.08a &=& \dfrac{(a+1)}{24} \\[2mm] 1.92a &=& a + 1 \\[2mm] a &=& \dfrac{1}{0.92} \\[2mm] a &\approx& 1.09. \end{array}$$

The dosage is the same if the age is 1.1 yr.

73. Solve for a and b.

$$-3a + b = 9$$
$$2a + b = -1.$$

Multiply the second equation by -1 and add to the first equation.

$$
\begin{array}{rcr}
-3a + b & = & 9 \\
-2a - b & = & 1 \\
\hline
-5a & = & 10 \\
a & = & -2
\end{array}
$$

Substituting $a = -2$ into $2a + b = -1$, one finds $b = 3$. An equation of the line is $y = -2x + 3$.

75. Solve for a and b.

$$-2a + b = 3$$
$$4a + b = -7$$

Multiply the first equation by -1 and add to the second equation.

$$
\begin{array}{rcr}
2a - b & = & -3 \\
4a + b & = & -7 \\
\hline
6a & = & -10 \\
a & = & -\dfrac{5}{3}
\end{array}
$$

Substituting $a = -\dfrac{5}{3}$ into $-2a + b = 3$, one

gets $b = -\dfrac{1}{3}$. An equation of the line is

$y = -\dfrac{5}{3}x - \dfrac{1}{3}$.

For Thought

1. True

2. False, since $(1, 1, 0)$ does not satisfy $-x - y + z = 4$.

3. True, adding the first two equations gives $0 = 6$ which is false.

4. False, since $(2, 3, -1)$ does not satisfy $x - y - z = 8$.

5. True

6. True, adding the first two equations gives an identity. Also, multiplying the first by -2 and then adding to the third equation gives an identity.

$$
\begin{array}{rcr}
x - y + z & = & 1 \\
-x + y - z & = & -1 \\
\hline
0 & = & 0
\end{array}
$$

$$
\begin{array}{rcr}
-2x + 2y - 2z & = & -2 \\
2x - 2y + 2z & = & 2 \\
\hline
0 & = & 0
\end{array}
$$

7. True. The calculations in Exercise 6 above show system (c) is dependent.

8. True

9. True, if $x = 1$ then $(x + 2, x, x - 1) = (3, 1, 0)$.

10. False, the value is $0.05x + 0.10y + 0.25z$ dollars.

5.2 Exercises

1. Points on the plane are $(5, 0, 0)$, $(0, 5, 0)$, and $(0, 0, 5)$.

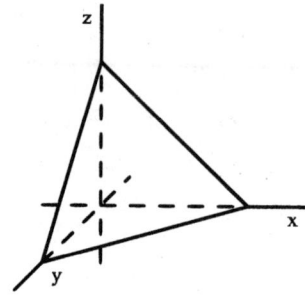

3. Points on the plane are $(3, 0, 0)$, $(0, 3, 0)$, and $(0, 0, -3)$.

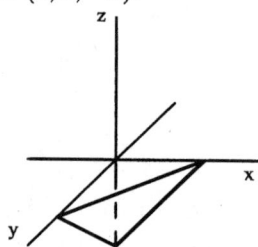

5. Note, $x = 1$, $y = 3$, and $z = 2$ satisfy all the three equations in the system . Yes, $(1, 3, 2)$ is a solution.

7. Note, $x = -1$, $y = 5$, and $z = 2$ does not satisfy $x - y - 2z = -8$. No, $(-1, 5, 2)$ is not a solution.

9. Add the first and second equations and add the first and third equations.

$$\begin{array}{rcr} x + y + z & = & 6 \\ 2x - 2y - z & = & -5 \\ \hline 3x - y & = & 1 \end{array}$$

$$\begin{array}{rcr} x + y + z & = & 6 \\ 3x + y - z & = & 2 \\ \hline 4x + 2y & = & 8 \end{array}$$

Divide $4x + 2y = 8$ by 2 and add to $3x - y = 1$.

$$\begin{array}{rcr} 2x + y & = & 4 \\ 3x - y & = & 1 \\ \hline 5x & = & 5 \end{array}$$

So $x = 1$. From $3x - y = 1$, $3 - y = 1$ and $y = 2$. From $x + y + z = 6$, $1 + 2 + z = 6$ and $z = 3$. The solution set is $\{(1, 2, 3)\}$.

11. Multiply the first equation by 2 and add to the second. Then multiply the first equation by -3 and add to the third equation.

$$\begin{array}{rcr} 6x + 4y + 2z & = & 2 \\ x + y - 2z & = & -4 \\ \hline 7x + 5y & = & -2 \end{array}$$

$$\begin{array}{rcr} -9x - 6y - 3z & = & -3 \\ 2x - 3y + 3z & = & 1 \\ \hline -7x - 9y & = & -2 \end{array}$$

Add $7x + 5y = -2$ to $-7x - 9y = -2$.

$$\begin{array}{rcr} 7x + 5y & = & -2 \\ -7x - 9y & = & -2 \\ \hline -4y & = & -4 \end{array}$$

So $y = 1$. From $7x + 5y = -2$, $7x + 5 = -2$ and $x = -1$. From $3x + 2y + z = 1$, $-3 + 2 + z = 1$ and $z = 2$. The solution set is $\{(-1, 1, 2)\}$.

13. Add first equation to third equation. Multiply second equation by 2 and add to the first.

$$\begin{array}{rcr} 2x + y - 2z & = & -15 \\ x + 3y + 2z & = & -5 \\ \hline 3x + 4y & = & -20 \end{array}$$

$$\begin{array}{rcr} 8x - 4y + 2z & = & 30 \\ 2x + y - 2z & = & -15 \\ \hline 10x - 3y & = & 15 \end{array}$$

Multiply $3x + 4y = -20$ by 3 and $10x - 3y = 15$ by 4 and then add the equations.

$$\begin{array}{rcr} 9x + 12y & = & -60 \\ 40x - 12y & = & 60 \\ \hline 49x & = & 0 \end{array}$$

So $x = 0$. From $3x + 4y = -20$, $4y = -20$ and $y = -5$. From $2x + y - 2z = -15$, we get $-5 - 2z = -15$ and $z = 5$. The solution set is $\{(0, -5, 5)\}$.

15. Adding the two equations, we have $4x - 4z = -20$ or $z = x + 5$. From $x + 2y - 3z = -17$, we obtain

$$\begin{array}{rcr} x + 2y - 3(x + 5) & = & -17 \\ 2y - 2x - 15 & = & -17 \\ 2y & = & 2x - 2 \\ y & = & x - 1. \end{array}$$

The solution set is $\{(x, x - 1, x + 5) | x \text{ is any real number}\}$.

17. Adding the first two equations, we have $0 = 0$, which is an identity. Multiply the second equation by 2 and add to the third equation.

$$\begin{array}{rcr} -2x - 4y + 6z & = & -10 \\ 2x + 4y - 6z & = & 10 \\ \hline 0 & = & 0 \end{array}$$

The system is dependent and the solution set is $\{(x, y, z) | x + 2y - 3z = 5\}$.

19. Multiply first equation by -2 and add to the second equation.

$$\begin{array}{rcr} -2x + 4y - 6z & = & -10 \\ 2x - 4y + 6z & = & 3 \\ \hline 0 & = & -7 \end{array}$$

This is false, so the solution set is $\emptyset$.

21. Adding the two equations, we get $3x = 6$ or $x = 2$. Substitute into the two equations. Then

$$\begin{array}{rcr} 2 + y - z & = & 2 \\ y - z & = & 0 \end{array}$$

and

$$4 - y + z = 4$$
$$-y + z = 0.$$

In any case $y = z$. The solution set is $\{(2, y, y) | y \text{ is any real number}\}$.

23. If the second equation is multiplied by -1 and added to the first equation, the result is the third equation.

$$\begin{array}{rcr} x + y &=& 5 \\ -y + z &=& -2 \\ \hline x + z &=& 3 \end{array}$$

There ae infinitely number many solutions. Since $y = 5 - x$ and $z = 3 - x$, the solution set $\{(x, 5 - x, 3 - x) | x \text{ is any real number}\}$.

25. Multiply the first equation by 2 and add to the second equation.

$$\begin{array}{rcr} 2x - 2y + 2z &=& 14 \\ 2y - 3z &=& -13 \\ \hline 2x - z &=& 1 \end{array}$$

Multiply $2x - z = 1$ by -2 and add to the third equation.

$$\begin{array}{rcr} -4x + 2z &=& -2 \\ 3x - 2z &=& -3 \\ \hline -x &=& -5 \end{array}$$

So $x = 5$. From $2x - z = 1$, we get $10 - z = 1$ and $z = 9$. Since $2y - 3z = -13$, $2y - 27 = -13$ and $y = 7$. The solution set is $\{(5, 7, 9)\}$.

27. Multiply first equation by -2 and add to third equation. Then multiply first equation by -4 and add to the second.

$$\begin{array}{rcr} -2x - 2y - 4z &=& -15 \\ 5x + 2y + 5z &=& 21 \\ \hline 3x + z &=& 6 \end{array}$$

$$\begin{array}{rcr} -4x - 4y - 8z &=& -30 \\ 3x + 4y + z &=& 12 \\ \hline -x - 7z &=& -18 \end{array}$$

Multiply $-x - 7z = -18$ by 3 and add to $3x + z = 6$.

$$\begin{array}{rcr} -3x - 21z &=& -54 \\ 3x + z &=& 6 \\ \hline -20z &=& -48 \end{array}$$

So $z = 48/20 = 2.4$. From $3x + z = 6$, we get $3x + 2.4 = 6$ and $x = 1.2$. From $x + y + 2z = 7.5$, $1.2 + y + 4.8 = 7.5$ and $y = 1.5$. The solution set is $\{(1.2, 1.5, 2.4)\}$.

29. Multiply the first equation by -5 and add to 100 times the second one.

$$\begin{array}{rcr} -5x - 5y - 5z &=& -45,000 \\ 5x + 6y + 9z &=& 71,000 \\ \hline y + 4z &=& 26,000 \end{array}$$

Substitute $z = 3y$ into $y + 4z = 26,000$.

$$\begin{array}{rcl} y + 12y &=& 26,000 \\ 13y &=& 26,000 \\ y &=& 2,000 \end{array}$$

From $z = 3y$, we obtain $z = 6000$. From $x + y + z = 9,000$, we have $x + 2000 + 6000 = 9000$ and $x = 1000$. The solution set is $\{(1000, 2000, 6000)\}$.

31. Substitute $x = 2y - 1$ into $z = 2x - 3$ to get $z = 4y - 5$. Then substitute $z = 4y - 5$ into $y = 3z + 2$.

$$\begin{array}{rcl} y &=& 3(4y - 5) + 2 \\ y &=& 12y - 13 \\ -11y &=& -13 \\ y &=& 13/11 \end{array}$$

From $z = 4y - 5$, we obtain $z = 52/11 - 5 = -3/11$. Since $x = 2y - 1$, $x = 26/11 - 1 = 15/11$. The solution set is $\{(15/11, 13/11, -3/11)\}$.

33. Substitute $(-1, -2)$, $(2, 1)$, $(-2, 1)$ into $y = ax^2 + bx + c$. So

$$\begin{array}{rcr} a - b + c &=& -2 \\ 4a + 2b + c &=& 1 \\ 4a - 2b + c &=& 1. \end{array}$$

Multiply first equation by -1 and add to the second and third equations.

$$\begin{array}{rcr} -a + b - c &=& 2 \\ 4a + 2b + c &=& 1 \\ \hline 3a + 3b &=& 3 \end{array}$$

$$-a + b - c = 2$$
$$\underline{4a - 2b + c = 1}$$
$$3a - b = 3$$

Multiply $3a + 3b = 3$ by -1 and add to $3a - b = 3$.

$$-3a - 3b = -3$$
$$\underline{3a - b = 3}$$
$$-4b = 0$$

So $b = 0$. From $3a - b = 3$, we get $3a = 3$ and $a = 1$. From $a - b + c = -2$, $1 + c = -2$ and $c = -3$. Since the solution is $(a, b, c) = (1, 0, -3)$, the parabola is $y = x^2 - 3$.

35. Substitute $(0,0)$, $(1,3)$, $(2,2)$ into $y = ax^2 + bx + c$. Then

$$c = 0$$
$$a + b + c = 3$$
$$4a + 2b + c = 2.$$

Multiply second one by -2 and add to third equation.

$$-2a - 2b - 2c = -6$$
$$\underline{4a + 2b + c = 2}$$
$$2a - c = -4$$

Substituting $c = 0$ into $2a - c = -4$, we get $2a = -4$ and $a = -2$. From $a + b + c = 3$, $-2 + b = 3$ and $b = 5$. Since $(a, b, c) = (-2, 5, 0)$, the parabola is $y = -2x^2 + 5x$.

37. Substitute $(0,4)$, $(-2,0)$, $(-3,1)$ into $y = ax^2 + bx + c$. Then

$$c = 4$$
$$4a - 2b + c = 0$$
$$9a - 3b + c = 1.$$

Multiply second and third equations by 3 and -2, respectively, then add the equations.

$$12a - 6b + 3c = 0$$
$$\underline{-18a + 6b - 2c = -2}$$
$$-6a + c = -2$$

Substituting $c = 4$ into $-6a + c = -2$, $-6a + 4 = -2$ and $a = 1$. From $4a - 2b + c = 0$, $4 - 2b + 4 = 0$ and $b = 4$. Since $(a, b, c) = (1, 4, 4)$, the parabola is $y = x^2 + 4x + 4$.

39. By substituting the coordinates of the points $(1,0,0)$, $(0,1,0)$, and $(0,0,1)$ into $ax + by + cz = 1$, we get $a = 1$, $b = 1$, and $c = 1$. A linear equation satisfied by the ordered triples is $x + y + z = 1$.

41. By substituting the coordinates of the points $(1,1,1)$, $(0,2,0)$, and $(1,0,0)$ into $ax + by + cz = 1$, we get $a + b + c = 1$, $2b = 1$, and $a = 1$. Then $b - = \frac{1}{2}$ and $c = -\frac{1}{2}$. A linear equation satisfied by the ordered triples is

$$x + \frac{1}{2}y - \frac{1}{2}z = 1 \text{ or } 2x + y - z = 2$$

43. Let x, y, and z be the amounts invested in stocks, bonds, and a mutual fund. Then

$$x + y + z = 25,000$$
$$0.08x + 0.10y + 0.06z = 1,860$$
$$2y = z.$$

Multiply the first equation by -8 and add to 100 times the second.

$$-8x - 8y - 8z = -200,000$$
$$\underline{8x + 10y + 6z = 186,000}$$
$$2y - 2z = -14,000$$

Substitute $z = 2y$ into $2y - 2z = -14,000$.

$$2y - 4y = -14,000$$
$$-2y = -14,000$$
$$y = 7,000$$

Since $z = 2y$, we obtain $z = 14,000$. Since $x + y + z = 25,000$, $x = 4000$. Marita invested $\$4,000$ in stocks, $\$7000$ in bonds, and $\$14,000$ in a mutual fund.

45. Let x, y, and z be the prices last year of a hamburger, fries, and a Coke, respectively. So

$$x + y + z = 1.90$$
$$1.1x + 1.2y + 1.5z = 2.37$$
$$1.5z = 1.1x + 0.09.$$

Multiply first equation by -12 and add to 100 times the second equation.

$$-12x - 12y - 12z = -22.8$$
$$\underline{11x + 12y + 15z = 23.7}$$
$$-x + 3z = 0.9$$

Substitute $x = 3z - 0.9$ into $1.5z = 1.1x + 0.09$.

$$
\begin{aligned}
1.5z &= 1.1(3z - 0.9) + 0.09 \\
1.5z &= 3.3z - 0.99 + 0.09 \\
0.90 &= 1.8z \\
0.50 &= z
\end{aligned}
$$

Since $x = 3z - 0.9$, we find $x = 1.50 - 0.90 = 0.60$. From $x + y + z = 1.90$, we have $y = 0.80$. The prices last year of a hamburger, fries and a Coke are \$0.60, \$0.80, and \$0.50, respectively.

47. Let L_f and L_r be the weights on the left front tire and left rear tire, respectively. Let R_f and R_r be the weights on the right front tire and right rear tire, respectively. Since $1200(0.51) = 612$ and $1200(0.48) = 576$, we obtain

$$
\begin{aligned}
L_f + L_r &= 612 \\
R_r + L_r &= 576 \\
L_f, L_r, R_f, R_r &\geq 280.
\end{aligned}
$$

Three possible weight distributions are
$(L_r, L_f, R_r, R_f) = (280, 332, 296, 292)$,
$(L_r, L_f, R_r, R_f) = (285, 327, 291, 297)$,
and $(L_r, L_f, R_r, R_f) = (290, 322, 286, 302)$.

49. Let x, y, and z be the number of pennies, nickels, and dimes, respectively. Then

$$
\begin{aligned}
x + y + z &= 232 \\
y + z &= x \\
0.01x + 0.05y + 0.10z &= 10.36.
\end{aligned}
$$

Multiply first equation by -1 and add to 10 times the third equation. Also combine first two equations.

$$
\begin{array}{rcr}
-x - y - z &=& -232 \\
0.1x + 0.5y + z &=& 103.6 \\
\hline
-0.9x - 0.5y &=& -128.4
\end{array}
$$

$$
\begin{array}{rcr}
-x - y - z &=& -232 \\
-x + y + z &=& 0 \\
\hline
-2x &=& -232
\end{array}
$$

Then $x = 116$. Substituting into $-0.9x - 0.5y = -128.4$, we obtain

$$
\begin{aligned}
-0.9(116) - 0.5y &= -128.4 \\
-104.4 - 0.5y &= -128.4 \\
24 &= 0.5y \\
48 &= y.
\end{aligned}
$$

Since $x + y + z = 232$, $116 + 48 + z = 232$ and $z = 68$. Emma used 116 pennies, 48 nickels, and 68 dimes.

51. Let x, y, and z be the prices of a carton of milk, a cup of coffee, and a doughnut, respectively. So

$$
\begin{aligned}
3x + 4y + 7z &= 5.45 \\
4x + 2y + 8z &= 5.30 \\
2x + 5y + 6z &= 5.15.
\end{aligned}
$$

Multiply third equation by -2 and add to second equation. Also, multiply first equation by -4 and add to 3 times the second equation.

$$
\begin{array}{rcr}
-4x - 10y - 12z &=& -10.30 \\
4x + 2y + 8z &=& 5.30 \\
\hline
-8y - 4z &=& -5
\end{array}
$$

$$
\begin{array}{rcr}
-12x - 16y - 28z &=& -21.80 \\
12x + 6y + 24z &=& 15.90 \\
\hline
-10y - 4z &=& -5.90
\end{array}
$$

Multiply $-8y - 4z = -5$ by -1 and add to $-10y - 4z = -5.90$.

$$
\begin{array}{rcr}
8y + 4z &=& 5 \\
-10y - 4z &=& -5.90 \\
\hline
-2y &=& -0.90
\end{array}
$$

Then $y = 0.45$. Substitute into $-8y - 4z = -5$ to get $-3.60 - 4z = -5$ and $z = 0.35$. Since $3x + 4y + 7z = 5.45$, we have $3x + 1.80 + 2.45 = 5.45$ and $x = 0.40$.

Alphonse's bill was
$5(0.40) + 2(0.45) + 9(0.35) = \$6.05$.
His change is \$3.95.

53. A system of equations is

$$4x = 6y$$
$$2(x + y) = 15(8)$$
$$6(x + y + 15 + 10) = 10z.$$

Rewriting the first two equations, one obtains

$$2x - 3y = 0$$
$$x + y = 60.$$

Solving this smaller system, one finds $x = 36$ lb, $y = 24$ lb. Substituting into the third equation, one finds

$$z = \frac{6(x + y + 25)}{10} = \frac{6(60 + 25)}{10} = 51 \text{ lb.}$$

55.

a) Substitute $(0,0)$, $(10,40)$, $(20,70)$ into $y = ax^2 + bx + c$. Consequently, we obtain the following system of equations:

$$c = 0$$
$$100a + 10b + c = 40$$
$$400a + 20b + c = 70.$$

Multiply the second equation by -2 and add to third equation.

$$-200a - 20b - 2c = -80$$
$$\underline{400a + 20b + c = 70}$$
$$200a - c = -10$$

Substitute $c = 0$ into $200a - c = -10$. Then $a = -\frac{1}{20}$. From $100a + 10b + c = 40$, we obtain $-5 + 10b = 40$ and $b = \frac{9}{2}$.

The parabola is $y = -\frac{1}{20}x^2 + \frac{9}{2}x$.

b) Since $-b/(2a) = 45$, the maximum height is

$$-\frac{1}{20}(45)^2 + \frac{9}{2}(45) = 101.25 \text{ m.}$$

c) Since the zeros of $y = -\frac{x}{20}(x - 90)$ are $x = 0, 90$, the missile will strike 90 m from the origin.

For Thought

1. True, since the line $y = x$ passes through the center of the circle.

2. False, when a line is tangent to a circle it intersects the circle at only one point.

3. True, the parabola $y = 2x^2 - 4$ intersects the circle $x^2 + y^2 = 16$ at three points.

4. False, they intersect at $(\pm 1, 0)$.

5. False, they intersect at $(1, 1)$ and $(-1, -1)$.

6. False, since three noncollinear points determine a unique circle through the points.

7. True, since either leg can serve as base and the other leg as altitude.

8. True 9. True

10. False, two such numbers are $\frac{1}{2}\left(7 \pm \sqrt{45}\right)$.

5.3 Exercises

1. Since $x = -1$ and $y = 4$ satisfy both equations in the system , $(-1, 4)$ is a solution.

3. Since $x = 4$ and $y = -5$ does not satisfy $x - y = 1$, $(4, -5)$ is not a solution.

5. Since $y = x$ and $y = x^2$, we obtain

$$x = x^2$$
$$x - x^2 = 0$$
$$x(x - 1) = 0.$$

Then $x = 0, 1$ and the solution set is $\{(0,0), (1,1)\}$.

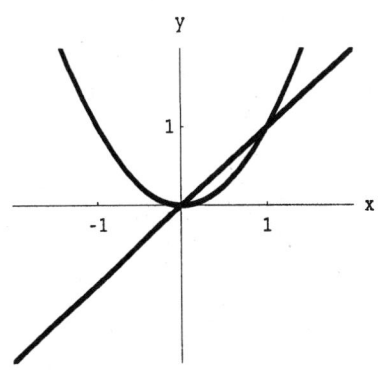

7. Substitute $y = x^2$ into $5x - y = 6$.

$$
\begin{aligned}
5x - x^2 &= 6 \\
x^2 - 5x &= -6 \\
x^2 - 5x + 6 &= 0 \\
(x - 3)(x - 2) &= 0
\end{aligned}
$$

If $x = 3, 2$ in $y = x^2$, then $y = 9, 4$.
The solution set is $\{(2, 4),\ (3, 9)\}$.

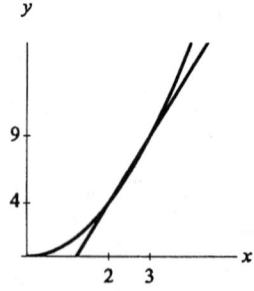

9. Substitute $y = |x|$ into $y = x + 3$.

$$
\begin{aligned}
|x| &= x + 3 \\
x = x + 3 \quad &\text{or} \quad -x = x + 3 \\
0 = 3 \quad &\text{or} \quad -2x = 3
\end{aligned}
$$

Then $x = -3/2$. Since $y = |x|$, $y = 3/2$.
The solution set is $\{(-3/2, 3/2)\}$.

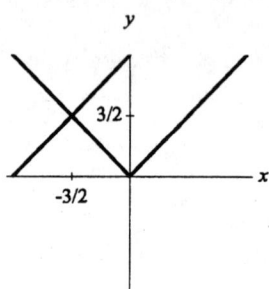

11. Substitute $y = x^2$ into $y = |x|$.

$$
\begin{aligned}
x^2 &= |x| \\
x^2 = x \quad &\text{or} \quad x^2 = -x \\
x^2 - x = 0 \quad &\text{or} \quad x^2 + x = 0 \\
x(x - 1) = 0 \quad &\text{or} \quad x(x + 1) = 0 \\
x = 0, -1 \quad &\text{or} \quad x = 0, -1
\end{aligned}
$$

Using $x = 0, \pm 1$ in $y = x^2$, one finds
$y = 0, 1$. The solution set is
$\{(0, 0), (-1, 1), (1, 1)\}$.

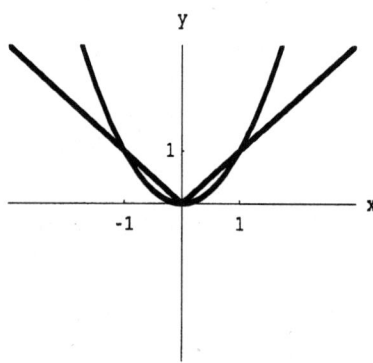

13. Substitute $y = \sqrt{x}$ into $y = 2x$ to obtain

$$
\begin{aligned}
\sqrt{x} &= 2x \\
x &= 4x^2 \\
x - 4x^2 &= 0 \\
x(1 - 4x) &= 0.
\end{aligned}
$$

Using $x = 0, 1/4$ in $y = 2x$, $y = 0, 1/2$.
The solution set is $\{(0, 0), (1/4, 1/2)\}$.

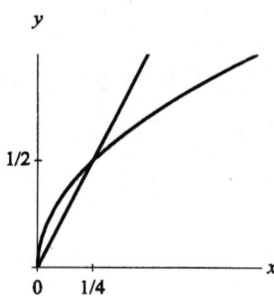

15. Substitute $y = x^3$ into $y = 4x$ to get

$$
\begin{aligned}
x^3 &= 4x \\
x^3 - 4x &= 0 \\
x(x^2 - 4) &= 0 \\
x &= 0, \pm 2.
\end{aligned}
$$

Substituting $x = 0, 2, -2$ into $y = 4x$,
we get $y = 0, 8, -8$.
The solution set is $\{(0, 0), (2, 8), (-2, -8)\}$.

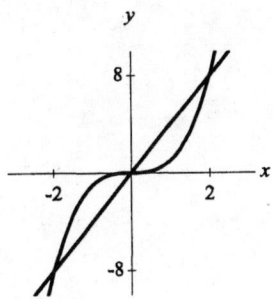

17. Substitute $y = x$ into $y = x^3 - x$.

$$
\begin{aligned}
x &= x^3 - x \\
2x - x^3 &= 0 \\
x(2 - x^2) &= 0 \\
x &= 0, \sqrt{2}, -\sqrt{2}
\end{aligned}
$$

Since $y = x$, the solution set is
$\{(0,0), (\sqrt{2}, \sqrt{2}), (-\sqrt{2}, -\sqrt{2})\}$.

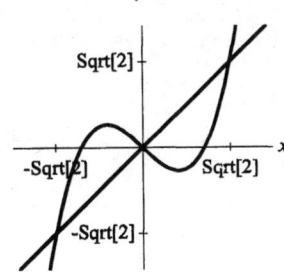

19. Substitute $y = x$ into $x^2 + y^2 = 1$.

$$
\begin{aligned}
x^2 + x^2 &= 1 \\
2x^2 &= 1 \\
x^2 &= \frac{1}{2} \\
x &= \pm \frac{\sqrt{2}}{2}
\end{aligned}
$$

Since $y = x$, the solution set is

$$
\left\{ \left(\frac{\sqrt{2}}{2}, \frac{\sqrt{2}}{2} \right), \left(-\frac{\sqrt{2}}{2}, -\frac{\sqrt{2}}{2} \right) \right\}.
$$

A sketch of the line and circle is shown on the next page.

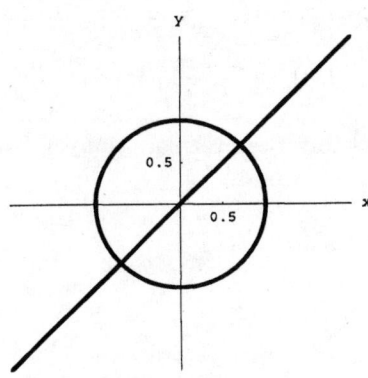

21. Substitute $y = -x - 4$ into $xy = 1$.

$$
\begin{aligned}
x(-x - 4) &= 1 \\
-x^2 - 4x &= 1 \\
x^2 + 4x &= -1 \\
x^2 + 4x + 4 &= 3 \\
(x + 2)^2 &= 3 \\
x &= -2 \pm \sqrt{3}
\end{aligned}
$$

Using $x = -2 + \sqrt{3}, -2 - \sqrt{3}$ in $y = -x - 4$, we find $y = -2 - \sqrt{3}, -2 + \sqrt{3}$.
The solution set is
$\left\{ (-2 + \sqrt{3}, -2 - \sqrt{3}), (-2 - \sqrt{3}, -2 + \sqrt{3}) \right\}$.

23. Substitute $x^2 = 2y^2 - 1$ into $2x^2 - y^2 = 1$.

$$
\begin{aligned}
2(2y^2 - 1) - y^2 &= 1 \\
3y^2 - 2 &= 1 \\
3y^2 &= 3 \\
y^2 &= 1 \\
y &= \pm 1
\end{aligned}
$$

Using $y = 1$ in $x^2 = 2y^2 - 1$, we get $x^2 = 1$ or $x = \pm 1$. Also, if $y = -1$ then $x = \pm 1$.
The solution set is
$\{(1,1), (1,-1), (-1,1), (-1,-1)\}$.

25. Since $2x - y = 1$, we obtain $y = 2x - 1$.
Then substitute into $xy - 2x = 2$.

$$
\begin{aligned}
x(2x - 1) - 2x &= 2 \\
2x^2 - 3x - 2 &= 0 \\
(2x + 1)(x - 2) &= 0 \\
x &= -\frac{1}{2}, 2
\end{aligned}
$$

Since $y = 2x - 1$, we get $y = 2 \cdot \left(-\dfrac{1}{2}\right) - 1 = -2$

and $x = 2(2) - 1 = 3$. The solution set

is $\left\{(2, 3), \left(-\dfrac{1}{2}, -2\right)\right\}$.

27. Multiply the first equation by 2 and add to the second.

$$\dfrac{6}{x} - \dfrac{2}{y} = \dfrac{26}{10}$$

$$\dfrac{1}{x} + \dfrac{2}{y} = \dfrac{9}{10}$$

$$\overline{\qquad\qquad\qquad}$$

$$\dfrac{7}{x} = \dfrac{35}{10}$$

So $70 = 35x$ and $x = 2$.

Substituting into $\dfrac{3}{x} - \dfrac{1}{y} = \dfrac{13}{10}$, we obtain

$$\dfrac{3}{2} - \dfrac{1}{y} = \dfrac{13}{10}$$

$$-\dfrac{1}{y} = \dfrac{13}{10} - \dfrac{15}{10}$$

$$-\dfrac{1}{y} = -\dfrac{2}{10}$$

$$y = 5.$$

The solution set is $\{(2, 5)\}$.

29. Substitute $y = 1 - x$ into $x^2 + xy - y^2 = -5$.

$$x^2 + x(1 - x) - (1 - x)^2 = -5$$
$$x^2 + x - x^2 - (1 - 2x + x^2) = -5$$
$$-x^2 + 3x - 1 = -5$$
$$x^2 - 3x + 1 = 5$$
$$x^2 - 3x - 4 = 0$$
$$(x - 4)(x + 1) = 0$$
$$x = 4, -1$$

Using $x = 4, -1$ in $y = 1 - x$, we get $y = -3, 2$.
The solution set is $\{(4, -3), (-1, 2)\}$.

31. Add the two given equations to obtain $xy = -2$. Substitute $y = -2/x$ into $x^2 + 2xy - 2y^2 = -11$.

$$x^2 + 2x\left(-\dfrac{2}{x}\right) - 2 \cdot \dfrac{4}{x^2} = -11$$

$$x^2 - 4 - \dfrac{8}{x^2} = -11$$
$$x^4 + 7x^2 - 8 = 0$$
$$(x^2 + 8)(x^2 - 1) = 0$$
$$x = \pm 1$$

Using $x = 1, -1$ in $y = -2/x$, $y = -2, 2$.
The solution set is $\{(1, -2), (-1, 2)\}$.

33. Substitute $y = 2^{x+1}$ into $y = 4^{-x}$.

$$2^{x+1} = (2^2)^{-x}$$
$$2^{x+1} = 2^{-2x}$$
$$x + 1 = -2x$$
$$3x = -1$$

Using $x = -1/3$ in $y = 2^{x+1}$, we get

$y = 2^{2/3}$. The solution set is $\left\{\left(-\dfrac{1}{3}, 2^{2/3}\right)\right\}$.

35. Substitute $y = \log_2(x)$ into $y = \log_4(x + 2)$ and use the base-changing formula.

$$\log_2(x) = \log_4(x + 2)$$
$$\log_2(x) = \dfrac{\log_2(x + 2)}{\log_2(4)}$$
$$\log_2(x) = \dfrac{\log_2(x + 2)}{2}$$
$$2 \cdot \log_2(x) = \log_2(x + 2)$$
$$2 \cdot \log_2(x) - \log_2(x + 2) = 0$$
$$\log_2\left(\dfrac{x^2}{x + 2}\right) = 0$$
$$\dfrac{x^2}{x + 2} = 1$$
$$x^2 = x + 2$$
$$x^2 - x - 2 = 0$$
$$(x - 2)(x + 1) = 0$$
$$x = 2, -1$$

But $x = -1$ is an extraneous root since $\log_2(-1)$ is undefined. Using $x = 2$ in $y = \log_2(x)$, we have $y = 1$.
The solution set is $\{(2, 1)\}$.

37. Substitute $y = \log_2(x + 2)$ into $y = 3 - \log_2(x)$.

$$\log_2(x + 2) = 3 - \log_2(x)$$
$$\log_2(x + 2) + \log_2(x) = 3$$

$$\log_2(x^2 + 2x) = 3$$
$$x^2 + 2x = 2^3$$
$$x^2 + 2x - 8 = 0$$
$$(x + 4)(x - 2) = 0$$
$$x = -4, 2$$

But $x = -4$ is an extraneous root since $\log_2(-4)$ is undefined. Using $x = 2$ in $y = \log_2(x + 2)$, we get $y = \log_2(4) = 2$. The solution set is $\{(2, 2)\}$.

39. Substitute $y = 3^x$ into $y = 2^x$.

$$3^x = 2^x$$
$$\log(3^x) = \log(2^x)$$
$$x \cdot \log(3) = x \cdot \log(2)$$
$$x [\log(3) - \log(2)] = 0$$
$$x = 0$$

Using $x = 0$ in $y = 3^x$, we obtain $y = 3^0 = 1$. The solution set is $\{(0, 1)\}$.

41. From the graphs, the solution set is $\{(2, 1), (0.3, -1.8)\}$.

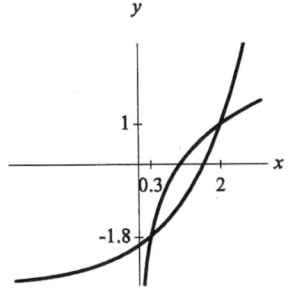

43. From the graphs, the solution set is $\{(1.9, 0.6), (0.1, -2.0)\}$.

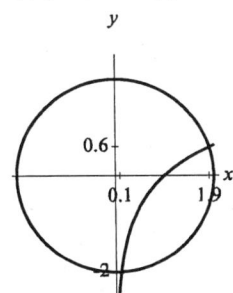

45. From the graphs, the solution set is $\{(2, 4), (-0.77, 0.59), (4, 16)\}$.

47. Let x and $6 - x$ be the two numbers.

$$x(6 - x) = -16$$
$$x^2 - 6x = 16$$
$$x^2 - 6x + 9 = 25$$
$$(x - 3)^2 = 25$$
$$x - 3 = \pm 5$$
$$x = 3 \pm 5$$
$$x = -2, 8$$

The numbers are -2 and 8.

49. Let x and y be the lengths of the legs of the triangle.

$$x^2 + y^2 = 15^2$$
$$\frac{1}{2}xy = 54$$

Substitute $y = \dfrac{108}{x}$ into $x^2 + y^2 = 225$ and use the quadratic formula.

$$x^2 + \frac{11664}{x^2} = 225$$
$$x^4 - 225x^2 + 11,664 = 0$$
$$x^2 = \frac{225 \pm \sqrt{225^2 - 4(11,664)}}{2}$$
$$x^2 = \frac{225 \pm 63}{2}$$
$$x^2 = 144, 81$$
$$x = 12, 9$$

Using $x = 12, 9$ in $y = \dfrac{108}{x}$, we get $y = 9, 12$. The sides of the right triangle are 9 m and 12 m.

51. If x is the length of the hypotenuse, then $\dfrac{x}{2}$ and $\dfrac{x\sqrt{3}}{2}$ are the lengths of the sides opposite the $30°$ and $60°$ angles.

$$x + \frac{x}{2} + \frac{x\sqrt{3}}{2} = 12$$
$$2x + x + \sqrt{3}x = 24$$
$$(3 + \sqrt{3})x = 24$$
$$x = \frac{24}{3 + \sqrt{3}} \cdot \frac{3 - \sqrt{3}}{3 - \sqrt{3}}$$
$$x = 12 - 4\sqrt{3}$$

Substituting $x = 12 - 4\sqrt{3}$ in $\dfrac{x}{2}$ and $\dfrac{x\sqrt{3}}{2}$, we find $6 - 2\sqrt{3}$ ft and $6\sqrt{3} - 6$ ft are the lengths of the two sides and the hypotenuse is $12 - 4\sqrt{3}$ ft.

53. The values of x and y must satisfy

$$6y = 6x$$
$$x(6 + y) = 7(4 + 12).$$

Since $x = y$ as seen from the first equation, upon substitution into the second equation one obtains

$$6x + x^2 = 112$$
$$x^2 + 6x - 112 = 0$$
$$(x + 14)(x - 8) = 0.$$

Then $x = 8$ in. and $y = 8$ oz. One must exclude the negative value $x = -14$.

55. Let x and y be the number of minutes it takes for pump A and pump B, respectively, to fill the vat.

$$\frac{1}{x} + \frac{1}{y} = \frac{1}{8}$$
$$\frac{1}{x} - \frac{1}{y} = \frac{1}{12}$$

Adding the two equations, we get

$$\frac{2}{x} = \frac{3}{24} + \frac{2}{24}$$
$$\frac{2}{x} = \frac{5}{24}$$
$$5x = 48$$
$$x = 9.6.$$

Substituting $x = \dfrac{48}{5}$ into $\dfrac{1}{x} + \dfrac{1}{y} = \dfrac{1}{8}$, we find

$$\frac{5}{48} + \frac{1}{y} = \frac{1}{8}$$
$$\frac{1}{y} = \frac{6}{48} - \frac{5}{48}$$
$$\frac{1}{y} = \frac{1}{48}$$
$$y = 48.$$

Pump A can fill the vat by itself in $x = 9.6$ min while Pump B will take $y = 48$ min.

57. Let x and y be two numbers satisfying

$$x + y = 6$$
$$xy = 10.$$

Substituting $y = 6 - x$ into $xy = 10$,

$$x(6 - x) = 10$$
$$6x - x^2 = 10$$
$$x^2 - 6x = -10$$
$$x^2 - 6x + 9 = -1$$
$$(x - 3)^2 = -1$$
$$x = 3 \pm i.$$

If $x = 3 + i$, then $y = 6 - (3 + i) = 3 - i$.
If $x = 3 - i$, then $y = 6 - (3 - i) = 3 + i$.
The two numbers are $3 + i$ and $3 - i$.

59. Let x and y be the length and width.

$$20xy = 36,000$$
$$40x + 40y + 2xy = 7200$$

Substitute $x = \dfrac{1800}{y}$ into $40x + 40y + 2xy = 7200$.

$$40 \cdot \frac{1800}{y} + 40y + 2 \cdot \frac{1800}{y} \cdot y = 7200$$
$$\frac{72,000}{y} + 40y + 3600 = 7200$$
$$\frac{72,000}{y} + 40y = 3600$$
$$40y^2 - 3600y + 72,000 = 0$$
$$y^2 - 90y + 1800 = 0$$
$$(y - 60)(y - 30) = 0$$

Using $y = 30, 60$ in $x = \dfrac{1800}{w}$, we get

$x = 60, 30$. The length is 60 ft and width is 30 ft.

61. From the graphs, the two models give the same population for $t = 0$ and $t = 9.65$ years. The exponential population model is twice the linear model when $t \approx 29.5$ years.

63. Let x be the time of the sunrise or the number of hours since midnight. Let S and B be Sally's and Bob's speeds. The number of miles Sally and Bob drove are $(16 - x)S$ and $(21 - x)B$, respectively. Since they met at noon, the distance between Sally's house and Bob's house is $(12 - x)S + (12 - x)B$; or equivalently $(12 - x)(S + B)$. Then

$$(16 - x)S = (12 - x)(S + B)$$
$$(16 - x)S = (16 - x)(S + B) - 4(S + B)$$
$$0 = (16 - x)B - 4(S + B)$$
$$(16 - x)B = 4(S + B).$$

Likewise,

$$(21 - x)B = (21 - x)(S + B) - 9(S + B)$$
$$(21 - x)S = 9(S + B).$$

Combining, one obtains
$$\dfrac{(16 - x)B}{(21 - x)S} = \dfrac{4(S + B)}{9(S + B)}. \text{ So, } \dfrac{(16 - x)9B}{(21 - x)4S} = 1.$$
Furthermore, since $(16 - x)S = (21 - x)B$ one finds $\dfrac{B}{S} = \dfrac{16 - x}{21 - x}$. Then

$$\dfrac{9(16 - x)}{4(21 - x)} \cdot \dfrac{B}{S} = 1$$

$$\dfrac{9}{4}\left(\dfrac{16 - x}{21 - x}\right)^2 = 1$$

$$\dfrac{16 - x}{21 - x} = \dfrac{2}{3} \text{ since } 16 - x > 0$$
$$48 - 3x = 42 - 2x$$
$$x = 6.$$

The sunrise was at 6:00 A.M.

For Thought

1. True, $\dfrac{1}{x} + \dfrac{3}{x + 1} = \dfrac{(x + 1) + 3x}{x(x + 1)} = \dfrac{4x + 1}{x(x + 1)}.$

2. True, $x + \dfrac{3x}{x^2 - 1} = \dfrac{x(x^2 - 1) + 3x}{x^2 - 1} = \dfrac{x^3 + 2x}{x^2 - 1}.$

3. False, by using long division we obtain
$$\dfrac{x^2}{x^2 - 9} = 1 + \dfrac{9}{x^2 - 9} = 1 + \dfrac{A}{x - 3} + \dfrac{B}{x + 3}.$$

4. True, since $\dfrac{1}{2} + \dfrac{1}{2^3} = \dfrac{2^2 + 1}{2^3} = \dfrac{5}{8}.$

5. False, since $\dfrac{3x - 1}{x^3 + x} = \dfrac{A}{x} + \dfrac{Bx + C}{x^2 + 1}.$

6. False, since $\dfrac{1}{x^2 - 1} = \dfrac{1/2}{x - 1} - \dfrac{1/2}{x + 1}.$

7. True, by using long division we get

$$x^2 + x - 2 \;\overline{\big)\; x^3 + 0x^2 + 0x + 1}$$
$$\underline{x^3 + x^2 - 2x}$$
$$-x^2 + 2x + 1$$
$$\underline{-x^2 - x + 2}$$
$$3x - 1$$

So $\dfrac{x^3 + 1}{x^2 + x - 2} = x - 1 + \dfrac{3x - 1}{x^2 + x - 2}.$

8. False, since $x^3 - 8 = (x - 2)(x^2 + 2x + 4).$

9. True, since $\dfrac{1}{x - 1} + \dfrac{1}{x^2 + x + 1} =$
$$\dfrac{(x^2 + x + 1) + (x - 1)}{x^3 - 1} = \dfrac{x^2 + 2x}{x^3 - 1}.$$

10. True, it is already in the form $\dfrac{Ax + B}{x^2 + 9}.$

5.4 Exercises

1. $\dfrac{3(x+1)+4(x-2)}{(x-2)(x+1)} = \dfrac{7x-5}{(x-2)(x+1)}$

3. $\dfrac{(x^2+2)-3(x-1)}{(x-1)(x^2+2)} = \dfrac{x^2-3x+5}{(x-1)(x^2+2)}$

5.
$$\dfrac{(2x+1)(x^2+3)+(x^3+2x+2)}{(x^2+3)^2} =$$
$$\dfrac{(2x^3+x^2+6x+3)+(x^3+2x+2)}{(x^2+3)^2} =$$
$$\dfrac{3x^3+x^2+8x+5}{(x^2+3)^2}$$

7.
$$\dfrac{(x-1)^2+(2x+3)(x-1)+(x^2+1)}{(x-1)^3} =$$
$$\dfrac{(x^2-2x+1)+(2x^2+x-3)+(x^2+1)}{(x-1)^3} =$$
$$\dfrac{4x^2-x-1}{(x-1)^3}$$

9. Multiply the equation by $(x-3)(x+3)$.

$$
\begin{aligned}
12 &= A(x+3)+B(x-3) \\
12 &= (A+B)x+(3A-3B)
\end{aligned}
$$

$A+B=0$ and $3A-3B=12$

Divide $3A-3B=12$ by 3 and add to $A+B=0$.

$$
\begin{aligned}
A-B &= 4 \\
A+B &= 0 \\
\hline
2A &= 4
\end{aligned}
$$

Using $A=2$ in $A+B=0$, $B=-2$.
Then $A=2$ and $B=-2$.

11. $\dfrac{5x-1}{(x+1)(x-2)} = \dfrac{A}{x+1}+\dfrac{B}{x-2}$

$$
\begin{aligned}
5x-1 &= A(x-2)+B(x+1) \\
5x-1 &= (A+B)x+(-2A+B)
\end{aligned}
$$

$A+B=5$ and $-2A+B=-1$

Multiply $-2A+B=-1$ by -1 and add to $A+B=5$.

$$
\begin{aligned}
2A-B &= 1 \\
A+B &= 5 \\
\hline
3A &= 6
\end{aligned}
$$

Using $A=2$ in $A+B=5$, $B=3$.

The answer is $\dfrac{2}{x+1}+\dfrac{3}{x-2}$.

13. $\dfrac{2x+5}{(x+4)(x+2)} = \dfrac{A}{x+4}+\dfrac{B}{x+2}$

$$
\begin{aligned}
2x+5 &= A(x+2)+B(x+4) \\
2x+5 &= (A+B)x+(2A+4B)
\end{aligned}
$$

$A+B=2$ and $2A+4B=5$

Multiply $A+B=2$ by -2 and add to $2A+4B=5$.

$$
\begin{aligned}
-2A-2B &= -4 \\
2A+4B &= 5 \\
\hline
2B &= 1
\end{aligned}
$$

Using $B=1/2$ in $A+B=2$, we find $A=3/2$.

The answer is $\dfrac{3/2}{x+4}+\dfrac{1/2}{x+2}$.

15. $\dfrac{2}{(x-3)(x+3)} = \dfrac{A}{x-3}+\dfrac{B}{x+3}$

$$
\begin{aligned}
2 &= A(x+3)+B(x-3) \\
2 &= (A+B)x+(3A-3B)
\end{aligned}
$$

$A+B=0$ and $3A-3B=2$

Multiply $A+B=0$ by 3 and add to $3A-3B=2$.

$$
\begin{aligned}
3A+3B &= 0 \\
3A-3B &= 2 \\
\hline
6A &= 2
\end{aligned}
$$

Using $A=1/3$ in $A+B=0$, we get $B=-1/3$.

The answer is $\dfrac{1/3}{x-3}+\dfrac{-1/3}{x+3}$.

17.

$$
\begin{aligned}
\dfrac{1}{x(x-1)} &= \dfrac{A}{x}+\dfrac{B}{x-1} \\
1 &= A(x-1)+Bx \\
1 &= (A+B)x-A
\end{aligned}
$$

$A+B=0$ and $-A=1$

Using $A=-1$ in $A+B=0$, we find $B=1$.

The answer is $\dfrac{-1}{x}+\dfrac{1}{x-1}$.

19. Multiplying the equation by $(x+3)^2(x-2)$, we obtain $x^2 + x - 31 =$

$$= A(x+3)(x-2) + B(x-2) + C(x+3)^2$$
$$= A(x^2 + x - 6) + B(x-2) + C(x^2 + 6x + 9)$$
$$= (A+C)x^2 + (A+B+6C)x + (-6A - 2B + 9C).$$

Equate the coefficients and solve the system.

$$A + C = 1$$
$$A + B + 6C = 1$$
$$-6A - 2B + 9C = -31$$

Multiply $A + B + 6C = 1$ by 2 and add to $-6A - 2B + 9C = -31$.

$$\begin{aligned} 2A + 2B + 12C &= 2 \\ -6A - 2B + 9C &= -31 \\ \hline -4A + 21C &= -29 \end{aligned}$$

Multiply $A + C = 1$ by 4 and add to $-4A + 21C = -29$.

$$\begin{aligned} 4A + 4C &= 4 \\ -4A + 21C &= -29 \\ \hline 25C &= -25 \end{aligned}$$

Using $C = -1$ in $A + C = 1$, we obtain $A = 2$. From $A + B + 6C = 1$, $2 + B - 6 = 1$ and $B = 5$. So $A = 2, B = 5$, and $C = -1$.

21. $\dfrac{4x - 1}{(x-1)^2(x+2)} = \dfrac{A}{x-1} + \dfrac{B}{(x-1)^2} + \dfrac{C}{x+2}$

$$\begin{aligned} 4x - 1 &= A(x-1)(x+2) + B(x+2) + C(x-1)^2 \\ 4x - 1 &= A(x^2 + x - 2) + B(x+2) + C(x^2 - 2x + 1) \\ 4x - 1 &= (A+C)x^2 + (A + B - 2C)x + \\ & \quad (-2A + 2B + C) \end{aligned}$$

If we equate the coefficients of x, we obtain

$$A + C = 0$$
$$A + B - 2C = 4$$
$$-2A + 2B + C = -1.$$

Solving the system, we get $A = 1$, $B = 1$, and $C = -1$. The answer is

$$\frac{1}{x-1} + \frac{1}{(x-1)^2} + \frac{-1}{x+2}.$$

23. $\dfrac{20 - 4x}{(x-2)^2(x+4)} = \dfrac{A}{x-2} + \dfrac{B}{(x-2)^2} + \dfrac{C}{x+4}$

$$\begin{aligned} 20 - 4x &= A(x-2)(x+4) + B(x+4) + \\ & \quad C(x-2)^2 \\ 20 - 4x &= A(x^2 + 2x - 8) + B(x+4) + \\ & \quad C(x^2 - 4x + 4) \\ 20 - 4x &= (A+C)x^2 + (2A + B - 4C)x + \\ & \quad (-8A + 4B + 4C) \end{aligned}$$

If we equate the coefficients of x, we obtain

$$A + C = 0$$
$$2A + B - 4C = -4$$
$$-8A + 4B + 4C = 20$$

Solving the system, we get $A = -1$, $B = 2$, and $C = 1$. The answer is

$$\frac{-1}{x-2} + \frac{2}{(x-2)^2} + \frac{1}{x+4}.$$

25. Note, $\dfrac{3x^2 + 3x - 2}{(x+1)^2(x-1)} = \dfrac{A}{x+1} + \dfrac{B}{(x+1)^2} + \dfrac{C}{x-1}$

$$\begin{aligned} 3x^2 + 3x - 2 &= A(x+1)(x-1) + B(x-1) + \\ & \quad C(x+1)^2 \\ 3x^2 + 3x - 2 &= A(x^2 - 1) + B(x-1) + \\ & \quad C(x^2 + 2x + 1) \\ 3x^2 + 3x - 2 &= (A+C)x^2 + (B + 2C)x + \\ & \quad (-A - B + C) \end{aligned}$$

If we equate the coefficients of x, we obtain

$$A + C = 3$$
$$B + 2C = 3$$
$$-A - B + C = -2.$$

Solving the system, we get $A = 2$, $B = 1$, and $C = 1$. The answer is $\dfrac{2}{x+1} + \dfrac{1}{(x+1)^2} + \dfrac{1}{x-1}$.

27. Multiplying the equation by $(x+1)(x^2+4)$, we get

$$\begin{aligned} x^2 - x - 7 &= A(x^2 + 4) + (Bx + C)(x+1) \\ x^2 - x - 7 &= (A+B)x^2 + (B+C)x + (4A + C). \end{aligned}$$

Equating the coefficients, we have

$$A + B = 1$$
$$B + C = -1$$
$$4A + C = -7.$$

Multiply $A + B = 1$ by -1 and add to $B + C = -1$.

$$-A - B = -1$$
$$\underline{B + C = -1}$$
$$-A + C = -2$$

Multiply $4A + C = -7$ by -1 and add to $-A + C = -2$.

$$-A + C = -2$$
$$\underline{-4A - C = 7}$$
$$-5A = 5$$

Using $A = -1$ in $A + B = 1$, $B = 2$.
Using $B = 2$ in $B + C = -1$, $C = -3$.
So $A = -1, B = 2$, and $C = -3$.

29. Note, $\dfrac{5x^2 + 5x}{(x + 2)(x^2 + 1)} = \dfrac{A}{x + 2} + \dfrac{Bx + C}{x^2 + 1}$.

$$5x^2 + 5x = A(x^2 + 1) + (Bx + C)(x + 2)$$
$$5x^2 + 5x = (A + B)x^2 + (2B + C)x + (A + 2C)$$

If we equate the coefficients of x, we obtain

$$A + B = 5$$
$$2B + C = 5$$
$$A + 2C = 0$$

Solving the system, we obtain $A = 2$, $B = 3$, and $C = -1$. The answer is $\dfrac{2}{x + 2} + \dfrac{3x - 1}{x^2 + 1}$.

31. Note,

$$\frac{x^2 - 2}{(x + 1)(x^2 + x + 1)} = \frac{A}{x + 1} + \frac{Bx + C}{x^2 + x + 1}.$$

$$x^2 - 2 = A(x^2 + x + 1) + (Bx + C)(x + 1)$$
$$x^2 - 2 = (A + B)x^2 + (A + B + C)x + (A + C)$$

If we equate the coefficients of x, we obtain

$$A + B = 1$$
$$A + B + C = 0$$
$$A + C = -2.$$

Solving the system, we obtain $A = -1$, $B = 2$, and $C = -1$. The answer is

$$\frac{-1}{x + 1} + \frac{2x - 1}{x^2 + x + 1}.$$

33. $\dfrac{-2x - 7}{(x + 2)^2} = \dfrac{A}{x + 2} + \dfrac{B}{(x + 2)^2}$

$$-2x - 7 = A(x + 2) + B$$
$$-2x - 7 = Ax + (2A + B)$$
$$A = -2 \quad \text{and} \quad 2A + B = -7$$

Using $A = -2$ in $2A + B = -7$, we get $B = -3$. The answer is

$$\frac{-2}{x + 2} + \frac{-3}{(x + 2)^2}.$$

35. Note that $x^3 + x^2 + x + 1 = x^2(x + 1) + (x + 1) = (x^2 + 1)(x + 1)$. Then we obtain

$$\frac{6x^2 - x + 1}{(x^2 + 1)(x + 1)} = \frac{A}{x + 1} + \frac{Bx + C}{x^2 + 1}$$

$$6x^2 - x + 1 = A(x^2 + 1) + (Bx + C)(x + 1)$$
$$6x^2 - x + 1 = (A + B)x^2 + (B + C)x + (A + C).$$

Equating the coefficients, we get

$$A + B = 6$$
$$B + C = -1$$
$$A + C = 1.$$

Multiply $A + B = 6$ by -1 and add to $B + C = -1$.

$$-A - B = -6$$
$$\underline{B + C = -1}$$
$$-A + C = -7$$

Adding $-A + C = -7$ and $A + C = 1$, $2C = -6$. Using $C = -3$ in $B + C = -1$ and $A + C = 1$, we obtain $B = 2$ and $A = 4$.

The answer is $\dfrac{4}{x + 1} + \dfrac{2x - 3}{x^2 + 1}$.

37. Note,

$$\frac{3x^3 - x^2 + 19x - 9}{(x^2 + 9)^2} = \frac{Ax + B}{x^2 + 9} + \frac{Cx + D}{(x^2 + 9)^2}.$$

So $3x^3 - x^2 + 19x - 9 =$
$(Ax + B)(x^2 + 9) + (Cx + D) =$
$Ax^3 + Bx^2 + (9A + C)x + (9B + D).$
Then $A = 3$ and $B = -1$. Since $9A + C = 19$
and $9B + D = -9$, we get $C = -8$ and $D = 0$.
The answer is $\dfrac{3x - 1}{x^2 + 9} + \dfrac{-8x}{(x^2 + 9)^2}.$

39. Observe that

$$\frac{3x^2 + 17x + 14}{(x - 2)(x^2 + 2x + 4)} = \frac{A}{x - 2} + \frac{Bx + C}{x^2 + 2x + 4}.$$

Then $3x^2 + 17x + 14 =$
$A(x^2 + 2x + 4) + (Bx + C)(x - 2) =$
$(A + B)x^2 + (2A - 2B + C)x + (4A - 2C)$

Equating the coefficients, we obtain

$$\begin{aligned} A + B &= 3 \\ 2A - 2B + C &= 17 \\ 4A - 2C &= 14. \end{aligned}$$

Multiply $A + B = 3$ by 2 and add to
$2A - 2B + C = 17$.

$$\begin{aligned} 2A + 2B &= 6 \\ 2A - 2B + C &= 17 \\ \hline 4A + C &= 23 \end{aligned}$$

Multiplying $4A - 2C = 14$ by -1 and
adding to $4A + C = 23$, $3C = 9$. So $C = 3$
and from $4A - 2C = 14$, $A = 5$. Using these
values in $2A - 2B + C = 17$, we get
$B = -2$. The answer is $\dfrac{5}{x - 2} + \dfrac{-2x + 3}{x^2 + 2x + 4}.$

41. Divide $2x^3 + x^2 + 3x - 2$ by $x^2 - 1$ by long
division.

$$\begin{array}{r} 2x + 1 \\ x^2 - 1 \overline{\smash{\big)}\,2x^3 + x^2 + 3x - 2} \\ \underline{2x^3 + 0x^2 - 2x} \\ x^2 + 5x - 2 \\ \underline{x^2 + 0x - 1} \\ 5x - 1 \end{array}$$

Then $\dfrac{2x^3 + x^2 + 3x - 2}{x^2 - 1} = 2x + 1 + \dfrac{5x - 1}{x^2 - 1}.$

Decompose $\dfrac{5x - 1}{x^2 - 1} = \dfrac{A}{x - 1} + \dfrac{B}{x + 1}.$

$$\begin{aligned} 5x - 1 &= A(x + 1) + B(x - 1) \\ 5x - 1 &= (A + B)x + (A - B). \end{aligned}$$
So $A + B = 5$ and $A - B = -1$.

Adding $A + B = 5$ and $A - B = -1$, $2A = 4$.
Using $A = 2$ in $A + B = 5$, we find $B = 3$.
The answer is $2x + 1 + \dfrac{2}{x - 1} + \dfrac{3}{x + 1}.$

43. Since $\dfrac{3x^3 - 2x^2 + x - 2}{(x^2 + x + 1)^2} =$

$\dfrac{Ax + B}{x^2 + x + 1} + \dfrac{Cx + D}{(x^2 + x + 1)^2}$, we get

$3x^3 - 2x^2 + x - 2 = (Ax + B)(x^2 + x + 1) +$
$(Cx + D)$
$\quad = Ax^3 + (A+B)x^2 + (A+B+C)x + (B+D).$

Equating the coefficients, we find $A = 3$.
Since $A + B = -2$, we get $B = -5$.
From $A + B + C = 1$ and $B + D = -2$,
we have $C = 3$ and $D = 3$.
The answer is $\dfrac{3x - 5}{x^2 + x + 1} + \dfrac{3x + 3}{(x^2 + x + 1)^2}.$

45. Since $\dfrac{3x^3 + 4x^2 - 12x + 16}{(x - 2)(x + 2)(x^2 + 4)} =$

$\dfrac{A}{x - 2} + \dfrac{B}{x + 2} + \dfrac{Cx + D}{x^2 + 4}$, we obtain

$3x^3 + 4x^2 - 12x + 16 =$
$\quad = A(x + 2)(x^2 + 4) + B(x - 2)(x^2 + 4) +$
$\quad\quad (Cx + D)(x^2 - 4)$
$\quad = (A + B + C)x^3 + (2A - 2B + D)x^2 +$
$\quad\quad (4A + 4B - 4C)x + (8A - 8B - 4D).$

Equating the coefficients, we get

$$\begin{aligned} A + B + C &= 3 \\ 2A - 2B + D &= 4 \\ 4A + 4B - 4C &= -12 \\ 8A - 8B - 4D &= 16. \end{aligned}$$

Multiply first equation by -4 and add to the
third. Multiply second equation by -4 and
add to the fourth. Also multiply first equation

by 2 and add to the second.

$$-4A - 4B - 4C = -12$$
$$\underline{4A + 4B - 4C = -12}$$
$$-8C = -24$$

$$-8A + 8B - 4D = -16$$
$$\underline{8A - 8B - 4D = 16}$$
$$-8D = 0$$

$$2A + 2B + 2C = 6$$
$$\underline{2A - 2B + D = 4}$$
$$4A + 2C + D = 10$$

So $C = 3$ and $D = 0$. From
$4A + 2C + D = 10$, we find $A = 1$ and
from $2A - 2B + D = 4$, we get $B = -1$.

The answer is $\dfrac{1}{x - 2} + \dfrac{-1}{x + 2} + \dfrac{3x}{x^2 + 4}$.

47.
$$\frac{5x^3 + x^2 + x - 3}{x^3(x - 1)} = \frac{A}{x} + \frac{B}{x^2} + \frac{C}{x^3} + \frac{D}{x - 1},$$

$$5x^3 + x^2 + x - 3 =$$
$$= Ax^2(x - 1) + Bx(x - 1) + C(x - 1) + Dx^3$$
$$= (A + D)x^3 + (-A + B)x^2 + (-B + C)x - C$$

Equating the coefficients, we get $C = 3$.
From $-B + C = 1$, we find $B = 2$.
From $-A + B = 1$, we obtain $A = 1$.
From $A + D = 5$, we have $D = 4$.

The answer is $\dfrac{1}{x} + \dfrac{2}{x^2} + \dfrac{3}{x^3} + \dfrac{4}{x - 1}$.

49. Note,
$$\frac{6x^2 - 28x + 33}{(x - 2)^2(x - 3)} = \frac{A}{x - 2} + \frac{B}{(x - 2)^2} + \frac{C}{x - 3}.$$

Then $6x^2 - 28x + 33 =$
$$= A(x - 2)(x - 3) + B(x - 3) + C(x - 2)^2$$
$$= (A + C)x^2 + (-5A + B - 4C)x +$$
$$(6A - 3B + 4C).$$

Equating the coefficients, we obtain

$$A + C = 6$$
$$-5A + B - 4C = -28$$
$$6A - 3B + 4C = 33.$$

Multiply second equation by 3 and add to the third.

$$-15A + 3B - 12C = -84$$
$$\underline{6A - 3B + 4C = 33}$$
$$-9A - 8C = -51$$

Multiply $A + C = 6$ by 8 and add to
$-9A - 8C = -51$

$$-9A - 8C = -51$$
$$\underline{8A + 8C = 48}$$
$$-A = -3$$

Using $A = 3$ in $A + C = 6$, we find $C = 3$.
From $6A - 3B + 4C = 33$, we obtain $B = -1$.

The answer is $\dfrac{3}{x - 2} + \dfrac{-1}{(x - 2)^2} + \dfrac{3}{x - 3}$.

51. Use synthetic division to factor the denominator.

$$\begin{array}{r|rrrr} -5 & 1 & 4 & -11 & -30 \\ & & -5 & 5 & 30 \\ \hline & 1 & -1 & -6 & 0 \end{array}$$

$$x^3 + 4x^2 - 11x - 30 = (x + 5)(x^2 - x - 6)$$
$$= (x + 5)(x + 2)(x - 3)$$

Decomposing, $\dfrac{9x^2 + 21x - 24}{(x + 5)(x + 2)(x - 3)} =$

$$= \frac{A}{x + 5} + \frac{B}{x + 2} + \frac{C}{x - 3}.$$

Then $9x^2 + 21x - 24 =$
$$= A(x + 2)(x - 3) + B(x + 5)(x - 3) +$$
$$C(x + 5)(x + 2)$$

and substituting $x = -2, 3, -5$, we obtain

$$-30 = -15B$$
$$2 = B$$

$$120 = 40C$$
$$3 = C$$

$$96 = -24A$$
$$4 = A.$$

The answer is $\dfrac{4}{x + 5} + \dfrac{2}{x + 2} + \dfrac{3}{x - 3}$.

53. Note that $x^3 - 3x^2 + 3x - 1 = (x - 1)^3$.

Then $\dfrac{x^2 - 2}{(x - 1)^3} = \dfrac{A}{x - 1} + \dfrac{B}{(x - 1)^2} + \dfrac{C}{(x - 1)^3}$.

$$x^2 - 2 = A(x - 1)^2 + B(x - 1) + C$$
$$= Ax^2 + (-2A + B)x + (A - B + C)$$

Then $A = 1$. Since $-2A + B = 0$, $B = 2$.
Since $A - B + C = -2$, we find $1 - 2 + C = -2$
and $C = -1$.

The answer is $\dfrac{1}{x - 1} + \dfrac{2}{(x - 1)^2} + \dfrac{-1}{(x - 1)^3}$.

55.

$$\frac{x}{(ax + b)^2} = \frac{A}{ax + b} + \frac{B}{(ax + b)^2}$$
$$x = A(ax + b) + B$$
$$= aAx + (bA + B)$$

So $aA = 1$ and $A = 1/a$. Since $bA + B = 0$,

we obtain $\dfrac{b}{a} + B = 0$ and $B = -b/a$.

The answer is $\dfrac{1/a}{ax + b} + \dfrac{-b/a}{(ax + b)^2}$.

57. Since $\dfrac{x + c}{x(ax + b)} = \dfrac{A}{x} + \dfrac{B}{ax + b}$, we have

$$x + c = A(ax + b) + Bx$$
$$= (aA + B)x + bA.$$

So $bA = c$ and $A = \dfrac{c}{b}$. Since $aA + B = 1$,

we have $\dfrac{ac}{b} + B = 1$ and $B = 1 - \dfrac{ac}{b}$.

Answer is $\dfrac{c/b}{x} + \dfrac{1 - ac/b}{ax + b}$.

59. Since $\dfrac{1}{x^2(ax + b)} = \dfrac{A}{x} + \dfrac{B}{x^2} + \dfrac{C}{ax + b}$, we get

$1 = Ax(ax + b) + B(ax + b) + Cx^2$
$1 = (aA + C)x^2 + (bA + aB)x + bB.$

So $bB = 1$ and $B = \dfrac{1}{b}$. Since $bA + aB = 0$,

we obtain $bA + \dfrac{a}{b} = 0$ and $A = -\dfrac{a}{b^2}$.

Since $aA + C = 0$, $-\dfrac{a^2}{b^2} + C = 0$ and $C = \dfrac{a^2}{b^2}$.

Answer is $\dfrac{-a/b^2}{x} + \dfrac{1/b}{x^2} + \dfrac{a^2/b^2}{ax + b}$.

For Thought

1. False, since $3 > 1 + 2$ is false.

2. False   **3.** True

4. False, because $x^2 + y^2 > 5$ is the region outside of a circle of radius $\sqrt{5}$.

5. True, since $(-2, 1)$ satisfies both equations in system (a).

6. True

7. False, $(-2, 0)$ does not satisfy $y < x + 2$.

8. False, $(-1, 2)$ lies on the line $y - 3x = 5$.

9. True

10. True

5.5 Exercises

1. c   **3.** d

5. $y < 2x$

7. $x + y > 3$

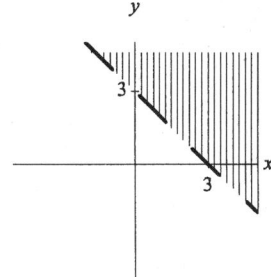

9. $2x - y \le 4$

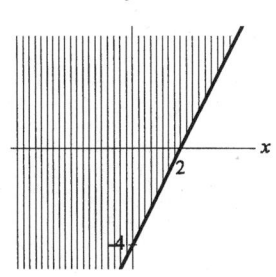

11. $y < -3x - 4$

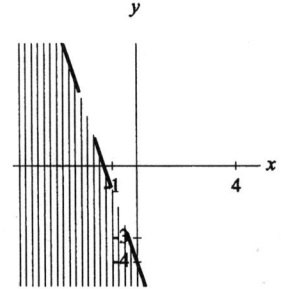

13. $x - 3 \geq 0$

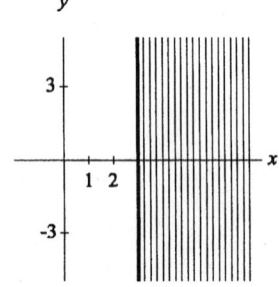

15. $20x - 30y \leq 6000$

17. $y < 3$

19. $y > -x^2$

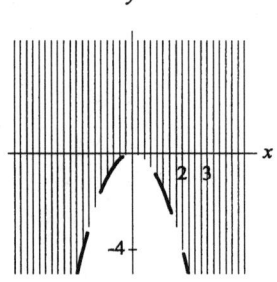

21. $x^2 + y^2 \geq 1$

23. $x > |y|$

25. $x \geq y^2$

27. $y \geq x^3$

29. $y > 2^x$

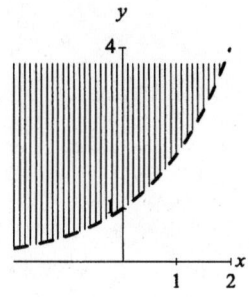

31. $y > x - 4$, $y < -x - 2$

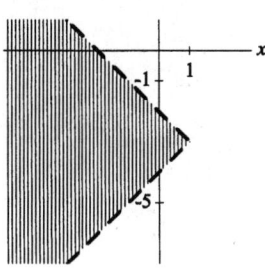

33. $3x - 4y \leq 12$, $x + y \geq -3$

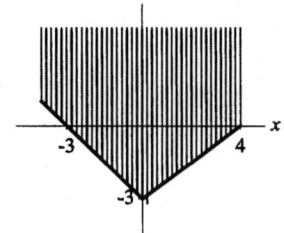

35. $3x - y < 4$, $y < 3x + 5$

37. No solution since the graphs of $y + x < 0$ and $y > 3 - x$ do not overlap.

39. $x + y < 5$, $y \geq 2$

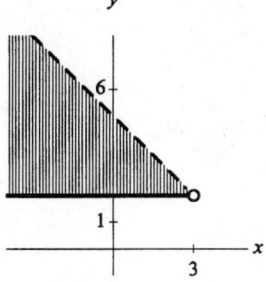

41. $y < x - 3$, $x \leq 4$

43. $y > x^2 - 3$, $y < x + 1$

45. $x^2 + y^2 \geq 4,\ x^2 + y^2 \leq 16$

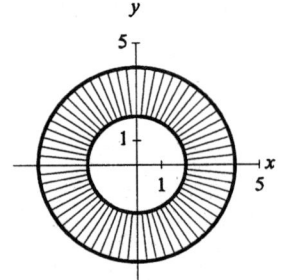

47. $(x-3)^2 + y^2 \leq 25,\ (x+3)^2 + y^2 \leq 25$

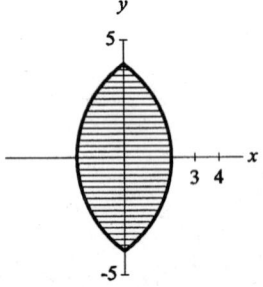

49. $x^2 + y^2 > 4,\ |x| \leq 4$

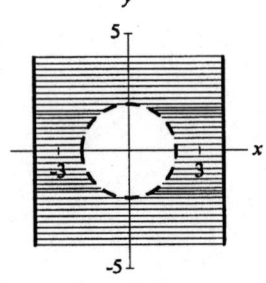

51. $y > |2x| - 4,\ y \leq \sqrt{4 - x^2}$

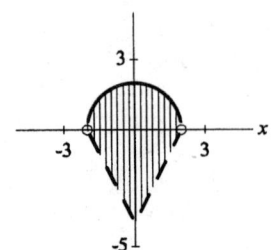

53. $|x - 1| < 2,\ |y - 1| < 4$

55. $x \geq 0,\ y \geq 0,\ x + y \leq 4$

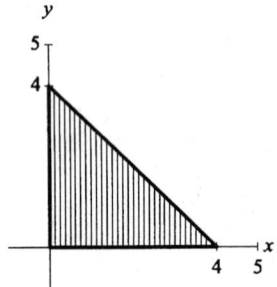

57. $x \geq 0,\ y \geq 0,\ x + y \geq 4,\ y \geq -2x + 6$

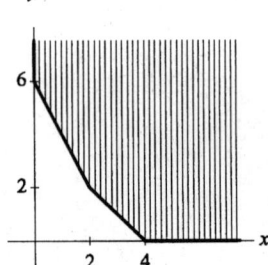

59. $x^2 + y^2 \geq 9,\ x^2 + y^2 \leq 25,\ y \geq |x|$

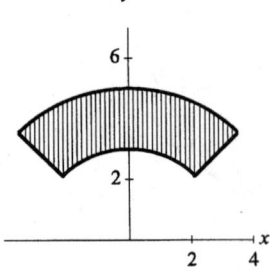

61. $y > (x-1)^3$, $y > 1$, $x + y > -2$

63. $y > 2^x$, $y < 6 - x^2$, $x + y > 0$

65. $x \geq 0$, $y \geq 0$, $y \leq -\dfrac{2}{3}x + 5$, $y \leq -3x + 12$

67. $x \geq 0$, $y \geq 0$, $y \geq -\dfrac{1}{2}x + 3$, $y \geq -\dfrac{3}{2}x + 5$

69. System is

$$
\begin{aligned}
|x| &< 2 \\
|y| &< 2.
\end{aligned}
$$

71. Since a circle of radius 9 with center at the origin is given by $x^2 + y^2 = 81$, the system is

$$
\begin{aligned}
x^2 + y^2 &< 81 \\
x &> 0 \\
y &> 0.
\end{aligned}
$$

73. $(-1.17,\ 1.84)$ is a solution.

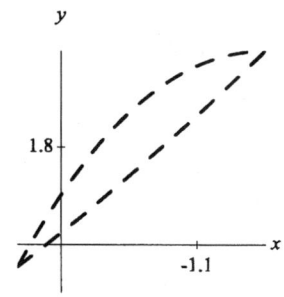

75. $(150,\ 22.4)$ is a solution.

77. Let w and h be the width and height, respectively. Then $50 + 2w + 2h \leq 130$. The system is

$$
\begin{aligned}
w + h &\leq 40 \\
w, h &\geq 0.
\end{aligned}
$$

79. Let x and y be the number of mid-size and full-size cars, respectively. Divide $10,000x + 15,000y \leq 1,500,000$ by 1000. The system is

$$
\begin{aligned}
x + y &\leq 110 \\
x + 1.5y &\leq 150 \\
x &\geq 0 \\
y &\geq 0.
\end{aligned}
$$

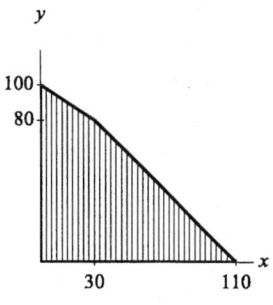

81. Let x and y be the number of $50 tickets and $100 tickets, respectively. Simplifying $y \leq 0.2(x + y)$, we get $0.8y \leq 0.2x$. Then $y \leq \frac{1}{4}x$. The system is

$$
\begin{aligned}
y &\leq \frac{1}{4}x \\
x + y &\leq 500 \\
x \geq 0, y &\geq 0.
\end{aligned}
$$

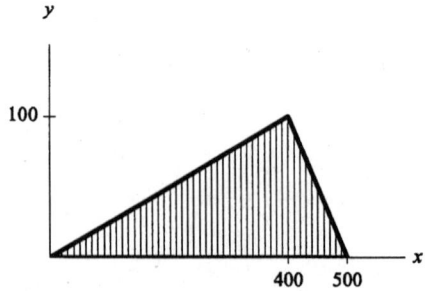

For Thought

1. False, $x \geq 0$ include points on the x-axis and the first and fourth quadrants.

2. False, $y \geq 2$ include points on or above the line $y = 2$.   **3.**   False

4. False, since x-intercept is $(6, 0)$ and y-intercept is $(0, 4)$.   **5.**   True   **6.**   True   **7.** False

8. True, since $R(1, 3) = 30(1) + 15(3) = 75$.

9. False, since $C(0, 5) = 7(0) + 9(5) + 3 = 48$.

10. True

5.6 Exercises

1. Vertices are $(0, 0), (0, 4), (4, 0)$

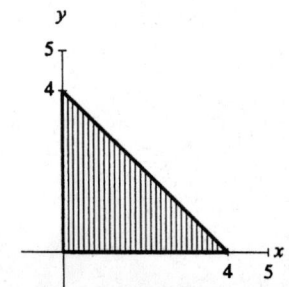

3. Vertices are $(0, 0), (1, 3), (1, 0), (0, 3)$

5. Vertices are $(0, 0), (2, 2), (0, 4), (3, 0)$

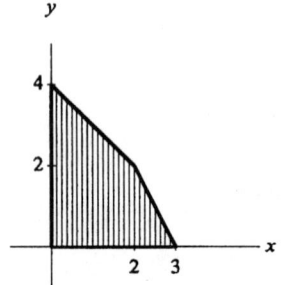

7. Vertices are $(3, 0), (1, 2), (0, 4)$

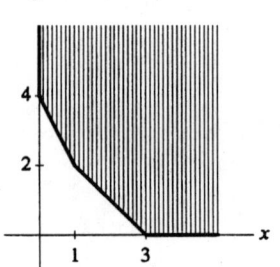

9. Vertices are $(1, 3), (4, 0), (0, 6)$

11. Vertices are $(1,5),(6,0),(0,8)$

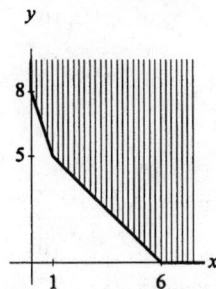

13. The values of $T(x,y) = 2x + 3y$ at the vertices are $T(0,0) = 0$, $T(0,4) = 12$, $T(3,3) = 15$, and $T(5,0) = 10$. The maximum value is 15.

15. The values of $H(x,y) = 2x + 2y$ at the vertices are $T(0,6) = 12$, $T(2,2) = 8$, and $T(5,0) = 10$. The minimum value is 8.

17. The values of $P(x,y) = 5x + 9y$ at the vertices are $P(0,0) = 0$, $P(6,0) = 30$, $P(0,3) = 27$. Maximum value is 30.

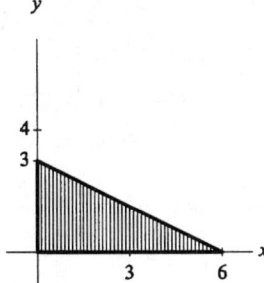

19. The values of $C(x,y) = 3x + 2y$ at the vertices are $C(0,4) = 8$ and $C(4,0) = 12$. The minimum value is 8.

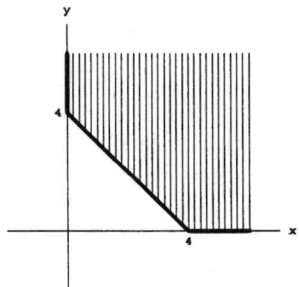

21. The values of $C(x,y) = 10x + 20y$ at the vertices are $C(0,8) = 160$, $C(5,3) = 110$, and $C(10,0) = 100$. Minimum value is 100.

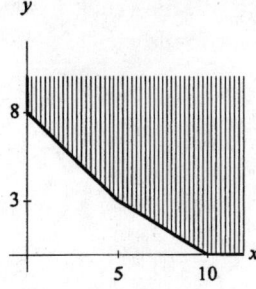

23. Let x and y be the number of bird houses and mailboxes, respectively.

$$\text{Maximize } 12x + 20y$$
$$\begin{aligned} \text{subject to } 3x + 4y &\leq 48 \\ x + 2y &\leq 20 \\ x, y &\geq 0 \end{aligned}$$

The values of $R(x,y) = 12x + 20y$ at the vertices are $R(0,0) = 0$, $R(0,10) = 200$, $R(8,6) = 216$, and $R(16,0) = 192$. To maximize revenue, they must sell 8 bird houses and 6 mailboxes.

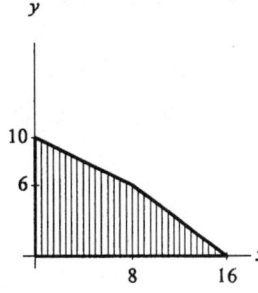

25. Let x and y be the number of bird houses and mailboxes, respectively. The values of $R(x,y) = 18x + 20y$ at the vertices are $R(0,0) = 0$, $R(0,10) = 200$, $R(8,6) = 264$, and $R(16,0) = 288$. To maximize revenue, they must sell 16 bird houses and 0 mailboxes.

27. Let x and y be the number of small and large truck loads, respectively.

$$\text{Minimize } 70x + 60y$$
$$\begin{aligned} \text{subject to } 12x + 20y &\geq 120 \\ x + y &\geq 8 \\ x, y &\geq 0 \end{aligned}$$

The values of $C(x, y) = 70x + 60y$ at the vertices are $C(0, 8) = 480$, $C(10, 0) = 700$, and $C(5, 3) = 530$, To minimize costs, they must make 8 large truck loads and 0 small truck loads.

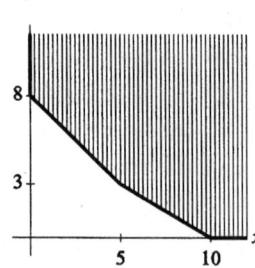

29. Let x and y be the number of small and large truck loads, respectively. The values of $C(x, y) = 70x + 75y$ at the vertices are $C(0, 8) = 600$, $C(10, 0) = 700$, and $C(5, 3) = 575$.

To minimize costs, they must make 5 small truck loads and 3 large truck loads.

Review Exercises

1. The solution set is $\{(3, 5)\}$.

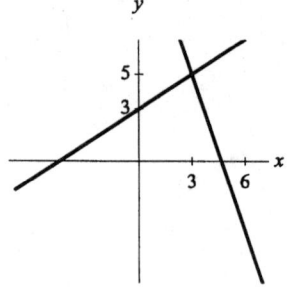

3. The solution set is $\{(-1, 3)\}$

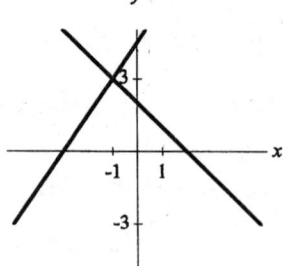

5. Substitute $y = x$ into $3x - 5y = 19$.

$$3x - 5x = 19$$

$$-2x = 19$$
$$x = -19/2$$

Independent and the solution set is $\{(-19/2, -19/2)\}$.

7. Multiply $4x - 3y = 6$ by 2 and $3x + 2y = 9$ by 3, then add the equations.

$$
\begin{aligned}
8x - 6y &= 12 \\
9x + 6y &= 27 \\
\hline
17x &= 39
\end{aligned}
$$

Substitute $x = 39/17$ into $4x - 3y = 6$.

$$\frac{156}{17} - 3y = 6$$
$$\frac{54}{17} = 3y$$
$$\frac{18}{17} = y$$

Independent and the solution set is $\{(39/17, 18/17)\}$.

9. Substitute $y = -3x + 1$ into $6x + 2y = 2$.

$$
\begin{aligned}
6x + 2(-3x + 1) &= 2 \\
6x - 6x + 2 &= 2 \\
2 &= 2
\end{aligned}
$$

Dependent and the solution set is $\{(x, y) \mid y = -3x + 1\}$.

11. Multiply $3x - 4y = 12$ by 2 and add to the second equation.

$$
\begin{aligned}
6x - 8y &= 24 \\
-6x + 8y &= 9 \\
\hline
0 &= 33
\end{aligned}
$$

Inconsistent and the solution set is $\emptyset$.

13. Add the first and second equations. Multiply $2x + y + z = 1$ by -3 and add to the third equation.

$$
\begin{aligned}
x + y - z &= 8 \\
2x + y + z &= 1 \\
\hline
3x + 2y &= 9
\end{aligned}
$$

$$
\begin{aligned}
-6x - 3y - 3z &= -3 \\
x + 2y + 3z &= -5 \\
\hline
-5x - y &= -8
\end{aligned}
$$

Multiply $-5x - y = -8$ by 2 and add to $3x + 2y = 9$.

$$\begin{array}{rcl} -10x - 2y & = & -16 \\ 3x + 2y & = & 9 \\ \hline -7x & = & -7 \end{array}$$

Using $x = 1$ in $3x + 2y = 9$, $3 + 2y = 9$ or $y = 3$. From $x + y - z = 8$, $1 + 3 - z = 8$ or $z = -4$. The Solution set is $\{(1, 3, -4)\}$.

15. Multiply first equation by -2 and add to the second equation. Multiply first equation by -2 and add to the third one.

$$\begin{array}{rcl} -2x - 2y - 2z & = & -2 \\ 2x - y + 2z & = & 2 \\ \hline -3y & = & 0 \\ y & = & 0 \end{array}$$

$$\begin{array}{rcl} -2x - 2y - 2z & = & -2 \\ 2x + 2y + 2z & = & 2 \\ \hline 0 & = & 0 \end{array}$$

Using $y = 0$ in $x + y + z = 1$, $x + z = 1$ and $z = 1 - x$. The solution set is $\{(x, 0, 1 - x) \mid x \text{ is any real number}\}$.

17. Multiply first equation by -1 and add to the third equation.

$$\begin{array}{rcl} -x - y - z & = & -1 \\ x + y + z & = & 4 \\ \hline 0 & = & 3 \end{array}$$

Inconsistent and the solution set is $\emptyset$.

19. Substitute $x = y^2$ into $x^2 + y^2 = 4$ and use the quadratic formula.

$$\begin{array}{rcl} y^4 + y^2 & = & 4 \\ y^4 + y^2 - 4 & = & 0 \\ y^2 & = & \dfrac{-1 + \sqrt{17}}{2} \\ y & = & \pm\sqrt{\dfrac{-1 + \sqrt{17}}{2}} \end{array}$$

Thus, $x = y^2 = \dfrac{-1 + \sqrt{17}}{2}$.
The solution set is

$$\left\{ \left(\frac{-1 + \sqrt{17}}{2}, \pm\sqrt{\frac{-1 + \sqrt{17}}{2}} \right) \right\}.$$

21. Substitute $y = x^2$ into $y = |x|$.

$$\begin{array}{rcl} x^2 & = & \sqrt{x^2} \\ x^4 & = & x^2 \\ x^2(x^2 - 1) & = & 0 \\ x & = & 0, \pm 1 \end{array}$$

Using $x = 0, 1, -1$ in $y = x^2$, we get $y = 0, 1, 1$, respectively. The solution set is $\{(0, 0), (1, 1), (-1, 1)\}$.

23. Note, $\dfrac{7x - 7}{(x - 3)(x + 4)} = \dfrac{A}{x - 3} + \dfrac{B}{x + 4}$.
Then

$$\begin{array}{rcl} 7x - 7 & = & A(x + 4) + B(x - 3) \\ 7x - 7 & = & (A + B)x + (4A - 3B). \end{array}$$

Equating the coefficients, we obtain

$$\begin{array}{rcl} A + B & = & 7 \\ 4A - 3B & = & -7. \end{array}$$

The solution of this system is $A = 2$, $B = 5$.
Answer is $\dfrac{2}{x - 3} + \dfrac{5}{x + 4}$.

25. Factoring the denominator, we obtain

$$\begin{array}{rcl} x^3 - 3x^2 + 4x - 12 & = & x^2(x - 3) + 4(x - 3) \\ & = & (x^2 + 4)(x - 3), \end{array}$$

and so $\dfrac{7x^2 - 7x + 23}{(x - 3)(x^2 + 4)} = \dfrac{A}{x - 3} + \dfrac{Bx + C}{x^2 + 4}$.
Then $7x^2 - 7x + 23 =$
$A(x^2 + 4) + (Bx + C)(x - 3) =$
$= (A + B)x^2 + (-3B + C)x + (4A - 3C)$.
Equating the coefficients, we have

$$\begin{array}{rcl} A + B & = & 7 \\ -3B + C & = & -7 \\ 4A - 3C & = & 23. \end{array}$$

The solution of this system is $A = 5$, $B = 2$, and $C = -1$. The answer is $\dfrac{5}{x - 3} + \dfrac{2x - 1}{x^2 + 4}$.

27. $x^2 + (y-3)^2 < 9$

29. $x \le (y-1)^2$

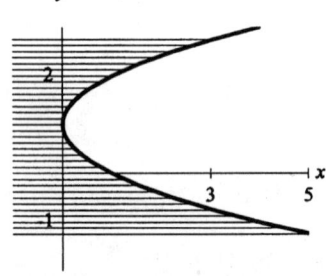

31. $2x - 3y \ge 6$, $x \le 2$

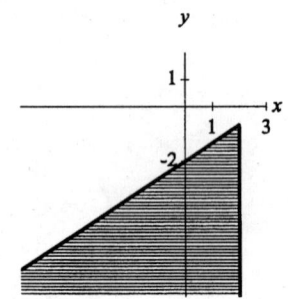

33. $y \ge 2x^2 - 6$, $x^2 + y^2 \le 9$

35. $x \ge 0$, $y \ge 1$, $x + 2y \le 10$, $3x + 4y \le 24$

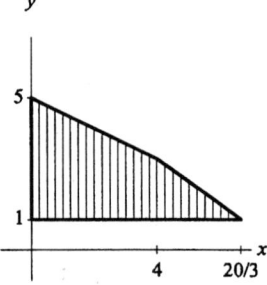

37. $x \ge 0$, $y \ge 0$, $x + 6y \ge 60$, $x + y \ge 35$

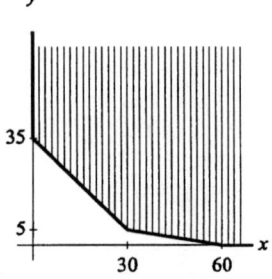

39. Substitute $(-2, 3)$ and $(4, -1)$ into $y = mx + b$.

$$\begin{aligned} -2m + b &= 3 \\ 4m + b &= -1 \end{aligned}$$

Multiply first equation by -1 and add to the second equation.

$$\begin{aligned} 2m - b &= -3 \\ \underline{4m + b} &= \underline{-1} \\ 6m &= -4 \end{aligned}$$

Using $m = -2/3$ in $-2m + b = 3$, $4/3 + b = 3$ and $b = 5/3$. Equation of line is $y = -\dfrac{2}{3}x + \dfrac{5}{3}$.

41. Substitute $(1, 4)$, $(3, 20)$, and $(-2, 25)$ into $y = ax^2 + bx + c$.

$$\begin{aligned} a + b + c &= 4 \\ 9a + 3b + c &= 20 \\ 4a - 2b + c &= 25 \end{aligned}$$

The solution of the above system is $a = 3$, $b = -4$, $c = 5$. The parabola is given by $y = 3x^2 - 4x + 5$.

43. Let x and y be the number of tacos and burritos, respectively.

$$\begin{aligned} x + 2y &= 181 \\ 2x + 3y &= 300 \end{aligned}$$

Solving the above system, we find $x = 57$ tacos and $y = 62$ burritos.

45. Let x, y and z be the selling price of a daisy, carnation, and a rose, respectively. Then

$$\begin{aligned} 5x + 3y + 2z &= 3.05 \\ 3x + y + 4z &= 2.75 \\ 4x + 2y + z &= 2.10. \end{aligned}$$

Solving the above system, we find $x = 0.30$, $y = 0.25$, and $z = 0.40$. Esther's economy special sells for $x + y + z = \$0.95$.

47. The values of $C(x, y) = 0.42x + 0.84y$ at the vertices are $C(0, 35) = 29.4$, $C(30, 5) = 16.8$, and $C(60, 0) = 25.2$. The minimum value is 16.8.

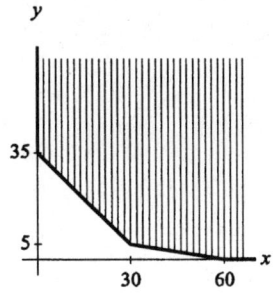

49. Let x and y be the number of barrels of oil obtained through the pipeline and barges, respectively.

Minimize $20x + 18y$

$$\begin{aligned} \text{subject to } x + y &\geq 12{,}000{,}000 \\ x &\leq 12{,}000{,}000 \\ x &\geq 6{,}000{,}000 \\ y &\leq 8{,}000{,}000 \\ x, y &\geq 0 \end{aligned}$$

The values of $C(x, y) = 20x + 18y$ at the vertices are
$C(12 \text{ million}, 0) = 240$ million,
$C(12 \text{ million}, 8 \text{ million}) = 384$ million,

$C(6 \text{ million}, 8 \text{ million}) = 264$ million, and
$C(6 \text{ million}, 6 \text{ million}) = 228$ million.
To minimize cost, purchase 6 million barrels from each source.

Chapter 8 Test

1. Solution set is $\{(-3, 4)\}$.

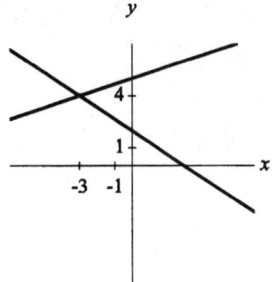

2. Substitute $y = 4 - 2x$ into $3x - 4y = 9$.

$$\begin{aligned} 3x - 4(4 - 2x) &= 9 \\ 3x - 16 + 8x &= 9 \\ 11x &= 25 \\ x &= 25/11 \end{aligned}$$

Using $x = 25/11$ in $y = 4 - 2x$, we have $y = 4 - 50/11 = -6/11$. Solution set is $\{(25/11, -6/11)\}$.

3. Multiply $10x - 3y = 22$ by 2 and $7x + 2y = 40$ by 3. Then add the equations.

$$\begin{aligned} 20x - 6y &= 44 \\ \underline{21x + 6y} &= \underline{120} \\ 41x &= 164 \\ x &= 4 \end{aligned}$$

Using $x = 4$ in $10x - 3y = 22$, we get $40 - 3y = 22$ and $y = 6$. The solution set is $\{(4, 6)\}$.

4. Substitute $x = 6 - y$ into $3x + 3y = 4$.

$$
\begin{aligned}
3(6 - y) + 3y &= 4 \\
18 - 3y + 3y &= 4 \\
18 &= 4
\end{aligned}
$$

The system is inconsistent.

5. Substitute $y = \dfrac{1}{2}x + 3$ into $x - 2y = -6$.

$$
\begin{aligned}
x - 2\left(\frac{1}{2}x + 3\right) &= -6 \\
x - x - 6 &= -6 \\
-6 &= -6
\end{aligned}
$$

The system is dependent.

6. Substitute $y = 2x - 1$ into $y = 3x + 20$.

$$
\begin{aligned}
2x - 1 &= 3x + 20 \\
-21 &= x
\end{aligned}
$$

Using $x = -21$ in $y = 2x - 1$, we get $y = -43$. The system is independent.

7. Substitute $y = -x + 2$ into $y = -x + 5$.

$$
\begin{aligned}
-x + 2 &= -x + 5 \\
2 &= 5
\end{aligned}
$$

The system is inconsistent.

8. Add the two equations.

$$
\begin{aligned}
2x - y + z &= 4 \\
-x + 2y - z &= 6 \\
\hline
x + y &= 10
\end{aligned}
$$

Using $y = 10 - x$ in $2x - y + z = 4$, we find $2x - (10 - x) + z = 4$ and $z = 14 - 3x$. The solution set is $\{(x, 10 - x, 14 - 3x) \mid x \text{ is any real number}\}$.

9. Add the first two equations. Also, multiply second equation by 3 and add to the third.

$$
\begin{aligned}
x - 2y - z &= 2 \\
2x + 3y + z &= -1 \\
\hline
3x + y &= 1
\end{aligned}
$$

$$
\begin{aligned}
6x + 9y + 3z &= -3 \\
3x - y - 3z &= -4 \\
\hline
9x + 8y &= -7
\end{aligned}
$$

Multiply $3x + y = 1$ by -3 and add to $9x + 8y = -7$.

$$
\begin{aligned}
-9x - 3y &= -3 \\
9x + 8y &= -7 \\
\hline
5y &= -10
\end{aligned}
$$

Using $y = -2$ in $3x + y = 1$, we get $3x - 2 = 1$ or $x = 1$. From $x - 2y - z = 2$, we have $1 + 4 - z = 2$ or $z = 3$. The solution set is $\{(1, -2, 3)\}$.

10. Add the second and third equations.

$$
\begin{aligned}
x + y - z &= 4 \\
-x - y + z &= 2 \\
\hline
0 &= 6
\end{aligned}
$$

Inconsistent and the solution set is $\emptyset$.

11. Multiply $x^2 + y^2 = 16$ by -1 and add to $x^2 - 4y^2 = 16$.

$$
\begin{aligned}
-x^2 - y^2 &= -16 \\
x^2 - 4y^2 &= 16 \\
\hline
-5y^2 &= 0 \\
y &= 0
\end{aligned}
$$

Using $y = 0$ in $x^2 + y^2 = 16$, we find $x^2 = 16$ and $x = \pm 4$. Solution set is $\{(4, 0), (-4, 0)\}$.

12. Substitute $y = x^2 - 5x$ into $x + y = -2$.

$$
\begin{aligned}
x + (x^2 - 5x) &= -2 \\
x^2 - 4x &= -2 \\
x^2 - 4x + 4 &= -2 + 4 \\
(x - 2)^2 &= 2 \\
x &= 2 \pm \sqrt{2}
\end{aligned}
$$

Using $x = 2 + \sqrt{2}$ and $x = 2 - \sqrt{2}$ in $y = -2 - x$, we have $y = -4 - \sqrt{2}$ and $y = -4 + \sqrt{2}$, respectively. The solution set is

$$
\{(2 + \sqrt{2}, -4 - \sqrt{2}), (2 - \sqrt{2}, -4 + \sqrt{2})\}.
$$

13.

$$
\begin{aligned}
\frac{2x + 10}{(x - 4)(x + 2)} &= \frac{A}{x - 4} + \frac{B}{x + 2} \\
2x + 10 &= A(x + 2) + B(x - 4) \\
2x + 10 &= (A + B)x + (2A - 4B)
\end{aligned}
$$

Equating the coefficients, we obtain

$$A + B = 2$$
$$2A - 4B = 10.$$

The solution of this system is $A = 3, B = -1.$
Answer is $\dfrac{3}{x-4} + \dfrac{-1}{x+2}.$

14. Note, $\dfrac{4x^2 + x - 2}{x^2(x-1)} = \dfrac{A}{x} + \dfrac{B}{x^2} + \dfrac{C}{x-1}.$ Then

$$
\begin{aligned}
4x^2 + x - 2 &= Ax(x-1) + B(x-1) + Cx^2 \\
&= (A+C)x^2 + (-A+B)x - B.
\end{aligned}
$$

Equating the coefficients, we have

$$
\begin{aligned}
A + C &= 4 \\
-A + B &= 1 \\
-B &= -2.
\end{aligned}
$$

The solution of this system is $B = 2, A = 1,$
and $C = 3.$ The answer is $\dfrac{1}{x} + \dfrac{2}{x^2} + \dfrac{3}{x-1}.$

15. $2x - y < 8$

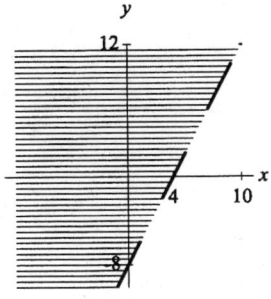

16. $x + y \le 5, \ x - y < 0$

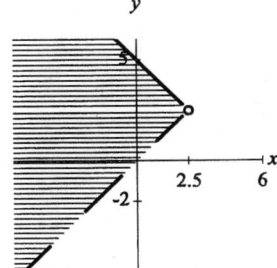

17. $x^2 + y^2 \le 9, \ y \le 1 - x^2$

18. Let x and y be the number of male and female students, respectively. Then

$$
\begin{aligned}
\tfrac{1}{3}x + \tfrac{1}{4}y &= 15 \\
x + y &= 52.
\end{aligned}
$$

Multiply the first equation by 12 and multiply the second equation by -3. Then add the resulting equations.

$$
\begin{array}{rcr}
4x + 3y &=& 180 \\
-3x - 3y &=& -156 \\
\hline
x &=& 24
\end{array}
$$

The solution of this system is $x = 24$ males and $y = 28$ females.

19. Let x and y be the number of TV commercials and newspaper ads, respectively. The linear program is given below.

Maximize $14,000x + 6000y$
  subject to $9000x + 3000y \le 99,000$
    $x + y \le 23$
    $x, y \ge 0$

The values of $N(x, y) = 14,000x + 6000y$ at the vertices are $N(0, 0) = 0,$ $N(0, 23) = 138,000,$ $N(5, 18) = 178,000,$ and $N(11, 0) = 154,000.$ To obtain maximum audience exposure, the hospital must have 5 TV commercials and 18 newspaper ads.

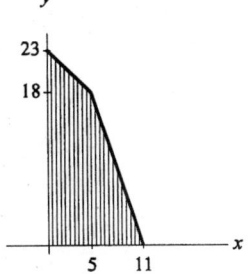

Tying It All Together

1. Multiply the equation by $24(x+5)$.

$$\begin{aligned} 24(x-2) &= 11(x+5) \\ 24x - 48 &= 11x + 55 \\ 13x &= 103 \end{aligned}$$

The solution set is $\{103/13\}$.

2. Multiply the equation by $24x(x+5)$.

$$\begin{aligned} 24(x+5) + 24x(x-2) &= 11x(x+5) \\ 24x + 120 + 24x^2 - 48x &= 11x^2 + 55x \\ 13x^2 - 79x + 120 &= 0 \\ (13x - 40)(x-3) &= 0 \end{aligned}$$

The solution set is $\{40/13, 3\}$.

3.

$$\begin{aligned} 5 - 3x - 6 - 2x + 4 &= 7 \\ 3 - 5x &= 7 \\ -5x &= 4 \end{aligned}$$

The solution set is $\{-4/5\}$.

4. We will solve an equivalent statement without absolute values.

$$\begin{aligned} 3 - 2x = 5 \quad &\text{or} \quad 3 - 2x = -5 \\ -2x = 2 \quad &\text{or} \quad -2x = -8 \\ x = -1 \quad &\text{or} \quad x = 4 \end{aligned}$$

The solution set is $\{4, -1\}$.

5. Square both sides of the equation.

$$\begin{aligned} 3 - 2x &= 25 \\ -2x &= 22 \end{aligned}$$

The solution set is $\{-11\}$.

6. Isolate x^2 on one side and take the square roots.

$$\begin{aligned} 3x^2 &= 4 \\ x^2 &= \frac{4}{3} \\ x &= \pm \frac{2}{\sqrt{3}} \\ x &= \pm \frac{2}{\sqrt{3}} \cdot \frac{\sqrt{3}}{\sqrt{3}} \end{aligned}$$

The solution set is $\left\{ \pm \frac{2\sqrt{3}}{3} \right\}$.

7. Multiply equation by x^2.

$$\begin{aligned} (x-2)^2 &= x^2 \\ x^2 - 4x + 4 &= x^2 \\ -4x &= -4 \end{aligned}$$

The solution set is $\{1\}$.

8. Since $2^{x-1} = 9$, we obtain $x - 1 = \log_2(9)$ by using the definition of a logarithm. The solution set is $\{1 + \log_2(9)\}$.

9. Write left-hand side as a single logarithm.

$$\begin{aligned} \log((x+1)(x+4)) &= 1 \\ x^2 + 5x + 4 &= 10^1 \\ x^2 + 5x - 6 &= 0 \\ (x+6)(x-1) &= 0 \\ x &= -6, 1 \end{aligned}$$

But $\log(x+1)$ is undefined when $x = -6$. The solution set is $\{1\}$.

10. Raise both sides of equation to the power $-3/2$.

$$\begin{aligned} x^{-2/3} &= \frac{1}{4} \\ x &= \pm \left(\frac{1}{4} \right)^{-3/2} \\ x &= \pm (4)^{3/2} \\ x &= \pm (4^{1/2})^3 \\ x &= \pm (2)^3 \end{aligned}$$

The solution set is $\{\pm 8\}$.

11. Use the quadratic formula.

$$\begin{aligned} x^2 - 3x - 6 &= 0 \\ x &= \frac{3 \pm \sqrt{33}}{2} \end{aligned}$$

The solution set is $\left\{ \frac{3 \pm \sqrt{33}}{2} \right\}$.

12. By using the square root property, we obtain

$$\begin{aligned} (x-3)^2 &= \frac{1}{2} \\ x - 3 &= \pm \frac{\sqrt{2}}{2} \\ x &= \frac{6}{2} \pm \frac{\sqrt{2}}{2}. \end{aligned}$$

The solution set is $\left\{\dfrac{6 \pm \sqrt{2}}{2}\right\}$.

13. Since $3 - 2x > 0$, we get $3 > 2x$ and $x < 3/2$. The solution set is $(-\infty, 3/2)$ and

its graph is
$$\xleftarrow{\hspace{1cm}} \overset{3/2}{\longrightarrow} \xrightarrow{\hspace{1cm}}$$

14. Since we must exclude $x = \dfrac{3}{2}$, the solution set is $(-\infty, 1.5) \cup (1.5, \infty)$ and the graph is
$$\xleftarrow{\hspace{2cm}} \overset{1.5}{\times} \xrightarrow{\hspace{2cm}}$$

15. The sign graph of $(x-3)(x+3) \geq 0$ is shown below.

```
- - - - - - - - 0 + + + +
- - - - 0 + + + + + + + +
←――――――――――――――――→
        -3        3
```
So the solution set is $(-\infty, -3] \cup [3, \infty)$

and the graph is
$$\xleftarrow{\hspace{1cm}} \overset{-3}{\rule{0pt}{0pt}} \quad \overset{3}{\rule{0pt}{0pt}} \xrightarrow{\hspace{1cm}}$$

16. Note that $x^2 + 2x - 8 \leq 27$ is equivalent to $x^2 + 2x - 35 \leq 0$. The sign graph of $(x+7)(x-5) \leq 0$ is

```
- - - - - - - - 0 + + + +
- - - - 0 + + + + + + + +
←――――――――――――――――→
        -7        5
```

So the solution set is $[-7, 5]$ and the

graph is
$$\xleftarrow{\hspace{1cm}} \overset{-7}{[} \quad \overset{5}{]} \xrightarrow{\hspace{1cm}}$$

17. $3 - 2x > y$

18. $|3 - 2x| > y$

19. $x^2 \geq 9$

20. $(x-2)(x+4) \leq y$

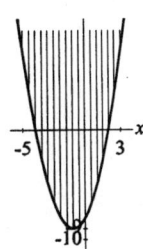

For Thought

1. False, the augmented matrix is a 2×3 matrix.

2. False, the required matrix is $\begin{bmatrix} 1 & -1 & | & 4 \\ 3 & 1 & | & 5 \end{bmatrix}$.

3. True   **4.**   True

5. True, since the row operation done is $R_1 + R_2 \to R_2$.

6. True, since the row operation done is $-R_1 + R_2 \to R_2$.

7. False, since it corresponds to
$$x = 2$$
$$y = 7.$$

8. True, since $0 \cdot x + 0 \cdot y = 7$ has no solution.

9. False, the system is dependent.   **10.**   True

6.1 Exercises

1. 1×3 **3.** 1×1 **5.** 3×2

7.
$$\begin{bmatrix} 1 & -2 & | & 4 \\ 3 & 2 & | & -5 \end{bmatrix}$$

9.
$$\begin{bmatrix} 1 & -1 & -1 & | & 4 \\ 1 & 3 & -1 & | & 1 \\ 0 & 2 & -5 & | & -6 \end{bmatrix}$$

11.
$$\begin{bmatrix} 1 & 3 & -1 & | & 5 \\ 1 & 0 & 1 & | & 0 \end{bmatrix}$$

13.
$$3x + 4y = -2$$
$$3x - 5y = 0$$

15.
$$5x = 6$$
$$-4x + 2z = -1$$
$$4x + 4y = 7$$

17.
$$x - y + 2z = 1$$
$$y + 4z = 3$$

19. Interchange R_1 and R_2
$$\begin{bmatrix} 1 & 2 & | & 0 \\ -2 & 4 & | & 1 \end{bmatrix}$$

21. Multiply $\frac{1}{2}$ to R_1
$$\begin{bmatrix} 1 & 4 & | & 1 \\ 0 & 3 & | & 6 \end{bmatrix}$$

23. Multiply 3 to R_1 then add the product to R_2. This is the new R_2.
$$\begin{bmatrix} 1 & -2 & | & 1 \\ -3 & 5 & | & 0 \end{bmatrix} =$$
$$\begin{bmatrix} 1 & -2 & | & 1 \\ 3 \cdot 1 + (-3) & 3 \cdot (-2) + 5 & | & 3 \cdot 1 + 0 \end{bmatrix} =$$
$$\begin{bmatrix} 1 & -2 & | & 1 \\ 0 & -1 & | & 3 \end{bmatrix}$$

25. From the augmented matrix
$$\begin{bmatrix} 2 & 4 & | & 14 \\ 5 & 4 & | & 5 \end{bmatrix}$$
the system is
$$2x + 4y = 14$$
$$5x + 4y = 5.$$

From the final augmented matrix
$$\begin{bmatrix} 1 & 0 & | & -3 \\ 0 & 1 & | & 5 \end{bmatrix}$$
we obtain that the solution set is $\{(-3, 5)\}$.

The row operations are $\frac{1}{2}R_1 \to R_1$,

$-5R_1 + R_2 \to R_2$, $-\frac{1}{6}R_2 \to R_2$,

and $-2R_2 + R_1 \to R_1$.

27. On $\begin{bmatrix} 1 & 1 & | & 5 \\ -2 & 1 & | & -1 \end{bmatrix}$ use $2R_1 + R_2 \to R_2$ to get

$\begin{bmatrix} 1 & 1 & | & 5 \\ 0 & 3 & | & 9 \end{bmatrix}$, use $\frac{1}{3}R_2 \to R_2$ to get

$\begin{bmatrix} 1 & 1 & | & 5 \\ 0 & 1 & | & 3 \end{bmatrix}$, use $-R_2 + R_1 \to R_1$ to get

$\begin{bmatrix} 1 & 0 & | & 2 \\ 0 & 1 & | & 3 \end{bmatrix}$, solution set is $\{(2,3)\}$, independent.

29. On $\begin{bmatrix} 2 & 2 & | & 8 \\ -3 & -1 & | & -6 \end{bmatrix}$ use $\frac{1}{2}R_1 \to R_1$ to get

$\begin{bmatrix} 1 & 1 & | & 4 \\ -3 & -1 & | & -6 \end{bmatrix}$, use $3R_1 + R_2 \to R_2$ to get

$\begin{bmatrix} 1 & 1 & | & 4 \\ 0 & 2 & | & 6 \end{bmatrix}$, use $\frac{1}{2}R_2 \to R_2$ to get

$\begin{bmatrix} 1 & 1 & | & 4 \\ 0 & 1 & | & 3 \end{bmatrix}$, use $-1R_2 + R_1 \to R_1$ to get

$\begin{bmatrix} 1 & 0 & | & 1 \\ 0 & 1 & | & 3 \end{bmatrix}$, solution set is $\{(1,3)\}$, independent.

31. On $\begin{bmatrix} 2 & -1 & | & 3 \\ 3 & 2 & | & 15 \end{bmatrix}$ use $-R_1 + R_2 \to R_1$ to get

$\begin{bmatrix} 1 & 3 & | & 12 \\ 3 & 2 & | & 15 \end{bmatrix}$, use $-3R_1 + R_2 \to R_2$ to get

$\begin{bmatrix} 1 & 3 & | & 12 \\ 0 & -7 & | & -21 \end{bmatrix}$, use $-\frac{1}{7}R_2 \to R_2$ to get

$\begin{bmatrix} 1 & 3 & | & 12 \\ 0 & 1 & | & 3 \end{bmatrix}$, use $-3R_2 + R_1 \to R_1$ to get

$\begin{bmatrix} 1 & 0 & | & 3 \\ 0 & 1 & | & 3 \end{bmatrix}$, solution set is $\{(3,3)\}$, independent.

33. On $\begin{bmatrix} 0.4 & -0.2 & | & 0 \\ 1 & 1.5 & | & 2 \end{bmatrix}$ use $5R_1 \to R_1$ and $2R_2 \to R_2$ to get

$\begin{bmatrix} 2 & -1 & | & 0 \\ 2 & 3 & | & 4 \end{bmatrix}$, use $R_1 + (-R_2) \to R_2$ to get

$\begin{bmatrix} 2 & -1 & | & 0 \\ 0 & -4 & | & -4 \end{bmatrix}$, use $-\frac{1}{4}R_2 \to R_2$ to get

$\begin{bmatrix} 2 & -1 & | & 0 \\ 0 & 1 & | & 1 \end{bmatrix}$, use $R_1 + R_2 \to R_1$ to get

$\begin{bmatrix} 2 & 0 & | & 1 \\ 0 & 1 & | & 1 \end{bmatrix}$, use $\frac{1}{2}R_1 \to R_1$ to get

$\begin{bmatrix} 1 & 0 & | & 0.5 \\ 0 & 1 & | & 1 \end{bmatrix}$, solution set is $\{(0.5,1)\}$, independent.

35. On $\begin{bmatrix} 3 & -5 & | & 7 \\ -3 & 5 & | & 4 \end{bmatrix}$ use $R_1 + R_2 \to R_2$ to get

$\begin{bmatrix} 3 & -5 & | & 7 \\ 0 & 0 & | & 11 \end{bmatrix}$, inconsistent system, and solution set is $\emptyset$.

37. On $\begin{bmatrix} 0.5 & 1.5 & | & 2 \\ 3 & 9 & | & 12 \end{bmatrix}$ use $2R_1 \to R_1$ to get

$\begin{bmatrix} 1 & 3 & | & 4 \\ 3 & 9 & | & 12 \end{bmatrix}$, use $-3R_1 + R_2 \to R_2$ to get

$\begin{bmatrix} 1 & 3 & | & 4 \\ 0 & 0 & | & 0 \end{bmatrix}$, dependent system, and solution set is $\{(u,v) \mid u + 3v = 4\}$.

39. Rewrite system as

$$2x + y = 4$$
$$x - y = 8.$$

On $\begin{bmatrix} 2 & 1 & | & 4 \\ 1 & -1 & | & 8 \end{bmatrix}$ use $R_1 + (-2R_2) \to R_2$ to get

$\begin{bmatrix} 2 & 1 & | & 4 \\ 0 & 3 & | & -12 \end{bmatrix}$, use $\frac{1}{3}R_2 \to R_2$ to get

$\begin{bmatrix} 2 & 1 & | & 4 \\ 0 & 1 & | & -4 \end{bmatrix}$, use $-1R_2 + R_1 \to R_1$ to get

$\begin{bmatrix} 2 & 0 & | & 8 \\ 0 & 1 & | & -4 \end{bmatrix}$, use $\frac{1}{2}R_1 \to R_1$ to get

$\begin{bmatrix} 1 & 0 & | & 4 \\ 0 & 1 & | & -4 \end{bmatrix}$, solution set is $\{(4,-4)\}$, independent.

41.

On $\begin{bmatrix} 1 & 1 & 1 & | & 6 \\ 1 & -1 & -1 & | & 0 \\ 0 & 2 & -1 & | & 3 \end{bmatrix}$ use $-1R_2 + R_1 \rightarrow R_2$

to get

$\begin{bmatrix} 1 & 1 & 1 & | & 6 \\ 0 & 2 & 2 & | & 6 \\ 0 & 2 & -1 & | & 3 \end{bmatrix}$, use $\frac{1}{2}R_2 \rightarrow R_2$ to get

$\begin{bmatrix} 1 & 1 & 1 & | & 6 \\ 0 & 1 & 1 & | & 3 \\ 0 & 2 & -1 & | & 3 \end{bmatrix}$, use $-1R_2 + R_1 \rightarrow R_1$ to

get

$\begin{bmatrix} 1 & 0 & 0 & | & 3 \\ 0 & 1 & 1 & | & 3 \\ 0 & 2 & -1 & | & 3 \end{bmatrix}$, use $-2R_2 + R_3 \rightarrow R_3$ to

get

$\begin{bmatrix} 1 & 0 & 0 & | & 3 \\ 0 & 1 & 1 & | & 3 \\ 0 & 0 & -3 & | & -3 \end{bmatrix}$, use $-\frac{1}{3}R_3 \rightarrow R_3$ to get

$\begin{bmatrix} 1 & 0 & 0 & | & 3 \\ 0 & 1 & 1 & | & 3 \\ 0 & 0 & 1 & | & 1 \end{bmatrix}$, use $-1R_3 + R_2 \rightarrow R_2$ to get

$\begin{bmatrix} 1 & 0 & 0 & | & 3 \\ 0 & 1 & 0 & | & 2 \\ 0 & 0 & 1 & | & 1 \end{bmatrix}$, solution set is $\{(3, 2, 1)\}$,

independent.

43.

Rewrite system as

$$2x + y - z = 2$$
$$x + 2y - z = 2$$
$$x - y + 2z = 2.$$

On $\begin{bmatrix} 2 & 1 & -1 & | & 2 \\ 1 & 2 & -1 & | & 2 \\ 1 & -1 & 2 & | & 2 \end{bmatrix}$, use $-1R_2 + R_1 \rightarrow R_1$

and $-1R_3 + R_2 \rightarrow R_3$ to get

$\begin{bmatrix} 1 & -1 & 0 & | & 0 \\ 1 & 2 & -1 & | & 2 \\ 0 & 3 & -3 & | & 0 \end{bmatrix}$, use $-1R_2 + R_1 \rightarrow R_2$

and $\frac{1}{3}R_3 \rightarrow R_3$ to get

$\begin{bmatrix} 1 & -1 & 0 & | & 0 \\ 0 & -3 & 1 & | & -2 \\ 0 & 1 & -1 & | & 0 \end{bmatrix}$, use $R_1 + R_3 \rightarrow R_1$

and $R_3 + R_2 \rightarrow R_2$ to get

$\begin{bmatrix} 1 & 0 & -1 & | & 0 \\ 0 & -2 & 0 & | & -2 \\ 0 & 1 & -1 & | & 0 \end{bmatrix}$, use $-\frac{1}{2}R_2 \rightarrow R_2$ to get

$\begin{bmatrix} 1 & 0 & -1 & | & 0 \\ 0 & 1 & 0 & | & 1 \\ 0 & 1 & -1 & | & 0 \end{bmatrix}$, use $-1R_3 + R_2 \rightarrow R_3$ to

get

$\begin{bmatrix} 1 & 0 & -1 & | & 0 \\ 0 & 1 & 0 & | & 1 \\ 0 & 0 & 1 & | & 1 \end{bmatrix}$, use $R_1 + R_3 \rightarrow R_1$ to get

$\begin{bmatrix} 1 & 0 & 0 & | & 1 \\ 0 & 1 & 0 & | & 1 \\ 0 & 0 & 1 & | & 1 \end{bmatrix}$, solution set is $\{(1, 1, 1)\}$,

independent.

45.

Interchange rows of $\begin{bmatrix} 2 & -2 & 1 & | & -2 \\ 1 & 1 & -3 & | & 3 \\ 1 & -3 & 1 & | & -5 \end{bmatrix}$ to

get

$\begin{bmatrix} 1 & 1 & -3 & | & 3 \\ 1 & -3 & 1 & | & -5 \\ 2 & -2 & 1 & | & -2 \end{bmatrix}$, use $-1R_2 + R_1 \rightarrow R_2$

and $-1R_3 + R_1 + R_2 \rightarrow R_3$ to get

$\begin{bmatrix} 1 & 1 & -3 & | & 3 \\ 0 & 4 & -4 & | & 8 \\ 0 & 0 & -3 & | & 0 \end{bmatrix}$, use $\frac{1}{4}R_2 \rightarrow R_2$

and $-\frac{1}{3}R_3 \rightarrow R_3$ to get

$\begin{bmatrix} 1 & 1 & -3 & | & 3 \\ 0 & 1 & -1 & | & 2 \\ 0 & 0 & 1 & | & 0 \end{bmatrix}$, use $R_2 + R_3 \rightarrow R_2$ to get

$\begin{bmatrix} 1 & 1 & -3 & | & 3 \\ 0 & 1 & 0 & | & 2 \\ 0 & 0 & 1 & | & 0 \end{bmatrix}$, use $-1R_2 + R_1 \rightarrow R_1$ to

get

$\begin{bmatrix} 1 & 0 & -3 & | & 1 \\ 0 & 1 & 0 & | & 2 \\ 0 & 0 & 1 & | & 0 \end{bmatrix}$, use $3R_3 + R_1 \rightarrow R_1$ to get

$\begin{bmatrix} 1 & 0 & 0 & | & 1 \\ 0 & 1 & 0 & | & 2 \\ 0 & 0 & 1 & | & 0 \end{bmatrix}$, solution set is $\{(1, 2, 0)\}$,

independent.

47.

Rewrite system as

$$\begin{aligned} x - 3y + z &= 0 \\ x - y - 3z &= 4 \\ x + y + 2z &= -1. \end{aligned}$$

On $\begin{bmatrix} 1 & -3 & 1 & | & 0 \\ 1 & -1 & -3 & | & 4 \\ 1 & 1 & 2 & | & -1 \end{bmatrix}$, use

$-1R_2 + R_1 \rightarrow R_2$ and $-1R_3 + R_1 \rightarrow R_3$ to get

$\begin{bmatrix} 1 & -3 & 1 & | & 0 \\ 0 & -2 & 4 & | & -4 \\ 0 & -4 & -1 & | & 1 \end{bmatrix}$, use $-\dfrac{1}{2}R_2 \rightarrow R_2$

and $-1R_3 \rightarrow R_3$ to get

$\begin{bmatrix} 1 & -3 & 1 & | & 0 \\ 0 & 1 & -2 & | & 2 \\ 0 & 4 & 1 & | & -1 \end{bmatrix}$, use $-1R_3 + R_1 \rightarrow R_1$

to get

$\begin{bmatrix} 1 & -7 & 0 & | & 1 \\ 0 & 1 & -2 & | & 2 \\ 0 & 4 & 1 & | & -1 \end{bmatrix}$, use $-4R_2 + R_3 \rightarrow R_3$

to get

$\begin{bmatrix} 1 & -7 & 0 & | & 1 \\ 0 & 1 & -2 & | & 2 \\ 0 & 0 & 9 & | & -9 \end{bmatrix}$, use $\dfrac{1}{9}R_3 \rightarrow R_3$ to get

$\begin{bmatrix} 1 & -7 & 0 & | & 1 \\ 0 & 1 & -2 & | & 2 \\ 0 & 0 & 1 & | & -1 \end{bmatrix}$, use $2R_3 + R_2 \rightarrow R_2$

to get

$\begin{bmatrix} 1 & -7 & 0 & | & 1 \\ 0 & 1 & 0 & | & 0 \\ 0 & 0 & 1 & | & -1 \end{bmatrix}$, use $R_1 + 7R_2 \rightarrow R_1$ to

get

$\begin{bmatrix} 1 & 0 & 0 & | & 1 \\ 0 & 1 & 0 & | & 0 \\ 0 & 0 & 1 & | & -1 \end{bmatrix}$, solution set is $\{(1, 0, -1)\}$,

independent.

49.

On $\begin{bmatrix} 1 & -2 & 3 & | & 1 \\ 2 & -4 & 6 & | & 2 \\ -3 & 6 & -9 & | & -3 \end{bmatrix}$, use $\dfrac{1}{2}R_2 \rightarrow R_2$

and $-\dfrac{1}{3}R_3 \rightarrow R_3$ to get

$\begin{bmatrix} 1 & -2 & 3 & | & 1 \\ 1 & -2 & 3 & | & 1 \\ 1 & -2 & 3 & | & 1 \end{bmatrix}$, use $-1R_1 + R_2 \rightarrow R_2$

and $-1R_1 + R_3 \rightarrow R_3$ to get

$\begin{bmatrix} 1 & -2 & 3 & | & 1 \\ 0 & 0 & 0 & | & 0 \\ 0 & 0 & 0 & | & 0 \end{bmatrix}$, dependent system,

and solution set is $\{(x, y, z) \mid x - 2y + 3z = 1\}$.

51.

On $\begin{bmatrix} 1 & -1 & 1 & | & 2 \\ 2 & 1 & -1 & | & 1 \\ 2 & -2 & 2 & | & 5 \end{bmatrix}$, use $-\dfrac{1}{2}R_3 + R_1 \rightarrow R_3$

to get $\begin{bmatrix} 1 & -1 & 1 & | & 2 \\ 2 & 1 & -1 & | & 1 \\ 0 & 0 & 0 & | & -1/2 \end{bmatrix}$, inconsistent,

and solution set is $\emptyset$.

53.

On $\begin{bmatrix} 1 & 1 & -1 & | & 3 \\ 3 & 1 & 1 & | & 7 \\ 1 & -1 & 3 & | & 1 \end{bmatrix}$, use $-1R_3 + R_1 \rightarrow R_3$

and $3R_1 + (-1R_2) \rightarrow R_2$ to get

$\begin{bmatrix} 1 & 1 & -1 & | & 3 \\ 0 & 2 & -4 & | & 2 \\ 0 & 2 & -4 & | & 2 \end{bmatrix}$, use $-1R_3 + R_2 \rightarrow R_3$ to

get

$\begin{bmatrix} 1 & 1 & -1 & | & 3 \\ 0 & 2 & -4 & | & 2 \\ 0 & 0 & 0 & | & 0 \end{bmatrix}$, use $\dfrac{1}{2}R_2 \rightarrow R_2$ to get

$\begin{bmatrix} 1 & 1 & -1 & | & 3 \\ 0 & 1 & -2 & | & 1 \\ 0 & 0 & 0 & | & 0 \end{bmatrix}$, use $-1R_2 + R_1 \rightarrow R_1$ to

get

$\begin{bmatrix} 1 & 0 & 1 & | & 2 \\ 0 & 1 & -2 & | & 1 \\ 0 & 0 & 0 & | & 0 \end{bmatrix}$. Substitute $z = 2 - x$

into $y = 1 + 2z$. Then $y = 1 + 2(2 - x) = 5 - 2x$.
The solution set is

$$\{(x, 5 - 2x, 2 - x) \mid x \text{ is any real number}\}$$

and the system is dependent.

55.

On $\begin{bmatrix} 2 & -1 & 3 & | & 1 \\ 1 & 1 & -1 & | & 4 \end{bmatrix}$, use $-2R_2 + R_1 \rightarrow R_2$

to get

$\begin{bmatrix} 2 & -1 & 3 & | & 1 \\ 0 & -3 & 5 & | & -7 \end{bmatrix}$, use $-3R_1 + R_2 \rightarrow R_1$

to get

$\begin{bmatrix} -6 & 0 & -4 & | & -10 \\ 0 & -3 & 5 & | & -7 \end{bmatrix}$, use $-\frac{1}{6}R_1 \rightarrow R_1$

and $-\frac{1}{3}R_2 \rightarrow R_2$ to get

$\begin{bmatrix} 1 & 0 & 2/3 & | & 5/3 \\ 0 & 1 & -5/3 & | & 7/3 \end{bmatrix}$. Note $x = \dfrac{5 - 2z}{3}$ and

$y = \dfrac{7 + 5z}{3}$. Solving for z, we get $z = \dfrac{5 - 3x}{2}$.

Then $y = \dfrac{7 + 5 \cdot \frac{5-3x}{2}}{3} = \dfrac{7 + 5 \cdot \frac{5-3x}{2}}{3} \cdot \dfrac{2}{2} =$

$\dfrac{39 - 15x}{6} = \dfrac{13 - 5x}{2}$. The solution set is

$\left\{ \left(x, \dfrac{13 - 5x}{2}, \dfrac{5 - 3x}{2} \right) \mid x \text{ is any real number} \right\}$

and the system is dependent.

57.

On $\begin{bmatrix} 1 & -1 & 1 & -1 & | & 2 \\ -1 & 2 & -1 & -1 & | & -1 \\ 2 & -1 & -1 & 1 & | & 4 \\ 1 & 3 & -2 & -3 & | & 6 \end{bmatrix}$,

use $2R_2 + R_3 \rightarrow R_3$ and $R_2 + R_4 \rightarrow R_4$ to get

$\begin{bmatrix} 1 & -1 & 1 & -1 & | & 2 \\ 0 & 1 & 0 & -2 & | & 1 \\ 0 & 3 & -3 & -1 & | & 2 \\ 0 & 5 & -3 & -4 & | & 5 \end{bmatrix}$, use

$-3R_2 + R_3 \rightarrow R_3$ and $-5R_2 + R_4 \rightarrow R_4$ to get

$\begin{bmatrix} 1 & -1 & 1 & -1 & | & 2 \\ 0 & 1 & 0 & -2 & | & 1 \\ 0 & 0 & -3 & 5 & | & -1 \\ 0 & 0 & -3 & 6 & | & 0 \end{bmatrix}$, use

$R_1 + R_2 \rightarrow R_1$ and $-\frac{1}{3}R_4 \rightarrow R_4$ to get

$\begin{bmatrix} 1 & 0 & 1 & -3 & | & 3 \\ 0 & 1 & 0 & -2 & | & 1 \\ 0 & 0 & -3 & 5 & | & -1 \\ 0 & 0 & 1 & -2 & | & 0 \end{bmatrix}$, use $R_3 \rightarrow R_4$

and $R_4 \rightarrow R_3$ to get

$\begin{bmatrix} 1 & 0 & 1 & -3 & | & 3 \\ 0 & 1 & 0 & -2 & | & 1 \\ 0 & 0 & 1 & -2 & | & 0 \\ 0 & 0 & -3 & 5 & | & -1 \end{bmatrix}$, use

$-3R_3 + (-1R_4) \rightarrow R_4$ to get

$\begin{bmatrix} 1 & 0 & 1 & -3 & | & 3 \\ 0 & 1 & 0 & -2 & | & 1 \\ 0 & 0 & 1 & -2 & | & 0 \\ 0 & 0 & 0 & 1 & | & 1 \end{bmatrix}$, use $-1R_3 + R_1 \rightarrow R_1$

to get

$\begin{bmatrix} 1 & 0 & 0 & -1 & | & 3 \\ 0 & 1 & 0 & -2 & | & 1 \\ 0 & 0 & 1 & -2 & | & 0 \\ 0 & 0 & 0 & 1 & | & 1 \end{bmatrix}$, use $2R_4 + R_3 \rightarrow R_3$ to

get

$\begin{bmatrix} 1 & 0 & 0 & -1 & | & 3 \\ 0 & 1 & 0 & -2 & | & 1 \\ 0 & 0 & 1 & 0 & | & 2 \\ 0 & 0 & 0 & 1 & | & 1 \end{bmatrix}$, use $2R_4 + R_2 \rightarrow R_2$ to

get

$\begin{bmatrix} 1 & 0 & 0 & -1 & | & 3 \\ 0 & 1 & 0 & 0 & | & 3 \\ 0 & 0 & 1 & 0 & | & 2 \\ 0 & 0 & 0 & 1 & | & 1 \end{bmatrix}$, use $R_4 + R_1 \rightarrow R_1$ to

get

$\begin{bmatrix} 1 & 0 & 0 & 0 & | & 4 \\ 0 & 1 & 0 & 0 & | & 3 \\ 0 & 0 & 1 & 0 & | & 2 \\ 0 & 0 & 0 & 1 & | & 1 \end{bmatrix}$, solution set is

$\{(4, 3, 2, 1)\}$, independent.

59. Let x and y be the number of hours Mike worked at Burgers and the Soap Opera, respectively. The augmented matrix is

$A = \begin{bmatrix} 1 & 1 & | & 60 \\ 8 & 9 & | & 502 \end{bmatrix}$. On A use

$-8R_1 + R_2 \rightarrow R_2$ to get

$\begin{bmatrix} 1 & 1 & | & 60 \\ 0 & 1 & | & 22 \end{bmatrix}$, use $-R_2 + R_1 \rightarrow R_1$ to get

$\begin{bmatrix} 1 & 0 & | & 38 \\ 0 & 1 & | & 22 \end{bmatrix}$. Mike worked $x = 38$ hours

at Burgers and $y = 22$ hours at Soap Opera.

61. Let x, y, and z be the amounts invested in a mutual fund, in treasury bills, and in bonds, respectively. Augmented matrix is

$$A = \begin{bmatrix} 1 & 1 & 1 & | & 40,000 \\ 0.08 & 0.09 & 0.12 & | & 3,660 \\ 1 & -1 & -1 & | & 0 \end{bmatrix}.$$

On A use $100R_2 \to R_2$ to get

$$\begin{bmatrix} 1 & 1 & 1 & | & 40,000 \\ 8 & 9 & 12 & | & 366,000 \\ 1 & -1 & -1 & | & 0 \end{bmatrix}, \text{ use}$$

$-8R_1 + R_2 \to R_2$ to get

$$\begin{bmatrix} 1 & 1 & 1 & | & 40,000 \\ 0 & 1 & 4 & | & 46,000 \\ 1 & -1 & -1 & | & 0 \end{bmatrix}, \text{ use}$$

$-1R_3 + R_1 \to R_3$ and $-1R_2 + R_1 \to R_1$ to get

$$\begin{bmatrix} 1 & 0 & -3 & | & -6,000 \\ 0 & 1 & 4 & | & 46,000 \\ 0 & 2 & 2 & | & 40,000 \end{bmatrix}, \text{ use } \frac{1}{2}R_3 \to R_3 \text{ to}$$

get

$$\begin{bmatrix} 1 & 0 & -3 & | & -6,000 \\ 0 & 1 & 4 & | & 46,000 \\ 0 & 1 & 1 & | & 20,000 \end{bmatrix}, \text{ use}$$

$-1R_3 + R_2 \to R_3$ to get

$$\begin{bmatrix} 1 & 0 & -3 & | & -6,000 \\ 0 & 1 & 4 & | & 46,000 \\ 0 & 0 & 3 & | & 26,000 \end{bmatrix}, \text{ use}$$

$-1R_3 + R_2 \to R_2$ and $\frac{1}{3}R_3 \to R_3$ to get

$$\begin{bmatrix} 1 & 0 & -3 & | & -6,000 \\ 0 & 1 & 1 & | & 20,000 \\ 0 & 0 & 1 & | & 8,666.67 \end{bmatrix}, \text{ use } 3R_3 + R_1 \to R_1$$

and $-1R_3 + R_2 \to R_2$ to get

$$\begin{bmatrix} 1 & 0 & 0 & | & 20,000 \\ 0 & 1 & 0 & | & 11,333.33 \\ 0 & 0 & 1 & | & 8,666.67 \end{bmatrix}. \text{ Investments were}$$

$x = \$20,000$ in a mutual fund, $y = \$11,333.33$ in treasury bills, and $z = \$8,666.67$ in bonds.

63. The augmented matrix is

$$A = \begin{bmatrix} -1 & -1 & 1 & | & 4 \\ 1 & 1 & 1 & | & 2 \\ 8 & 2 & 1 & | & 7 \end{bmatrix}.$$

On A use $R_1 \to R_2$ and $R_2 \to R_1$ to get

$$\begin{bmatrix} 1 & 1 & 1 & | & 2 \\ -1 & -1 & 1 & | & 4 \\ 8 & 2 & 1 & | & 7 \end{bmatrix}, \text{ use } 8R_1 + (-1R_3) \to R_3$$

and $-\frac{8}{3}R_2 + \left(-\frac{1}{3}R_3\right) \to R_2$ to get

$$\begin{bmatrix} 1 & 1 & 1 & | & 2 \\ 0 & 2 & -3 & | & -13 \\ 0 & 6 & 7 & | & 9 \end{bmatrix}, \text{ use } -3R_2 + R_3 \to R_3$$

to get

$$\begin{bmatrix} 1 & 1 & 1 & | & 2 \\ 0 & 2 & -3 & | & -13 \\ 0 & 0 & 16 & | & 48 \end{bmatrix}, \text{ use } \frac{1}{2}R_2 \to R_2$$

and $\frac{1}{16}R_3 \to R_3$ to get

$$\begin{bmatrix} 1 & 1 & 1 & | & 2 \\ 0 & 1 & -3/2 & | & -13/2 \\ 0 & 0 & 1 & | & 3 \end{bmatrix}, \text{ use}$$

$\frac{3}{2}R_3 + R_2 \to R_2$ to get

$$\begin{bmatrix} 1 & 1 & 1 & | & 2 \\ 0 & 1 & 0 & | & -2 \\ 0 & 0 & 1 & | & 3 \end{bmatrix}, \text{ use } -1R_3 + R_1 \to R_1 \text{ to}$$

get

$$\begin{bmatrix} 1 & 1 & 0 & | & -1 \\ 0 & 1 & 0 & | & -2 \\ 0 & 0 & 1 & | & 3 \end{bmatrix}, \text{ use } -1R_2 + R_1 \to R_1 \text{ to}$$

get

$$\begin{bmatrix} 1 & 0 & 0 & | & 1 \\ 0 & 1 & 0 & | & -2 \\ 0 & 0 & 1 & | & 3 \end{bmatrix}. \text{ Then } a = 1, b = -2,$$

$c = 3$, and the equation is $y = x^3 - 2x + 3$.

65. Since the number of cars entering M.L. King Dr. and Washington St. is 750 and $x + y$ is the number of cars leaving the intersection of M.L. King Dr. and Washington St. then $x + y = 750$.

On the intersection of M.L. King Dr. and JFK Blvd., the number of cars entering this intersection is $450 + x$ and the number of cars leaving is $700 + z$. So $450 + x = 700 + z$.

Simplifying, one gets $y = 750 - x$ and $z = x - 250$; and since y and z are nonnegative,

$250 \leq x \leq 750$. The values of x, y, and z that realizes this traffic flow must satisfy

$$
\begin{aligned}
y &= 750 - x \\
z &= x - 250 \\
250 \leq x &\leq 750
\end{aligned}
$$

If $z = 50$, then $50 = x - 250$ or $x = 300$ and $y = 750 - 300 = 450$.

For Thought

1. True

2. False, since the orders of matrices A and C are different.

3. False, $A + B = \begin{bmatrix} 2 \\ 6 \end{bmatrix}$.

4. True, $C + D = \begin{bmatrix} 1-3 & 1+5 \\ 3+1 & 3-2 \end{bmatrix} =$
   $= \begin{bmatrix} -2 & 6 \\ 4 & 1 \end{bmatrix}$.

5. True, $A - B = \begin{bmatrix} 1-1 \\ 3-3 \end{bmatrix} = \begin{bmatrix} 0 \\ 0 \end{bmatrix}$.

6. False, $3B = 3\begin{bmatrix} 1 \\ 3 \end{bmatrix} = \begin{bmatrix} 3 \\ 9 \end{bmatrix}$.

7. False, $-A = -\begin{bmatrix} 1 \\ 3 \end{bmatrix} = \begin{bmatrix} -1 \\ -3 \end{bmatrix}$.

8. False, matrices of different orders cannot be added.   9.   False, matrices of different orders cannot be subtracted.

10. False, $C + 2D = \begin{bmatrix} 1 & 1 \\ 3 & 3 \end{bmatrix} + \begin{bmatrix} -6 & 10 \\ 2 & -4 \end{bmatrix} =$
    $= \begin{bmatrix} -5 & 11 \\ 5 & -1 \end{bmatrix}$.

6.2 Exercises

1. $x = 2$, $y = 5$

3. Since $2x = 6$ and $4y = 16$, $x = 3$ and $y = 4$. Also $3z = z + y$ and so $z = y/2 = 4/2 = 2$. Then $x = 3$, $y = 4$, and $z = 2$.

5. $\begin{bmatrix} 3+2 \\ 5+1 \end{bmatrix} = \begin{bmatrix} 5 \\ 6 \end{bmatrix}$

7. $\begin{bmatrix} -0.5+2 & -0.03+1 \\ 2-0.05 & -0.33+1 \end{bmatrix} = \begin{bmatrix} 1.5 & 0.97 \\ 1.95 & 0.67 \end{bmatrix}$

9. $\begin{bmatrix} 2+1 & -3-1 & 4+1 \\ 4+0 & -6+1 & 8-1 \\ 6+0 & -3+0 & 1+1 \end{bmatrix} = \begin{bmatrix} 3 & -4 & 5 \\ 4 & -5 & 7 \\ 6 & -3 & 2 \end{bmatrix}$

11. $-A = -\begin{bmatrix} 1 & -4 \\ -5 & 6 \end{bmatrix} = \begin{bmatrix} -1 & 4 \\ 5 & -6 \end{bmatrix}$
    and $A + (-A) = \begin{bmatrix} 0 & 0 \\ 0 & 0 \end{bmatrix}$

13. $-A = -\begin{bmatrix} 3 & 0 & -1 \\ 8 & -2 & 1 \\ -3 & 6 & 3 \end{bmatrix} =$
    $\begin{bmatrix} -3 & 0 & 1 \\ -8 & 2 & -1 \\ 3 & -6 & -3 \end{bmatrix}$ and
    $A + (-A) = \begin{bmatrix} 0 & 0 & 0 \\ 0 & 0 & 0 \\ 0 & 0 & 0 \end{bmatrix}$

15. $B - A = \begin{bmatrix} -1+4 & -2-1 \\ 7-3 & 4-0 \end{bmatrix} = \begin{bmatrix} 3 & -3 \\ 4 & 4 \end{bmatrix}$

17. $B - C = \begin{bmatrix} -1+3 & -2+4 \\ 7-2 & 4+5 \end{bmatrix} = \begin{bmatrix} 2 & 2 \\ 5 & 9 \end{bmatrix}$

19. $B - E$ is undefined since B and E have different sizes

21. $3A = 3\begin{bmatrix} -4 & 1 \\ 3 & 0 \end{bmatrix} = \begin{bmatrix} -12 & 3 \\ 9 & 0 \end{bmatrix}$

23. $-1D = -1\begin{bmatrix} -4 \\ 5 \end{bmatrix} = \begin{bmatrix} 4 \\ -5 \end{bmatrix}$

25.
$3A + 3C = \begin{bmatrix} -12 & 3 \\ 9 & 0 \end{bmatrix} + \begin{bmatrix} -9 & -12 \\ 6 & -15 \end{bmatrix} =$
$\begin{bmatrix} -21 & -9 \\ 15 & -15 \end{bmatrix}$

27.

$$2A - B = \begin{bmatrix} -8 & 2 \\ 6 & 0 \end{bmatrix} - \begin{bmatrix} -1 & -2 \\ 7 & 4 \end{bmatrix} =$$

$$\begin{bmatrix} -7 & 4 \\ -1 & -4 \end{bmatrix}$$

29.

$$2D - 3E = \begin{bmatrix} -8 \\ 10 \end{bmatrix} - \begin{bmatrix} -3 \\ 6 \end{bmatrix} = \begin{bmatrix} -5 \\ 4 \end{bmatrix}$$

31. $D+A$ is undefined since D and A have different sizes

33.

$$(A + B) + C = \begin{bmatrix} -5 & -1 \\ 10 & 4 \end{bmatrix} + \begin{bmatrix} -3 & -4 \\ 2 & -5 \end{bmatrix} =$$

$$\begin{bmatrix} -8 & -5 \\ 12 & -1 \end{bmatrix}$$

35. $\begin{bmatrix} 0.2+0.2 & 0.1+0.05 \\ 0.4+0.3 & 0.3+0.8 \end{bmatrix} = \begin{bmatrix} 0.4 & 0.15 \\ 0.7 & 1.1 \end{bmatrix}$

37. $\begin{bmatrix} 1/2 & 3/2 \\ 3 & -12 \end{bmatrix}$

39. $\begin{bmatrix} -2 & 4 \\ 6 & 8 \end{bmatrix} - \begin{bmatrix} -12 & 4 \\ 8 & -8 \end{bmatrix} = \begin{bmatrix} 10 & 0 \\ -2 & 16 \end{bmatrix}$

41. Undefined since we cannot add matrices with different sizes

43. $\begin{bmatrix} -1 & 13 \\ -9 & 3 \\ 6 & -2 \end{bmatrix}$

45. $\begin{bmatrix} 3\sqrt{2} & 2 & 3\sqrt{3} \end{bmatrix}$

47.

$$\begin{bmatrix} 2a \\ 2b \end{bmatrix} + \begin{bmatrix} 6a \\ 12b \end{bmatrix} + \begin{bmatrix} 5a \\ -15b \end{bmatrix} = \begin{bmatrix} 13a \\ -b \end{bmatrix}$$

49.

$$\begin{bmatrix} -0.4x - 0.6x & 0.4y - 0.9y \\ 0.8x - 1.5x & 3.2y + 0.3y \end{bmatrix} =$$

$$\begin{bmatrix} -x & -0.5y \\ -0.7x & 3.5y \end{bmatrix}$$

51.

$$\begin{bmatrix} 2x & 2y & 2z \\ -2x & 4y & 6z \\ 2x & -2y & -6z \end{bmatrix} - \begin{bmatrix} -x & 0 & 3z \\ 4x & y & -z \\ 2x & 5y & z \end{bmatrix} =$$

$$\begin{bmatrix} 3x & 2y & -z \\ -6x & 3y & 7z \\ 0 & -7y & -7z \end{bmatrix}$$

53. Equate the corresponding entries.

$$\begin{aligned} x + y &= 5 \\ x - y &= 1 \end{aligned}$$

Adding the two equations, we get $2x = 6$. Substitute $x = 3$ into $x+y = 5$. Then $3+y = 5$ and $y = 2$. Solution set is $\{(3,2)\}$.

55. Equate the corresponding entries.

$$\begin{aligned} 2x + 3y &= 7 \\ x - 4y &= -13 \end{aligned}$$

Multiply second equation by -2 and add to the first one.

$$\begin{aligned} 2x + 3y &= 7 \\ \underline{-2x + 8y} &= \underline{26} \\ 11y &= 33 \\ y &= 3 \end{aligned}$$

Substitute $y = 3$ into $x - 4y = -13$. Then $x - 12 = -13$ and $x = -1$. The solution set is $\{(-1,3)\}$.

57. Equate the corresponding entries.

$$\begin{aligned} x + y + z &= 8 \\ x - y - z &= -7 \\ x - y + z &= 2 \end{aligned}$$

Adding the first and second equations, $2x = 1$ and $x = 0.5$. Multiply second equation by -1 and add to the third.

$$\begin{aligned} -x + y + z &= 7 \\ \underline{x - y + z} &= \underline{2} \\ 2z &= 9 \\ z &= 4.5 \end{aligned}$$

Substitute $x = 0.5$ and $z = 4.5$ into $x + y + z = 8$. Then $y + 5 = 8$ and $y = 3$. The solution set is $\{(0.5, 3, 4.5)\}$.

59. The matrices for January, February and March are, respectively,

$$J = \begin{bmatrix} 120 \\ 30 \\ 40 \end{bmatrix}, F = \begin{bmatrix} 130 \\ 70 \\ 50 \end{bmatrix}, \text{ and}$$

$$M = \begin{bmatrix} 140 \\ 60 \\ 45 \end{bmatrix}. \text{ The sum}$$

$$J + F + M = \begin{bmatrix} \$390 \\ \$160 \\ \$135 \end{bmatrix} \text{ represents the}$$

total expenses on food, clothing and utilities for the three months.

61. The supply matrix for the first week is

$$S = \begin{bmatrix} 40 & 80 \\ 30 & 90 \\ 80 & 200 \end{bmatrix}. \text{ Next week's supply matrix}$$

$$\text{is } S + 0.5S = \begin{bmatrix} 40 & 80 \\ 30 & 90 \\ 80 & 200 \end{bmatrix} + \begin{bmatrix} 20 & 40 \\ 15 & 45 \\ 40 & 100 \end{bmatrix}$$

$$= \begin{bmatrix} 60 & 120 \\ 45 & 135 \\ 120 & 300 \end{bmatrix}.$$

63. yes, yes

65. yes, yes

67. yes, yes

69. $\begin{bmatrix} 0 & 0 \\ 0 & 0 \end{bmatrix}$

For Thought

1. True **2.** True **3.** False, they cannot be multiplied since the number of columns of A is not the same as the number of rows of C.

4. False, the order of CA is 2×1.

5. True

6. True, $BC = [7 \cdot 2 + 9 \cdot 4 \quad 7 \cdot 3 + 9 \cdot 5] =$
$= [14 + 36 \quad 21 + 45] = [50 \quad 66].$

7. True, $AB = \begin{bmatrix} 1 \\ 6 \end{bmatrix} [7 \quad 9] = \begin{bmatrix} 1 \cdot 7 & 1 \cdot 9 \\ 6 \cdot 7 & 6 \cdot 9 \end{bmatrix} =$
$= \begin{bmatrix} 7 & 9 \\ 42 & 54 \end{bmatrix}.$

8. True, $\begin{bmatrix} 2 & 3 \\ 4 & 5 \end{bmatrix} \begin{bmatrix} 2 & -1 \\ 0 & 3 \end{bmatrix} =$
$\begin{bmatrix} 4+0 & -2+9 \\ 8+0 & -4+15 \end{bmatrix} = \begin{bmatrix} 4 & 7 \\ 8 & 11 \end{bmatrix}.$

9. True, $BA = [7 \quad 9] \begin{bmatrix} 1 \\ 6 \end{bmatrix} = [7 + 54] = [61].$

10. False, since $EC = \begin{bmatrix} 2 & -1 \\ 0 & 3 \end{bmatrix} \begin{bmatrix} 2 & 3 \\ 4 & 5 \end{bmatrix} =$
$= \begin{bmatrix} 4-4 & 6-5 \\ 0+12 & 0+15 \end{bmatrix} = \begin{bmatrix} 0 & 1 \\ 12 & 15 \end{bmatrix}$ and from Exercise 8 one sees $EC \neq CE.$

6.3 Exercises

1. 3×5 **3.** 1×1 **5.** 5×5

7. 3×3 **9.** undefined

11. $[-3(4) + 2(1)] = [-10]$

13. $\begin{bmatrix} 1(1) + 3(3) \\ 2(1) + (-4)(3) \end{bmatrix} = \begin{bmatrix} 10 \\ -10 \end{bmatrix}$

15. $\begin{bmatrix} 5(1) + 1(3) & 5(2) + 1(1) \\ 2(1) + 1(3) & 2(2) + 1(1) \end{bmatrix} = \begin{bmatrix} 8 & 11 \\ 5 & 5 \end{bmatrix}$

17. $\begin{bmatrix} 3(5) & 3(6) \\ 1(5) & 1(6) \end{bmatrix} = \begin{bmatrix} 15 & 18 \\ 5 & 6 \end{bmatrix}$

19. $AB = \begin{bmatrix} 1(1)+3(-1) & 1(0)+3(1) & 1(1)+3(0) \\ 2(1)+4(-1) & 2(0)+4(1) & 2(1)+4(0) \\ 5(1)+6(-1) & 5(0)+6(1) & 5(1)+6(0) \end{bmatrix}$
$= \begin{bmatrix} -2 & 3 & 1 \\ -2 & 4 & 2 \\ -1 & 6 & 5 \end{bmatrix}$ and

$BA = \begin{bmatrix} 1(1)+0(2)+1(5) & 1(3)+0(4)+1(6) \\ -1(1)+1(2)+0(5) & -1(3)+1(4)+0(6) \end{bmatrix}$
$= \begin{bmatrix} 6 & 9 \\ 1 & 1 \end{bmatrix}$

21. $AB =$

$$\begin{bmatrix} 1(1)+2(0)+3(0) & 1(1)+2(1)+3(0) & 1(1)+2(1)+3(1) \\ 2(1)+1(0)+3(0) & 2(1)+1(1)+3(0) & 2(1)+1(1)+3(1) \\ 3(1)+2(0)+1(0) & 3(1)+2(1)+1(0) & 3(1)+2(1)+1(1) \end{bmatrix}$$

$$= \begin{bmatrix} 1 & 3 & 6 \\ 2 & 3 & 6 \\ 3 & 5 & 6 \end{bmatrix} \text{ and}$$

$BA =$

$$\begin{bmatrix} 1(1)+1(2)+1(3) & 1(2)+1(1)+1(2) & 1(3)+1(3)+1(1) \\ 0(1)+1(2)+1(3) & 0(2)+1(1)+1(2) & 0(3)+1(3)+1(1) \\ 0(1)+0(2)+1(3) & 0(2)+0(1)+1(2) & 0(3)+0(3)+1(1) \end{bmatrix}$$

$$= \begin{bmatrix} 6 & 5 & 7 \\ 5 & 3 & 4 \\ 3 & 2 & 1 \end{bmatrix}$$

23.

$$AB = \begin{bmatrix} 2\cdot 2 & 2\cdot 3 & 2\cdot 4 \\ -3\cdot 2 & -3\cdot 3 & -3\cdot 4 \\ 1\cdot 2 & 1\cdot 3 & 1\cdot 4 \end{bmatrix} =$$

$$\begin{bmatrix} 4 & 6 & 8 \\ -6 & -9 & -12 \\ 2 & 3 & 4 \end{bmatrix}$$

25. $\begin{bmatrix} 2+0+0 & 2+3+0 & 2+3+4 \end{bmatrix} =$

$\begin{bmatrix} 2 & 5 & 9 \end{bmatrix}$

27. Undefined

29.

$$EC = \begin{bmatrix} 2+4+1 & 3+5+0 \\ 0+4+1 & 0+5+0 \\ 0+0+1 & 0+0+0 \end{bmatrix} = \begin{bmatrix} 7 & 8 \\ 5 & 5 \\ 1 & 0 \end{bmatrix}$$

31.

$$DC = \begin{bmatrix} 4-4+1 & 6-5+0 \\ 0+12+2 & 0+15+0 \end{bmatrix} =$$

$$\begin{bmatrix} 1 & 1 \\ 14 & 15 \end{bmatrix}$$

33. Undefined

35.

$$EA = \begin{bmatrix} 2-3+1 \\ 0-3+1 \\ 0+0+1 \end{bmatrix} = \begin{bmatrix} 0 \\ -2 \\ 1 \end{bmatrix}$$

37. Undefined

39. We will use the answer in Exercise 23.

$AB + 2E =$

$$\begin{bmatrix} 4 & 6 & 8 \\ -6 & -9 & -12 \\ 2 & 3 & 4 \end{bmatrix} + \begin{bmatrix} 2 & 2 & 2 \\ 0 & 2 & 2 \\ 0 & 0 & 2 \end{bmatrix}$$

$$= \begin{bmatrix} 6 & 8 & 10 \\ -6 & -7 & -10 \\ 2 & 3 & 6 \end{bmatrix}$$

41. $A^2 = \begin{bmatrix} 1 & 0 \\ 1 & 1 \end{bmatrix} \begin{bmatrix} 1 & 0 \\ 1 & 1 \end{bmatrix} =$

$$\begin{bmatrix} 1(1)+0(1) & 1(0)+0(1) \\ 1(1)+1(1) & 1(0)+1(1) \end{bmatrix} = \begin{bmatrix} 1 & 0 \\ 2 & 1 \end{bmatrix}$$

43. We will use the answer in Exercise 42.

$$A^4 = A^3 A = \begin{bmatrix} 1 & 0 \\ 3 & 1 \end{bmatrix} \begin{bmatrix} 1 & 0 \\ 1 & 1 \end{bmatrix} =$$

$$\begin{bmatrix} 1(1)+0(1) & 1(0)+0(1) \\ 3(1)+1(1) & 3(0)+1(1) \end{bmatrix} = \begin{bmatrix} 1 & 0 \\ 4 & 1 \end{bmatrix}$$

45.

$$\begin{bmatrix} 2\cdot 1+0\cdot 0 & 2\cdot 1+0\cdot 1 \\ 3\cdot 1+1\cdot 0 & 3\cdot 1+1\cdot 1 \end{bmatrix} = \begin{bmatrix} 2 & 2 \\ 3 & 4 \end{bmatrix}$$

47.

$$\begin{bmatrix} 7\cdot 3+4\cdot(-5) & 7\cdot(-4)+4\cdot 7 \\ 5\cdot 3+3\cdot(-5) & 5\cdot(-4)+3\cdot 7 \end{bmatrix} =$$

$$\begin{bmatrix} 1 & 0 \\ 0 & 1 \end{bmatrix}$$

49.

$$\begin{bmatrix} -0.5\cdot 1+4\cdot 0 & -0.5\cdot 0+4\cdot 1 \\ 9\cdot 1+0.7\cdot 0 & 9\cdot 0+0.7\cdot 1 \end{bmatrix} =$$

$$\begin{bmatrix} -0.5 & 4 \\ 9 & 0.7 \end{bmatrix}$$

51. $\begin{bmatrix} -2a+6a & -6b+3b \end{bmatrix} = \begin{bmatrix} 4a & -3b \end{bmatrix}$

53.

$$\begin{bmatrix} -2a+0\cdot 1 & 5a+0\cdot 4 & 3a+0\cdot 6 \\ 0\cdot(-2)+b & 0\cdot 5+4b & 0\cdot 3+6b \end{bmatrix} =$$

$$\begin{bmatrix} -2a & 5a & 3a \\ b & 4b & 6b \end{bmatrix}$$

55. $\begin{bmatrix} 1\cdot 1+2\cdot 0+3\cdot 1 & 1\cdot 0+2\cdot 1+3\cdot 0 & 1\cdot 1+2\cdot 1+3\cdot 1 \end{bmatrix}$
$= \begin{bmatrix} 4 & 2 & 6 \end{bmatrix}$

57. $[-1 \cdot (-5) + 0 \cdot 1 + 3 \cdot 4] = [17]$

59.
$$\begin{bmatrix} x^2 & xy \\ xy & y^2 \end{bmatrix}$$

61.
$$\begin{bmatrix} -1 \cdot \sqrt{2} + 2 \cdot 0 + 3\sqrt{2} \\ 3 \cdot \sqrt{2} + 4 \cdot 0 + 4\sqrt{2} \end{bmatrix} = \begin{bmatrix} 2\sqrt{2} \\ 7\sqrt{2} \end{bmatrix}$$

63.
$$\begin{bmatrix} (1/2) \cdot (-8) + (1/3) \cdot (-5) & 6 + (1/3) \cdot 15 \\ (1/4) \cdot (-8) + (1/5) \cdot (-5) & 3 + (1/5) \cdot 15 \end{bmatrix}$$
$$= \begin{bmatrix} -4 - (5/3) & 6+5 \\ -2-1 & 3+3 \end{bmatrix} = \begin{bmatrix} -17/3 & 11 \\ -3 & 6 \end{bmatrix}$$

65. Undefined

67.
$$\begin{bmatrix} 9+0-7 & 8+0-8 & 10+0-4 \\ 0+3+0 & 0+5+0 & 0+2+0 \\ 9+3+7 & 8+5+8 & 10+2+4 \end{bmatrix} =$$
$$\begin{bmatrix} 2 & 0 & 6 \\ 3 & 5 & 2 \\ 19 & 21 & 16 \end{bmatrix}$$

69.
$$\begin{bmatrix} 1-0.6-0.6 & 1.5+1.2-1 \\ 0.8+0.4+1.8 & 1.2-0.8+3 \\ 0.4+0.6-2.4 & 0.6-1.2-4 \end{bmatrix} =$$
$$\begin{bmatrix} -0.2 & 1.7 \\ 3 & 3.4 \\ -1.4 & -4.6 \end{bmatrix}$$

71. System of equations is
$$\begin{aligned} 2x - 3y &= 0 \\ x + 2y &= 7. \end{aligned}$$

Multiply second equation by -2 and add to the first one.
$$\begin{aligned} 2x - 3y &= 0 \\ -2x - 4y &= -14 \\ \hline -7y &= -14 \end{aligned}$$

Substitute $y = 2$ into $x + 2y = 7$. Then $x + 4 = 7$ and $x = 3$. Solution set is $\{(3,2)\}$.

73. System of equations is
$$\begin{aligned} 2x + 3y &= 5 \\ 4x + 6y &= 9 \end{aligned}$$

Multiply first equation by -2 and add to the second one.
$$\begin{aligned} -4x - 6y &= -10 \\ 4x + 6y &= 9 \\ \hline 0 &= -1 \end{aligned}$$

Inconsistent and solution set is $\emptyset$.

75. System of equations is
$$\begin{aligned} x + y + z &= 4 \\ y + z &= 5 \\ z &= 6. \end{aligned}$$

Substitute $z = 6$ into $y + z = 5$ to get $y + 6 = 5$ and $y = -1$. From $x + y + z = 4$, we have $x - 1 + 6 = 4$ and $x = -1$. Solution set is $\{(-1, -1, 6)\}$.

77.
$$\begin{bmatrix} 2 & 3 \\ 4 & -1 \end{bmatrix} \begin{bmatrix} x \\ y \end{bmatrix} = \begin{bmatrix} 9 \\ 6 \end{bmatrix}$$

79.
$$\begin{bmatrix} 1 & 2 & -1 \\ 3 & -1 & 3 \\ 2 & 1 & -4 \end{bmatrix} \begin{bmatrix} x \\ y \\ z \end{bmatrix} = \begin{bmatrix} 3 \\ 1 \\ 0 \end{bmatrix}$$

81.
$$A = \begin{bmatrix} \$24{,}000 & \$40{,}000 \\ \$38{,}000 & \$70{,}000 \end{bmatrix}, Q = \begin{bmatrix} 4 \\ 7 \end{bmatrix}, \text{ and}$$

matrix product $AQ = \begin{bmatrix} \$376{,}000 \\ \$642{,}000 \end{bmatrix}$

represents the costs for labor and material for building 4 economy houses and 7 deluxe models.

83. False

85. True

87. True

For Thought

1. True, $AB = \begin{bmatrix} 10-9 & -6+6 \\ 15-15 & -9+10 \end{bmatrix} =$

$= \begin{bmatrix} 1 & 0 \\ 0 & 1 \end{bmatrix} = \begin{bmatrix} 10-9 & 15-15 \\ -6+6 & -9+10 \end{bmatrix} = BA.$

2. True, $AB = BA = I$ by Exercise 1.

3. True, by Exercise 2 the inverse of B is A.

4. False, AC is undefined, although CA is defined. **5.** True

6. False, a non-square matrix has no inverse.

7. False, the coefficient matrix is $\begin{bmatrix} 2 & 3 \\ 3 & 1 \end{bmatrix}$.

8. True, since $A^{-1} = B$ then $A^{-1}D =$

$= \begin{bmatrix} 5 & -3 \\ -3 & 2 \end{bmatrix} \begin{bmatrix} 11 \\ 19 \end{bmatrix} = \begin{bmatrix} -2 \\ 5 \end{bmatrix}.$

9. False, $(-2, 5)$ does not satisfy $3x + y = 19$.

10. False, the solution is $\begin{bmatrix} 2 & 3 \\ 3 & 1 \end{bmatrix}^{-1} \begin{bmatrix} 3 \\ -7 \end{bmatrix}.$

6.4 Exercises

1.

$AI = \begin{bmatrix} 1 & 3 \\ 4 & 6 \end{bmatrix} \begin{bmatrix} 1 & 0 \\ 0 & 1 \end{bmatrix}$

$= \begin{bmatrix} 1(1)+3(0) & 1(0)+3(1) \\ 4(1)+6(0) & 4(0)+6(1) \end{bmatrix}$

$= \begin{bmatrix} 1 & 3 \\ 4 & 6 \end{bmatrix}$

$= A$

and

$IA = \begin{bmatrix} 1 & 0 \\ 0 & 1 \end{bmatrix} \begin{bmatrix} 1 & 3 \\ 4 & 6 \end{bmatrix}$

$= \begin{bmatrix} 1(1)+0(4) & 1(3)+0(6) \\ 0(1)+1(4) & 0(3)+1(6) \end{bmatrix}$

$= \begin{bmatrix} 1 & 3 \\ 4 & 6 \end{bmatrix}$

$= A$

3. $AI = \begin{bmatrix} 3 & 2 & 1 \\ 5 & 6 & 2 \\ 7 & 8 & 3 \end{bmatrix} \begin{bmatrix} 1 & 0 & 0 \\ 0 & 1 & 0 \\ 0 & 0 & 1 \end{bmatrix} =$

$\begin{bmatrix} 3(1)+2(0)+1(0) & 3(0)+2(1)+1(0) & 3(0)+2(0)+1(1) \\ 5(1)+6(0)+2(0) & 5(0)+6(1)+2(0) & 5(0)+6(0)+2(1) \\ 7(1)+8(0)+3(0) & 7(0)+8(1)+3(0) & 7(0)+8(0)+3(1) \end{bmatrix}$

$= \begin{bmatrix} 3 & 2 & 1 \\ 5 & 6 & 2 \\ 7 & 8 & 3 \end{bmatrix} = A$

and

$IA = \begin{bmatrix} 1 & 0 & 0 \\ 0 & 1 & 0 \\ 0 & 0 & 1 \end{bmatrix} \begin{bmatrix} 3 & 2 & 1 \\ 5 & 6 & 2 \\ 7 & 8 & 3 \end{bmatrix} =$

$\begin{bmatrix} 1(3)+0(5)+0(7) & 1(2)+0(6)+0(8) & 1(1)+0(2)+0(3) \\ 0(3)+1(5)+0(7) & 0(2)+1(6)+0(8) & 0(1)+1(2)+0(3) \\ 0(3)+0(5)+1(7) & 0(2)+0(6)+1(8) & 0(1)+0(2)+1(3) \end{bmatrix}$

$= \begin{bmatrix} 3 & 2 & 1 \\ 5 & 6 & 2 \\ 7 & 8 & 3 \end{bmatrix} = A$

5.

$I \begin{bmatrix} -3 & 5 \\ 12 & 6 \end{bmatrix} = \begin{bmatrix} -3 & 5 \\ 12 & 6 \end{bmatrix}$

7.

$\begin{bmatrix} -8+9 & -6+6 \\ 12-12 & 9-8 \end{bmatrix} = \begin{bmatrix} 1 & 0 \\ 0 & 1 \end{bmatrix}$

9.

$\begin{bmatrix} 5-4 & -4+4 \\ 5-5 & -4+5 \end{bmatrix} = \begin{bmatrix} 1 & 0 \\ 0 & 1 \end{bmatrix}$

11.

$\begin{bmatrix} 3 & 5 & 1 \\ 4 & 5 & 7 \\ 4 & 9 & 2 \end{bmatrix} I = \begin{bmatrix} 3 & 5 & 1 \\ 4 & 5 & 7 \\ 4 & 9 & 2 \end{bmatrix}$

13.

$\begin{bmatrix} 0+0+1 & 1+0-1 & -3+0+3 \\ 0+0+0 & 1+0+0 & -3+3+0 \\ 0+0+0 & 0+0+0 & 0+1+0 \end{bmatrix} =$

$$\begin{bmatrix} 1 & 0 & 0 \\ 0 & 1 & 0 \\ 0 & 0 & 1 \end{bmatrix}$$

15.
$$\begin{bmatrix} 0.5+0.5+0 & -0.5+0.5+0 & 0.5-0.5+0 \\ 0+0.5-0.5 & 0+0.5+0.5 & 0-0.5+0.5 \\ 0.5+0-0.5 & -0.5+0+0.5 & 0.5+0+0.5 \end{bmatrix} =$$

$$\begin{bmatrix} 1 & 0 & 0 \\ 0 & 1 & 0 \\ 0 & 0 & 1 \end{bmatrix}$$

17.

Yes, since $\begin{bmatrix} 3 & 1 \\ 11 & 4 \end{bmatrix} \begin{bmatrix} 4 & -1 \\ -11 & 3 \end{bmatrix} =$

$\begin{bmatrix} 12-11 & -3+3 \\ 44-44 & -11+12 \end{bmatrix} = \begin{bmatrix} 1 & 0 \\ 0 & 1 \end{bmatrix}$ and

similarly $\begin{bmatrix} 4 & -1 \\ -11 & 3 \end{bmatrix} \begin{bmatrix} 3 & 1 \\ 11 & 4 \end{bmatrix} = I.$

19.

No, since $\begin{bmatrix} 1/2 & -1 \\ 3 & -12 \end{bmatrix} \begin{bmatrix} 4 & 2 \\ 1 & 1 \end{bmatrix} =$

$\begin{bmatrix} 2-1 & 1-1 \\ 12-12 & 6-12 \end{bmatrix} = \begin{bmatrix} 1 & 0 \\ 0 & -6 \end{bmatrix} \neq I.$

21. No, since only square matrices may have inverses.

23.

On $\begin{bmatrix} 1 & 4 & | & 1 & 0 \\ 0 & 2 & | & 0 & 1 \end{bmatrix}$, use $-2R_2 + R_1 \to R_1$ to get

$\begin{bmatrix} 1 & 0 & | & 1 & -2 \\ 0 & 2 & | & 0 & 1 \end{bmatrix}$, use $\frac{1}{2}R_2 \to R_2$ to get

$\begin{bmatrix} 1 & 0 & | & 1 & -2 \\ 0 & 1 & | & 0 & 1/2 \end{bmatrix}.$ Then $A^{-1} = \begin{bmatrix} 1 & -2 \\ 0 & 1/2 \end{bmatrix}.$

25.

On $\begin{bmatrix} 1 & 6 & | & 1 & 0 \\ 1 & 9 & | & 0 & 1 \end{bmatrix}$, use $-1R_1 + R_2 \to R_2$ to get

$\begin{bmatrix} 1 & 6 & | & 1 & 0 \\ 0 & 3 & | & -1 & 1 \end{bmatrix}$, use $-2R_2 + R_1 \to R_1$ to get

$\begin{bmatrix} 1 & 0 & | & 3 & -2 \\ 0 & 3 & | & -1 & 1 \end{bmatrix}$, use $\frac{1}{3}R_2 \to R_2$ to get

$\begin{bmatrix} 1 & 0 & | & 3 & -2 \\ 0 & 1 & | & -1/3 & 1/3 \end{bmatrix}.$

Thus, $A^{-1} = \begin{bmatrix} 3 & -2 \\ -1/3 & 1/3 \end{bmatrix}.$

27.

On $\begin{bmatrix} -2 & -3 & | & 1 & 0 \\ 3 & 4 & | & 0 & 1 \end{bmatrix}$, use $R_2 + R_1 \to R_1$ to get

$\begin{bmatrix} 1 & 1 & | & 1 & 1 \\ 3 & 4 & | & 0 & 1 \end{bmatrix}$, use $-3R_1 + R_2 \to R_2$ to get

$\begin{bmatrix} 1 & 1 & | & 1 & 1 \\ 0 & 1 & | & -3 & -2 \end{bmatrix}$, use $-1R_2 + R_1 \to R_1$ to get

$\begin{bmatrix} 1 & 0 & | & 4 & 3 \\ 0 & 1 & | & -3 & -2 \end{bmatrix}.$ So $A^{-1} = \begin{bmatrix} 4 & 3 \\ -3 & -2 \end{bmatrix}.$

29.

On $\begin{bmatrix} 1 & -5 & | & 1 & 0 \\ -1 & 3 & | & 0 & 1 \end{bmatrix}$, use $R_1 + R_2 \to R_2$ to get

$\begin{bmatrix} 1 & -5 & | & 1 & 0 \\ 0 & -2 & | & 1 & 1 \end{bmatrix}$, use $-\frac{1}{2}R_2 \to R_2$ to get

$\begin{bmatrix} 1 & -5 & | & 1 & 0 \\ 0 & 1 & | & -1/2 & -1/2 \end{bmatrix}$, use $5R_2 + R_1 \to R_1$ to get

$\begin{bmatrix} 1 & 0 & | & -3/2 & -5/2 \\ 0 & 1 & | & -1/2 & -1/2 \end{bmatrix}.$

Then $A^{-1} = \begin{bmatrix} -3/2 & -5/2 \\ -1/2 & -1/2 \end{bmatrix}.$

31.

On $\begin{bmatrix} -1 & 5 & | & 1 & 0 \\ 2 & -10 & | & 0 & 1 \end{bmatrix}$, use $R_1 + R_2 \to R_1$

and $\frac{1}{2}R_2 \to R_2$ to get

$\begin{bmatrix} 1 & -5 & | & 1 & 1 \\ 1 & -5 & | & 0 & 1/2 \end{bmatrix}$, use $-1R_2 + R_1 \to R_1$ to get

$\begin{bmatrix} 0 & 0 & | & 1 & 1/2 \\ 1 & -5 & | & 0 & 1/2 \end{bmatrix}.$ So A has no inverse.

33.

On $\begin{bmatrix} 1 & 1 & 0 & | & 1 & 0 & 0 \\ 0 & -1 & -1 & | & 0 & 1 & 0 \\ 1 & 0 & -1 & | & 0 & 0 & 1 \end{bmatrix}$, use

$R_2 + R_1 \to R_1$ and $-1R_3 + R_1 \to R_3$ to get

$\begin{bmatrix} 1 & 0 & -1 & | & 1 & 1 & 0 \\ 0 & -1 & -1 & | & 0 & 1 & 0 \\ 0 & 1 & 1 & | & 1 & 0 & -1 \end{bmatrix}$, use

$R_2 + R_3 \to R_2$ to get

$\begin{bmatrix} 1 & 0 & -1 & | & 1 & 1 & 0 \\ 0 & 0 & 0 & | & 1 & 1 & -1 \\ 0 & 1 & 1 & | & 1 & 0 & 1 \end{bmatrix}$. So A^{-1} does not exist.

35.

On $\begin{bmatrix} 1 & 1 & 1 & | & 1 & 0 & 0 \\ 1 & -1 & -1 & | & 0 & 1 & 0 \\ 1 & -1 & 1 & | & 0 & 0 & 1 \end{bmatrix}$, use

$R_1 + R_2 \to R_2$ and $R_2 + (-1R_3) \to R_3$ to get

$\begin{bmatrix} 1 & 1 & 1 & | & 1 & 0 & 0 \\ 2 & 0 & 0 & | & 1 & 1 & 0 \\ 0 & 0 & -2 & | & 0 & 1 & -1 \end{bmatrix}$, use

$-2R_1 + R_2 \to R_2$ and $-\frac{1}{2}R_3 \to R_3$ to get

$\begin{bmatrix} 1 & 1 & 1 & | & 1 & 0 & 0 \\ 0 & -2 & -2 & | & -1 & 1 & 0 \\ 0 & 0 & 1 & | & 0 & -1/2 & 1/2 \end{bmatrix}$, use

$-\frac{1}{2}R_2 \to R_2$ to get

$\begin{bmatrix} 1 & 1 & 1 & | & 1 & 0 & 0 \\ 0 & 1 & 1 & | & 1/2 & -1/2 & 0 \\ 0 & 0 & 1 & | & 0 & -1/2 & 1/2 \end{bmatrix}$, use

$-1R_2 + R_1 \to R_1$ to get

$\begin{bmatrix} 1 & 0 & 0 & | & 1/2 & 1/2 & 0 \\ 0 & 1 & 1 & | & 1/2 & -1/2 & 0 \\ 0 & 0 & 1 & | & 0 & -1/2 & 1/2 \end{bmatrix}$, use

$-1R_3 + R_2 \to R_2$ to get

$\begin{bmatrix} 1 & 0 & 0 & | & 1/2 & 1/2 & 0 \\ 0 & 1 & 0 & | & 1/2 & 0 & -1/2 \\ 0 & 0 & 1 & | & 0 & -1/2 & 1/2 \end{bmatrix}$.

So $A^{-1} = \begin{bmatrix} 1/2 & 1/2 & 0 \\ 1/2 & 0 & -1/2 \\ 0 & -1/2 & 1/2 \end{bmatrix}$.

37.

On $\begin{bmatrix} 0 & 2 & 0 & | & 1 & 0 & 0 \\ 3 & 3 & 2 & | & 0 & 1 & 0 \\ 2 & 5 & 1 & | & 0 & 0 & 1 \end{bmatrix}$, use

$-1R_3 + R_2 \to R_3$ to get

$\begin{bmatrix} 0 & 2 & 0 & | & 1 & 0 & 0 \\ 3 & 3 & 2 & | & 0 & 1 & 0 \\ 1 & -2 & 1 & | & 0 & 1 & -1 \end{bmatrix}$, use

$R_3 \to R_1$ and $R_1 \to R_3$ to get

$\begin{bmatrix} 1 & -2 & 1 & | & 0 & 1 & -1 \\ 3 & 3 & 2 & | & 0 & 1 & 0 \\ 0 & 2 & 0 & | & 1 & 0 & 0 \end{bmatrix}$, use

$R_3 + R_1 \to R_1$ and $-1R_3 + R_2 \to R_2$ to get

$\begin{bmatrix} 1 & 0 & 1 & | & 1 & 1 & -1 \\ 3 & 1 & 2 & | & -1 & 1 & 0 \\ 0 & 2 & 0 & | & 1 & 0 & 0 \end{bmatrix}$, use

$-3R_1 + R_2 \to R_2$ to get

$\begin{bmatrix} 1 & 0 & 1 & | & 1 & 1 & -1 \\ 0 & 1 & -1 & | & -4 & -2 & 3 \\ 0 & 2 & 0 & | & 1 & 0 & 0 \end{bmatrix}$, use

$-2R_2 + R_3 \to R_3$ to get

$\begin{bmatrix} 1 & 0 & 1 & | & 1 & 1 & -1 \\ 0 & 1 & -1 & | & -4 & -2 & 3 \\ 0 & 0 & 2 & | & 9 & 4 & -6 \end{bmatrix}$, use

$\frac{1}{2}R_3 \to R_3$ to get

$\begin{bmatrix} 1 & 0 & 1 & | & 1 & 1 & -1 \\ 0 & 1 & -1 & | & -4 & -2 & 3 \\ 0 & 0 & 1 & | & 9/2 & 2 & -3 \end{bmatrix}$, use

$R_2 + R_3 \to R_2$ to get

$\begin{bmatrix} 1 & 0 & 1 & | & 1 & 1 & -1 \\ 0 & 1 & 0 & | & 1/2 & 0 & 0 \\ 0 & 0 & 1 & | & 9/2 & 2 & -3 \end{bmatrix}$, use

$-1R_3 + R_1 \to R_1$ to get

$\begin{bmatrix} 1 & 0 & 0 & | & -7/2 & -1 & 2 \\ 0 & 1 & 0 & | & 1/2 & 0 & 0 \\ 0 & 0 & 1 & | & 9/2 & 2 & -3 \end{bmatrix}$.

So $A^{-1} = \begin{bmatrix} -7/2 & -1 & 2 \\ 1/2 & 0 & 0 \\ 9/2 & 2 & -3 \end{bmatrix}$.

39.

On $\begin{bmatrix} 1 & 0 & 1 & | & 1 & 0 & 0 \\ 0 & 2 & 2 & | & 0 & 1 & 0 \\ 2 & 1 & 0 & | & 0 & 0 & 1 \end{bmatrix}$, use

$2R_1 + (-1R_3) \rightarrow R_3$ and $\frac{1}{2}R_2 \rightarrow R_2$ to get

$\begin{bmatrix} 1 & 0 & 1 & | & 1 & 0 & 0 \\ 0 & 1 & 1 & | & 0 & 1/2 & 0 \\ 0 & -1 & 2 & | & 2 & 0 & -1 \end{bmatrix}$, use

$R_2 + R_3 \rightarrow R_3$ to get

$\begin{bmatrix} 1 & 0 & 1 & | & 1 & 0 & 0 \\ 0 & 1 & 1 & | & 0 & 1/2 & 0 \\ 0 & 0 & 3 & | & 2 & 1/2 & -1 \end{bmatrix}$, use

$\frac{1}{3}R_3 \rightarrow R_3$ to get

$\begin{bmatrix} 1 & 0 & 1 & | & 1 & 0 & 0 \\ 0 & 1 & 1 & | & 0 & 1/2 & 0 \\ 0 & 0 & 1 & | & 2/3 & 1/6 & -1/3 \end{bmatrix}$, use

$-1R_3 + R_2 \rightarrow R_2$ and $-1R_3 + R_1 \rightarrow R_1$ to get

$\begin{bmatrix} 1 & 0 & 0 & | & 1/3 & -1/6 & 1/3 \\ 0 & 1 & 0 & | & -2/3 & 1/3 & 1/3 \\ 0 & 0 & 1 & | & 2/3 & 1/6 & -1/3 \end{bmatrix}$.

Then $A^{-1} = \begin{bmatrix} 1/3 & -1/6 & 1/3 \\ -2/3 & 1/3 & 1/3 \\ 2/3 & 1/6 & -1/3 \end{bmatrix}$.

41.

On $\begin{bmatrix} 0 & 4 & 2 & | & 1 & 0 & 0 \\ 0 & 3 & 2 & | & 0 & 1 & 0 \\ 1 & -1 & 1 & | & 0 & 0 & 1 \end{bmatrix}$, use

$R_1 \rightarrow R_3$ and $R_3 \rightarrow R_1$ to get

$\begin{bmatrix} 1 & -1 & 1 & | & 0 & 0 & 1 \\ 0 & 3 & 2 & | & 0 & 1 & 0 \\ 0 & 4 & 2 & | & 1 & 0 & 0 \end{bmatrix}$, use

$-1R_2 + R_3 \rightarrow R_2$ to get

$\begin{bmatrix} 1 & -1 & 1 & | & 0 & 0 & 1 \\ 0 & 1 & 0 & | & 1 & -1 & 0 \\ 0 & 4 & 2 & | & 1 & 0 & 0 \end{bmatrix}$, use

$R_1 + R_2 \rightarrow R_1$ to get

$\begin{bmatrix} 1 & 0 & 1 & | & 1 & -1 & 1 \\ 0 & 1 & 0 & | & 1 & -1 & 0 \\ 0 & 4 & 2 & | & 1 & 0 & 0 \end{bmatrix}$, use

$-4R_2 + R_3 \rightarrow R_3$ to get

$\begin{bmatrix} 1 & 0 & 1 & | & 1 & -1 & 1 \\ 0 & 1 & 0 & | & 1 & -1 & 0 \\ 0 & 0 & 2 & | & -3 & 4 & 0 \end{bmatrix}$, use

$\frac{1}{2}R_3 \rightarrow R_3$ to get

$\begin{bmatrix} 1 & 0 & 1 & | & 1 & -1 & 1 \\ 0 & 1 & 0 & | & 1 & -1 & 0 \\ 0 & 0 & 1 & | & -3/2 & 2 & 0 \end{bmatrix}$, use

$-1R_3 + R_1 \rightarrow R_1$ to get

$\begin{bmatrix} 1 & 0 & 0 & | & 5/2 & -3 & 1 \\ 0 & 1 & 0 & | & 1 & -1 & 0 \\ 0 & 0 & 1 & | & -3/2 & 2 & 0 \end{bmatrix}$.

Then $A^{-1} = \begin{bmatrix} -5/2 & -3 & 1 \\ 1 & -1 & 0 \\ -3/2 & 2 & 0 \end{bmatrix}$.

43.

On $\begin{bmatrix} 1 & 2 & 3 & 4 & | & 1 & 0 & 0 & 0 \\ 0 & 1 & 2 & 3 & | & 0 & 1 & 0 & 0 \\ 0 & 0 & 1 & 2 & | & 0 & 0 & 1 & 0 \\ 0 & 0 & 0 & 1 & | & 0 & 0 & 0 & 1 \end{bmatrix}$, use

$-2R_2 + R_1 \rightarrow R_1$ to get

$\begin{bmatrix} 1 & 0 & -1 & -2 & | & 1 & -2 & 0 & 0 \\ 0 & 1 & 2 & 3 & | & 0 & 1 & 0 & 0 \\ 0 & 0 & 1 & 2 & | & 0 & 0 & 1 & 0 \\ 0 & 0 & 0 & 1 & | & 0 & 0 & 0 & 1 \end{bmatrix}$, use

$-2R_4 + R_3 \rightarrow R_3$ to get

$\begin{bmatrix} 1 & 0 & -1 & -2 & | & 1 & -2 & 0 & 0 \\ 0 & 1 & 2 & 3 & | & 0 & 1 & 0 & 0 \\ 0 & 0 & 1 & 0 & | & 0 & 0 & 1 & -2 \\ 0 & 0 & 0 & 1 & | & 0 & 0 & 0 & 1 \end{bmatrix}$, use

$-2R_3 + R_2 \rightarrow R_2$ and $-2R_4 + R_3 \rightarrow R_3$ to get

$\begin{bmatrix} 1 & 0 & -1 & -2 & | & 1 & -2 & 0 & 0 \\ 0 & 1 & 0 & 3 & | & 0 & 1 & -2 & 4 \\ 0 & 0 & 1 & -2 & | & 0 & 0 & 1 & -4 \\ 0 & 0 & 0 & 1 & | & 0 & 0 & 0 & 1 \end{bmatrix}$, use

$R_1 + R_3 \rightarrow R_1$ to get

$\begin{bmatrix} 1 & 0 & 0 & -4 & | & 1 & -2 & 1 & -4 \\ 0 & 1 & 0 & 3 & | & 0 & 1 & -2 & 4 \\ 0 & 0 & 1 & -2 & | & 0 & 0 & 1 & -4 \\ 0 & 0 & 0 & 1 & | & 0 & 0 & 0 & 1 \end{bmatrix}$, use

$2R_4 + R_3 \rightarrow R_3$, $-3R_4 + R_2 \rightarrow R_2$ and $4R_4 + R_1 \rightarrow R_1$ to get

$$\left[\begin{array}{cccc|cccc} 1 & 0 & 0 & 0 & 1 & -2 & 1 & 0 \\ 0 & 1 & 0 & 0 & 0 & 1 & -2 & 1 \\ 0 & 0 & 1 & 0 & 0 & 0 & 1 & -2 \\ 0 & 0 & 0 & 1 & 0 & 0 & 0 & 1 \end{array}\right].$$

So $A^{-1} = \begin{bmatrix} 1 & -2 & 1 & 0 \\ 0 & 1 & -2 & 1 \\ 0 & 0 & 1 & -2 \\ 0 & 0 & 0 & 1 \end{bmatrix}$.

45.

Since the coefficient matrix is $A = \begin{bmatrix} 1 & 6 \\ 1 & 9 \end{bmatrix}$,

$$A^{-1}\begin{bmatrix} -3 \\ -6 \end{bmatrix} = \begin{bmatrix} 3 & -2 \\ -1/3 & 1/3 \end{bmatrix}\begin{bmatrix} -3 \\ -6 \end{bmatrix} =$$

$$= \begin{bmatrix} 3 \\ -1 \end{bmatrix}.$$ The solution set is $\{(3, -1)\}$.

47.

Since the coefficient matrix is $A = \begin{bmatrix} 1 & 6 \\ 1 & 9 \end{bmatrix}$,

$$A^{-1}\begin{bmatrix} 4 \\ 5 \end{bmatrix} = \begin{bmatrix} 3 & -2 \\ -1/3 & 1/3 \end{bmatrix}\begin{bmatrix} 4 \\ 5 \end{bmatrix} =$$

$$= \begin{bmatrix} 2 \\ 1/3 \end{bmatrix}.$$ The solution set is $\{(2, 1/3)\}$.

49.

Since the coefficient matrix is
$A = \begin{bmatrix} -2 & -3 \\ 3 & 4 \end{bmatrix}$, we get

$$A^{-1}\begin{bmatrix} 1 \\ -1 \end{bmatrix} = \begin{bmatrix} 4 & 3 \\ -3 & -2 \end{bmatrix}\begin{bmatrix} 1 \\ -1 \end{bmatrix} =$$

$$= \begin{bmatrix} 1 \\ -1 \end{bmatrix}.$$ The solution set is $\{(1, -1)\}$.

51.

Since the coefficient matrix is
$A = \begin{bmatrix} 1 & -5 \\ -1 & 3 \end{bmatrix}$, we obtain

$$A^{-1}\begin{bmatrix} -5 \\ 1 \end{bmatrix} = \begin{bmatrix} -3/2 & -5/2 \\ -1/2 & -1/2 \end{bmatrix}\begin{bmatrix} -5 \\ 1 \end{bmatrix} =$$

$$= \begin{bmatrix} 5 \\ 2 \end{bmatrix}.$$ The solution set is $\{(5, 2)\}$.

53.

Since coefficient matrix is

$A = \begin{bmatrix} 1 & 1 & 1 \\ 1 & -1 & -1 \\ 1 & -1 & 1 \end{bmatrix}$, we find $A^{-1}\begin{bmatrix} 3 \\ -1 \\ 5 \end{bmatrix} =$

$$= \begin{bmatrix} 1/2 & 1/2 & 0 \\ 1/2 & 0 & -1/2 \\ 0 & -1/2 & 1/2 \end{bmatrix}\begin{bmatrix} 3 \\ -1 \\ 5 \end{bmatrix} = \begin{bmatrix} 1 \\ -1 \\ 3 \end{bmatrix}.$$

The solution set is $\{(1, -1, 3)\}$.

55.

Since the coefficient matrix is

$A = \begin{bmatrix} 0 & 2 & 0 \\ 3 & 3 & 2 \\ 2 & 5 & 1 \end{bmatrix}$, we get $A^{-1}\begin{bmatrix} 6 \\ 16 \\ 19 \end{bmatrix} =$

$$= \begin{bmatrix} -7/2 & -1 & 2 \\ 1/2 & 0 & 0 \\ 9/2 & 2 & -3 \end{bmatrix}\begin{bmatrix} 6 \\ 16 \\ 19 \end{bmatrix} = \begin{bmatrix} 1 \\ 3 \\ 2 \end{bmatrix}.$$

The solution set is $\{(1, 3, 2)\}$.

57. The coefficient matrix of

$$0.3x + 0.1y = 3$$
$$2x + 4y = 7$$

is $A = \begin{bmatrix} 0.3 & 0.1 \\ 2 & 4 \end{bmatrix}$. Note $A^{-1}\begin{bmatrix} 3 \\ 7 \end{bmatrix} =$

$$\begin{bmatrix} 4 & -0.1 \\ -2 & 0.3 \end{bmatrix}\begin{bmatrix} 3 \\ 7 \end{bmatrix} = \begin{bmatrix} 11.3 \\ -3.9 \end{bmatrix}.$$

The solution set is $\{(11.3, -3.9)\}$.

59. Use the Gauss-Jordan method. On the

augmented matrix $\begin{bmatrix} 1 & -1 & 1 & 5 \\ 2 & -1 & 3 & 1 \\ 0 & 1 & 1 & -9 \end{bmatrix}$,

use $-1R_1 + R_2 \to R_1$ to get

$$\begin{bmatrix} 1 & 0 & 2 & -4 \\ 2 & -1 & 3 & 1 \\ 0 & 1 & 1 & -9 \end{bmatrix},$$ use

$-2R_1 + R_2 \to R_2$ to get

$$\begin{bmatrix} 1 & 0 & 2 & -4 \\ 0 & -1 & -1 & 9 \\ 0 & 1 & 1 & -9 \end{bmatrix},$$ use

$R_2 + R_3 \to R_3$ and $-1R_2 \to R_2$ to get

$$\begin{bmatrix} 1 & 0 & 2 & | & -4 \\ 0 & 1 & 1 & | & -9 \\ 0 & 0 & 0 & | & 0 \end{bmatrix}. \text{ Since } y+z=-9$$

and $x+2z=-4$, the solution set is

$\{(-2z-4,-z-9,z) \mid z \text{ is any real number }\}$.

61.

Note coefficient matrix is $A = \begin{bmatrix} 1 & 1 & 1 \\ 2 & 4 & 1 \\ 1 & 3 & 6 \end{bmatrix}$

and $A^{-1} \begin{bmatrix} 1 \\ 2 \\ 3 \end{bmatrix} =$

$$\begin{bmatrix} 7/4 & -1/4 & -1/4 \\ -11/12 & 5/12 & 1/12 \\ 1/6 & -1/6 & 1/6 \end{bmatrix} \begin{bmatrix} 1 \\ 2 \\ 3 \end{bmatrix} = \begin{bmatrix} 1/2 \\ 1/6 \\ 1/3 \end{bmatrix}.$$

The solution set is $\{(1/2, 1/6, 1/3)\}$.

63.

Note $A^{-1} = \begin{bmatrix} -55/6 & 5/2 & 5/3 \\ 35/12 & -5/4 & 5/6 \\ 40/3 & 0 & -10/3 \end{bmatrix}$

and $A^{-1} \begin{bmatrix} 27 \\ 9 \\ 16 \end{bmatrix} = \begin{bmatrix} -165 \\ 97.5 \\ 240 \end{bmatrix}.$

The solution set is $\{(-165, 97.5, 240)\}$.

65.

We get $A^{-1} = \begin{bmatrix} 68/133 & -36/133 & 8/19 \\ 10/19 & -12/19 & 6/19 \\ 127/133 & -122/133 & 6/19 \end{bmatrix}$

and $A^{-1} \begin{bmatrix} 16 \\ 24 \\ -8 \end{bmatrix} \approx \begin{bmatrix} -1.6842 \\ -9.2632 \\ -9.2632 \end{bmatrix}.$

Solution set is approximately

$$\{(-1.6842, -9.2632, -9.2632)\}.$$

67.

Since $AA^{-1} = \begin{bmatrix} a & 7 \\ 3 & b \end{bmatrix} \begin{bmatrix} -b & 7 \\ 3 & -a \end{bmatrix} =$

$= \begin{bmatrix} 21-ab & 0 \\ 0 & 21-ab \end{bmatrix}, 21-ab=1$ and

$ab = 20$. List of permissible pairs (a, b):

a	1	2	4	5	10	20
b	20	10	5	4	2	1

Matrices are $\begin{bmatrix} 1 & 7 \\ 3 & 20 \end{bmatrix}, \begin{bmatrix} 2 & 7 \\ 3 & 10 \end{bmatrix},$

$\begin{bmatrix} 4 & 7 \\ 3 & 5 \end{bmatrix}, \begin{bmatrix} 5 & 7 \\ 3 & 4 \end{bmatrix},$

$\begin{bmatrix} 10 & 7 \\ 3 & 2 \end{bmatrix},$ and $\begin{bmatrix} 20 & 7 \\ 3 & 1 \end{bmatrix}.$

69. Let x and y be the costs of a dozen eggs and a magazine before taxes. Then

$$0.08x + 0.05y = 0.15$$
$$x + y = 2.70 - 0.15.$$

If $A = \begin{bmatrix} .08 & .05 \\ 1 & 1 \end{bmatrix}$, then $A^{-1} \begin{bmatrix} .15 \\ 2.55 \end{bmatrix} =$

$= \begin{bmatrix} 100/3 & -5/3 \\ -100/3 & 8/3 \end{bmatrix} \begin{bmatrix} .15 \\ 2.55 \end{bmatrix} = \begin{bmatrix} .75 \\ 1.80 \end{bmatrix}.$

Eggs costs $0.75 a dozen, the magazine costs $1.80.

71. Let x and y be the costs of one load of plywood and a load insulation, respectively. So

$$4x + 6y = 2500$$
$$3x + 5y = 1950.$$

If $A = \begin{bmatrix} 4 & 6 \\ 3 & 5 \end{bmatrix}$ then $A^{-1} \begin{bmatrix} 2500 \\ 1950 \end{bmatrix} =$

$= \begin{bmatrix} 5/2 & -3 \\ -3/2 & 2 \end{bmatrix} \begin{bmatrix} 2500 \\ 1950 \end{bmatrix} = \begin{bmatrix} 400 \\ 150 \end{bmatrix}.$

One load of plywood costs $400 and a load of insulation costs $150 .

73.

One computes $A^{-1} = \begin{bmatrix} 2 & -1 \\ -5 & 3 \end{bmatrix}$. To decode the message, we find

$\begin{bmatrix} 2 & -1 \\ -5 & 3 \end{bmatrix} \begin{bmatrix} 36 \\ 65 \end{bmatrix} = \begin{bmatrix} 7 \\ 15 \end{bmatrix} = \begin{bmatrix} g \\ o \end{bmatrix},$

$\begin{bmatrix} 2 & -1 \\ -5 & 3 \end{bmatrix} \begin{bmatrix} 49 \\ 83 \end{bmatrix} = \begin{bmatrix} 15 \\ 4 \end{bmatrix} = \begin{bmatrix} o \\ d \end{bmatrix},$

$\begin{bmatrix} 2 & -1 \\ -5 & 3 \end{bmatrix} \begin{bmatrix} 12 \\ 24 \end{bmatrix} = \begin{bmatrix} 0 \\ 12 \end{bmatrix} = \begin{bmatrix} space \\ l \end{bmatrix},$

$\begin{bmatrix} 2 & -1 \\ -5 & 3 \end{bmatrix} \begin{bmatrix} 66 \\ 111 \end{bmatrix} = \begin{bmatrix} 21 \\ 3 \end{bmatrix} = \begin{bmatrix} u \\ c \end{bmatrix},$

$$\begin{bmatrix} 2 & -1 \\ -5 & 3 \end{bmatrix}\begin{bmatrix} 33 \\ 55 \end{bmatrix} = \begin{bmatrix} 11 \\ 0 \end{bmatrix} = \begin{bmatrix} k \\ space \end{bmatrix}.$$

The message is 'good luck.'

75. Let x and y be the amounts invested in the Asset Manager Fund and Magellan Fund, respectively. Since $0.86(60,000) = 51,600$, we get

$$\begin{aligned} x + y &= 60,000 \\ 0.76x + 0.90y &= 51,600. \end{aligned}$$

If $A = \begin{bmatrix} 1 & 1 \\ 0.76 & 0.90 \end{bmatrix}$, then

$$A^{-1} = \begin{bmatrix} 45/7 & -50/7 \\ -38/7 & 50/7 \end{bmatrix} \text{ and}$$

$$A^{-1}\begin{bmatrix} 60,000 \\ 51,600 \end{bmatrix} = \begin{bmatrix} 17,142.86 \\ 42,857.14 \end{bmatrix}. \text{ In the}$$

Asset Manager Fund, \$17,142.86 was invested and in the Magellan Fund \$42,857.14 was invested.

77. Let $x, y,$ and z be the prices of an animal totem, a trade-bead necklace, and a tribal mask, respectively. Then we obtain the system

$$\begin{aligned} 24x + 33y + 12z &= 202.23 \\ 19x + 40y + 22z &= 209.38 \\ 30x + 9y + 19z &= 167.66. \end{aligned}$$

The inverse of the coefficient matrix A is

$$A^{-1} = \frac{1}{11,007}\begin{bmatrix} 562 & -519 & 246 \\ 299 & 96 & -300 \\ 1029 & 774 & 333 \end{bmatrix}.$$

Since $A^{-1}\begin{bmatrix} 202.23 \\ 209.38 \\ 167.66 \end{bmatrix} \approx \begin{bmatrix} \$4.20 \\ \$2.75 \\ \$0.89 \end{bmatrix}$, an

animal totem costs \$4.20, a necklace \$2.75, and a tribal mask \$0.89.

For Thought

1. False, $|A| = 12 - (-5) = 17$.

2. True, $|A| \neq 0$.

3. True, $|B| = 4 \cdot 5 - (-2)(-10) = 0$.

4. False, $|B| = 0$.

5. True, since the determinant of the coefficient matrix A is nonzero.

6. True, in general $|LM| = |L||M|$ for any square matrices L and M of the same size.

7. False, the system is not linear.   **8.**   True

9. False, $\begin{vmatrix} 2 & 0.1 \\ 100 & 5 \end{vmatrix} = 2 \cdot 5 - 100(0.1) = 0$.

10. False, because a 2×2 matrix is not equal to the number 27.

6.5 Exercises

1.
$$\begin{vmatrix} 1 & 3 \\ 0 & 2 \end{vmatrix} = 1 \cdot 2 - 0 \cdot 3 = 2$$

3.
$$\begin{vmatrix} 3 & 4 \\ 2 & 9 \end{vmatrix} = 3(9) - 2(4) = 19$$

5.
$$\begin{vmatrix} -0.3 & -0.5 \\ -0.7 & 0.2 \end{vmatrix} = (-0.3)(0.2) - (-0.7)(-0.5)$$
$$= -0.41$$

7.
$$\begin{vmatrix} 1/8 & -3/8 \\ 2 & -1/4 \end{vmatrix} = (1/8)(-1/4) - (2)(-3/8)$$
$$= -1/32 + 3/4 = 23/32$$

9.
$$\begin{vmatrix} 0.02 & 0.4 \\ 1 & 20 \end{vmatrix} = (.02)(20) - (.4)(1) = 0$$

11.
$$\begin{vmatrix} 3 & -5 \\ -9 & 15 \end{vmatrix} = (3)(15) - (-9)(-5) = 0$$

13.
Note $D = \begin{vmatrix} 2 & 1 \\ 1 & 2 \end{vmatrix} = 3$, $D_x = \begin{vmatrix} 5 & 1 \\ 7 & 2 \end{vmatrix} = 3$,

and $D_y = \begin{vmatrix} 2 & 5 \\ 1 & 7 \end{vmatrix} = 9$.

Then $x = \dfrac{D_x}{D} = \dfrac{3}{3} = 1$ and $y = \dfrac{D_y}{D} = \dfrac{9}{3} = 3$.
Solution set is $\{(1,3)\}$.

15.

Note $D = \begin{vmatrix} 1 & -2 \\ 1 & 2 \end{vmatrix} = 4$, $D_x = \begin{vmatrix} 7 & -2 \\ -5 & 2 \end{vmatrix} = 4$, and $D_y = \begin{vmatrix} 1 & 7 \\ 1 & -5 \end{vmatrix} = -12$.

So $x = \dfrac{D_x}{D} = \dfrac{4}{4} = 1$ and $y = \dfrac{D_y}{D} = \dfrac{-12}{4} = -3$. The solution set is $\{(1, -3)\}$.

17.

Note $D = \begin{vmatrix} 2 & -1 \\ 1 & 3 \end{vmatrix} = 7$, $D_x = \begin{vmatrix} -11 & -1 \\ 12 & 3 \end{vmatrix} = -21$, and $D_y = \begin{vmatrix} 2 & -11 \\ 1 & 12 \end{vmatrix} = 35$. Then

$x = \dfrac{D_x}{D} = -\dfrac{21}{7} = -3$ and $y = \dfrac{D_y}{D} = \dfrac{35}{7} = 5$.
Solution set is $\{(-3, 5)\}$.

19. Rewrite system as

$$x - y = 6$$
$$x + y = 5.$$

Note $D = \begin{vmatrix} 1 & -1 \\ 1 & 1 \end{vmatrix} = 2$, $D_x = \begin{vmatrix} 6 & -1 \\ 5 & 1 \end{vmatrix} = 11$, and $D_y = \begin{vmatrix} 1 & 6 \\ 1 & 5 \end{vmatrix} = -1$.

So $x = \dfrac{D_x}{D} = \dfrac{11}{2}$ and $y = \dfrac{D_y}{D} = -\dfrac{1}{2}$.
Solution set is $\{(11/2, -1/2)\}$.

21.

Note $D = \begin{vmatrix} 1/2 & -1/3 \\ 1/4 & 1/2 \end{vmatrix} = 1/3$,

$D_x = \begin{vmatrix} 4 & -1/3 \\ 6 & 1/2 \end{vmatrix} = 4$, and

$D_y = \begin{vmatrix} 1/2 & 4 \\ 1/4 & 6 \end{vmatrix} = 2$. So $x = \dfrac{D_x}{D}$

$= \dfrac{4}{1/3} = 12$ and $y = \dfrac{D_y}{D} = \dfrac{2}{1/3} = 6$.
Solution set is $\{(12, 6)\}$.

23.

Note $D = \begin{vmatrix} 0.2 & 0.12 \\ 1 & 1 \end{vmatrix} = 0.08$,

$D_x = \begin{vmatrix} 148 & 0.12 \\ 900 & 1 \end{vmatrix} = 40$, and

$D_y = \begin{vmatrix} 0.2 & 148 \\ 1 & 900 \end{vmatrix} = 32$. Then

$x = \dfrac{D_x}{D} = \dfrac{40}{0.08} = 500$ and

$y = \dfrac{D_y}{D} = \dfrac{32}{0.08} = 400$.
Solution set is $\{(500, 400)\}$.

25. Cramer's rule does not apply since

$D = \begin{vmatrix} 3 & 1 \\ -6 & -2 \end{vmatrix} = 0$. Dividing the second

equation by -2, one gets the first equation.
Solution set is $\{(x, y) \mid 3x + y = 6\}$.

27. Cramer's Rule does not apply since the determinant D is zero.

Adding the two equations, one gets $0 = 19$. Inconsistent and the solution set is $\emptyset$.

29. We use Cramer's Rule on the system

$$x - y = 3$$
$$3x - y = -9.$$

Note, $D = \begin{vmatrix} 1 & -1 \\ 3 & -1 \end{vmatrix} = 2$,

$D_x = \begin{vmatrix} 3 & -1 \\ -9 & -1 \end{vmatrix} = -12$, and

$D_y = \begin{vmatrix} 1 & 3 \\ 3 & -9 \end{vmatrix} = -18$.

So $x = \dfrac{D_x}{D} = \dfrac{-12}{2} = -6$ and

$y = \dfrac{D_y}{D} = \dfrac{-18}{2} = -9$.
Solution set is $\{(-6, -9)\}$.

31. Note, $D = \begin{vmatrix} \sqrt{2} & \sqrt{3} \\ 3\sqrt{2} & -2\sqrt{3} \end{vmatrix} = -5\sqrt{6}$,

$D_x = \begin{vmatrix} 4 & \sqrt{3} \\ -3 & -2\sqrt{3} \end{vmatrix} = -5\sqrt{3}$, and

$D_y = \begin{vmatrix} \sqrt{2} & 4 \\ 3\sqrt{2} & -3 \end{vmatrix} = -15\sqrt{2}$.

So $x = \dfrac{D_x}{D} = \dfrac{-5\sqrt{3}}{-5\sqrt{6}} = \dfrac{1}{\sqrt{2}} = \dfrac{\sqrt{2}}{2}$

and $y = \dfrac{D_y}{D} = \dfrac{-15\sqrt{2}}{-5\sqrt{6}} = \dfrac{3}{\sqrt{3}} = \sqrt{3}$.

The solution set is $\{(\sqrt{2}/2, \sqrt{3})\}$.

33. Multiply second equation by -1 and add to the first one.

$$
\begin{aligned}
x^2 + y^2 &= 25 \\
\underline{-x^2 + y} &= \underline{-5} \\
y^2 + y &= 20 \\
y^2 + y - 20 &= 0 \\
(y + 5)(y - 4) &= 0
\end{aligned}
$$

If $y = -5$, then $x^2 = 0$ and $x = 0$.
If $y = 4$, then $x^2 = 9$ and $x = \pm 3$.
The solution set is $\{(0, -5), (\pm 3, 4)\}$.

35. Solving for x in the second equation, we find $x = 4 + 2y$. Substituting into the first equation, we obtain

$$
\begin{aligned}
4 + 2y - 2y &= y^2 \\
4 &= y^2 \\
\pm 2 &= y.
\end{aligned}
$$

Using $y = 2$ in $x = 4 + 2y$, we get $x = 8$.
Similarly, if $y = -2$ then $x = 0$.
Solution set is $\{(8, 2), (0, -2)\}$.

37.
Invertible, since $\begin{vmatrix} 4 & 0.5 \\ 2 & 3 \end{vmatrix} = 12 - 1 = 11 \neq 0$

39.
Not invertible, since $\begin{vmatrix} 3 & -4 \\ 9 & -12 \end{vmatrix} = -36 + 36 = 0$

41. Note, $D = \begin{vmatrix} 3.47 & 23.09 \\ 12.48 & 3.98 \end{vmatrix} = -274.3526$,

$D_x = \begin{vmatrix} 5978.95 & 23.09 \\ 2765.34 & 3.98 \end{vmatrix} = -40,055.4796$,

and

$D_y = \begin{vmatrix} 3.47 & 5978.95 \\ 12.48 & 2765.34 \end{vmatrix} = -65,021.5662$.

Then $x = \dfrac{D_x}{D} = 146$ and $y = \dfrac{D_y}{D} = 237$.
The solution set is $\{(146, 237)\}$.

43. Let x and y be the number of boys and girls, respectively. Then

$$
\begin{aligned}
0.44x + 0.35y &= 231 \\
x + y &= 615.
\end{aligned}
$$

Note, $D = \begin{vmatrix} 0.44 & 0.35 \\ 1 & 1 \end{vmatrix} = 0.09$,

$D_x = \begin{vmatrix} 231 & 0.35 \\ 615 & 1 \end{vmatrix} = 15.75$, and

$D_y = \begin{vmatrix} 0.44 & 231 \\ 1 & 615 \end{vmatrix} = 39.6$.

There were $x = \dfrac{D_x}{D} = \dfrac{15.75}{0.09} = 175$ boys

and $y = \dfrac{D_y}{D} = \dfrac{39.6}{0.09} = 440$ girls.

45. Let x and y be the measurements of the two acute angles. Then we obtain

$$
\begin{aligned}
x + y &= 90 \\
x - 2y &= 1.
\end{aligned}
$$

Note, $D = \begin{vmatrix} 1 & 1 \\ 1 & -2 \end{vmatrix} = -3$,

$D_x = \begin{vmatrix} 90 & 1 \\ 1 & -2 \end{vmatrix} = -181$, and

$D_y = \begin{vmatrix} 1 & 90 \\ 1 & 1 \end{vmatrix} = -89$.

The acute angles are $x = \dfrac{D_x}{D} = \dfrac{181}{3}$ degrees

and $y = \dfrac{D_y}{D} = \dfrac{89}{3}$ degrees.

47. Let x and y be the salaries of the president and vice-president, respectively. Then we have

$$
\begin{aligned}
x + y &= 400,000 \\
x - y &= 100,000.
\end{aligned}
$$

Note, $D = \begin{vmatrix} 1 & 1 \\ 1 & -1 \end{vmatrix} = -2$,

$D_x = \begin{vmatrix} 400,000 & 1 \\ 100,000 & -1 \end{vmatrix} = -500,000$, and

$D_y = \begin{vmatrix} 1 & 400,000 \\ 1 & 100,000 \end{vmatrix} = -300,000$.

Thus, $x = \dfrac{D_x}{D} = \dfrac{-500,000}{-2} = 250,000$ and

$y = \dfrac{D_y}{D} = \dfrac{-300,000}{-2} = 150,000.$

The president's salary is $250,000 and the vice-president's salary is $150,000.

49. Yes, $|MN| = |M||N|$ since $|M| = 2$, $|N| = 3$,

and $|MN| = \begin{vmatrix} 8 & 31 \\ 14 & 55 \end{vmatrix} = 6.$

51.

$\left| \begin{bmatrix} a & b \\ c & d \end{bmatrix} \begin{bmatrix} e & f \\ g & h \end{bmatrix} \right| =$

$\begin{vmatrix} ae+bg & af+bh \\ ce+dg & cf+dh \end{vmatrix} =$

$(ae+bg)(cf+dh) - (ce+dg)(af+bh) =$
$aecf + bgcf + aedh + bgdh - ceaf - dgaf -$
$cebh - dgbh = bgcf + aedh - dgaf - cebh =$
$ad(eh-gf) - bc(eh-gf) = (ad-bc)(eh-gf) =$

$= \begin{vmatrix} a & b \\ c & d \end{vmatrix} \begin{vmatrix} e & f \\ g & h \end{vmatrix}$

53.

No, since $|-2M| = \begin{vmatrix} -6 & -4 \\ -10 & -8 \end{vmatrix} = 8$

and $-2|M| = -4.$

For Thought

1. False, the sign array of A is used in evaluating $|A|$.

2. False, the last term should be $1 \cdot \begin{vmatrix} 3 & 4 \\ 0 & 0 \end{vmatrix}$.

3. True

4. True, $|A|$ was expanded about the third row.

5. False, $|A|$ can be expanded only about a row or column.   **6.** False, a minor is a 2×2 matrix only if it comes from a 3×3 matrix.

7. False, $x = \dfrac{D_x}{D}$.   **8.**   True

9. False, it can happen that $D = 0$ and there are infinitely many solutions.

10. False, Cramer's Rule applies only to a system of linear equations.

6.6 Exercises

1.

$\begin{vmatrix} 5 & -6 \\ 9 & -8 \end{vmatrix} = -40 + 54 = 14$

3.

$\begin{vmatrix} 4 & 5 \\ 7 & 9 \end{vmatrix} = 36 - 35 = 1$

5.

$\begin{vmatrix} 2 & 1 \\ 7 & -8 \end{vmatrix} = -16 - 7 = -23$

7.

$\begin{vmatrix} 2 & 1 \\ 4 & -6 \end{vmatrix} = -12 - 4 = -16$

9.

$1 \begin{vmatrix} 1 & -2 \\ -1 & 5 \end{vmatrix} - (-3) \begin{vmatrix} -4 & 0 \\ -1 & 5 \end{vmatrix} + 3 \begin{vmatrix} -4 & 0 \\ 1 & -2 \end{vmatrix}$

$= 1(3) - (-3)(-20) + 3(8) = -33$

11.

$3 \begin{vmatrix} 4 & -1 \\ 1 & -2 \end{vmatrix} - 0 \begin{vmatrix} -1 & 2 \\ 1 & -2 \end{vmatrix} + 5 \begin{vmatrix} -1 & 2 \\ 4 & -1 \end{vmatrix}$

$= 3(-7) - 0 + 5(-7) = -56$

13.

$-2 \begin{vmatrix} 0 & -1 \\ 2 & -7 \end{vmatrix} - (-3) \begin{vmatrix} 5 & 1 \\ 2 & -7 \end{vmatrix} + 0 \begin{vmatrix} 5 & 1 \\ 0 & -1 \end{vmatrix}$

$= -2(2) - (-3)(-37) + 0 = -115$

15.

$0.1 \begin{vmatrix} 20 & 6 \\ 90 & 8 \end{vmatrix} - 0.4 \begin{vmatrix} 30 & 1 \\ 90 & 8 \end{vmatrix} + 0.7 \begin{vmatrix} 30 & 1 \\ 20 & 6 \end{vmatrix}$

$= 0.1(-380) - 0.4(150) + 0.7(160) = 14$

17. Expanding about the second row,

$D = -(-2) \begin{vmatrix} 3 & 5 \\ 3 & -4 \end{vmatrix} = -(-2)(-27) = -54.$

19. Expanding about the first row,

$D = 1 \begin{vmatrix} 2 & 2 \\ 4 & 4 \end{vmatrix} - 1 \begin{vmatrix} 2 & 2 \\ 4 & 4 \end{vmatrix} + 1 \begin{vmatrix} 2 & 2 \\ 4 & 4 \end{vmatrix}$

$= 1(0) - 1(0) + 1(0) = 0.$

21. Expanding about the first row,

$$D = -(-1) \begin{vmatrix} 3 & 6 \\ -2 & -5 \end{vmatrix} = -(-1)(-3) = -3.$$

23. Expanding about the second column,

$$D = -9 \begin{vmatrix} 2 & 1 \\ 4 & 6 \end{vmatrix} = -9(8) = -72.$$

25. Expanding about the first row,

$$3 \begin{vmatrix} -3 & 2 & 0 \\ 3 & 1 & 2 \\ -4 & 1 & 3 \end{vmatrix} + 0 + 1 \begin{vmatrix} 2 & -3 & 0 \\ -2 & 3 & 2 \\ 2 & -4 & 3 \end{vmatrix} -$$

$$5 \begin{vmatrix} 2 & -3 & 2 \\ -2 & 3 & 1 \\ 2 & -4 & 1 \end{vmatrix} =$$

$$= 3(-37) + 1(4) - 5(6) = -137.$$

27. Expand the determinant about the second row.

$$-(1) \begin{vmatrix} -3 & 4 & 6 \\ 3 & 1 & -3 \\ 0 & 2 & 1 \end{vmatrix} + (-5) \begin{vmatrix} 2 & 4 & 6 \\ 1 & 1 & -3 \\ -2 & 2 & 1 \end{vmatrix} =$$

$$= -(1)(3) + (-5)(58) = -293.$$

29.

One finds $D = \begin{vmatrix} 1 & 1 & 1 \\ 1 & -1 & 1 \\ 2 & 1 & 1 \end{vmatrix} = 2,$

$$D_x = \begin{vmatrix} 6 & 1 & 1 \\ 2 & -1 & 1 \\ 7 & 1 & 1 \end{vmatrix} = 2, D_y = \begin{vmatrix} 1 & 6 & 1 \\ 1 & 2 & 1 \\ 2 & 7 & 1 \end{vmatrix} =$$

$$4, \text{ and } D_z = \begin{vmatrix} 1 & 1 & 6 \\ 1 & -1 & 2 \\ 2 & 1 & 7 \end{vmatrix} = 6.$$

Since $x = \dfrac{D_x}{D} = 2/2 = 1$, $y = \dfrac{D_y}{D} = 4/2 = 2$,

and $z = \dfrac{D_z}{D} = 6/2 = 3$,

the solution set is $\{(1, 2, 3)\}$.

31.

One finds $D = \begin{vmatrix} 1 & 2 & 0 \\ 1 & -3 & 1 \\ 2 & -1 & 0 \end{vmatrix} = 5,$

$$D_x = \begin{vmatrix} 8 & 2 & 0 \\ -2 & -3 & 1 \\ 1 & -1 & 0 \end{vmatrix} = 10,$$

$$D_y = \begin{vmatrix} 1 & 8 & 0 \\ 1 & -2 & 1 \\ 2 & 1 & 0 \end{vmatrix} = 15,$$

and $D_z = \begin{vmatrix} 1 & 2 & 8 \\ 1 & -3 & -2 \\ 2 & -1 & 1 \end{vmatrix} = 25.$

Since $x = \dfrac{D_x}{D} = 10/5 = 2$,

$$y = \dfrac{D_y}{D} = 15/5 = 3,$$

and $z = \dfrac{D_z}{D} = 25/5 = 5$,

the solution set is $\{(2, 3, 5)\}$.

33.

One finds $D = \begin{vmatrix} 2 & -3 & 1 \\ 1 & 4 & -1 \\ 3 & -1 & 2 \end{vmatrix} = 16,$

$$D_x = \begin{vmatrix} 1 & -3 & 1 \\ 0 & 4 & -1 \\ 0 & -1 & 2 \end{vmatrix} = 7,$$

$$D_y = \begin{vmatrix} 2 & 1 & 1 \\ 1 & 0 & -1 \\ 3 & 0 & 2 \end{vmatrix} = -5,$$

and $D_z = \begin{vmatrix} 2 & -3 & 1 \\ 1 & 4 & 0 \\ 3 & -1 & 0 \end{vmatrix} = -13.$

Then $x = \dfrac{D_x}{D} = 7/16$,

$$y = \dfrac{D_y}{D} = -5/16, \text{ and } z = \dfrac{D_z}{D} = -13/16.$$

The solution set is $\{(7/16, -5/16, -13/16)\}$.

35.

One finds $D = \begin{vmatrix} 1 & 1 & 1 \\ 2 & -1 & 3 \\ 3 & 1 & -1 \end{vmatrix} = 14,$

$$D_x = \begin{vmatrix} 2 & 1 & 1 \\ 0 & -1 & 3 \\ 0 & 1 & -1 \end{vmatrix} = -4,$$

$$D_y = \begin{vmatrix} 1 & 2 & 1 \\ 2 & 0 & 3 \\ 3 & 0 & -1 \end{vmatrix} = 22,$$

and $D_z = \begin{vmatrix} 1 & 1 & 2 \\ 2 & -1 & 0 \\ 3 & 1 & 0 \end{vmatrix} = 10.$

Then $x = \dfrac{D_x}{D} = -4/14 = -2/7,$

$y = \dfrac{D_y}{D} = 22/14 = 11/7,$ and

$z = \dfrac{D_z}{D} = 10/14 = 5/7.$

The solution set is $\{(-2/7, 11/7, 5/7)\}$.

37. This system is dependent since the sum of the first two equations is the third equation. Adding the first and third equations, we get $3x - 3z = 4$ or $z = x - 4/3$. Substituting into the first equation, we find

$$x + y - 2\left(x - \frac{4}{3}\right) = 1$$
$$y = 1 - \frac{8}{3} + x$$
$$y = x - \frac{5}{3}.$$

Solution set is

$$\left\{ \left(x, x - \frac{5}{3}, x - \frac{4}{3}\right) \mid x \text{ is any real number} \right\}.$$

39. Multiply the first equation by -2 and add to the third equation.

$$
\begin{array}{rcr}
-2x + 2y - 2z &=& -10 \\
2x - 2y + 2z &=& 16 \\
\hline
0 &=& 6
\end{array}
$$

Inconsistent and the solution set is $\emptyset$.

41. Let $x, y,$ and z be the ages of Jackie, Rochelle, and Alisha, respectively. Since $\dfrac{x+y}{2} = 33$, $\dfrac{y+z}{2} = 25,$ and $\dfrac{x+z}{2} = 19$, we obtain

$$
\begin{array}{rcr}
x + y &=& 66 \\
y + z &=& 50 \\
x + z &=& 38.
\end{array}
$$

One finds $D = \begin{vmatrix} 1 & 1 & 0 \\ 0 & 1 & 1 \\ 1 & 0 & 1 \end{vmatrix} = 2,$

$D_x = \begin{vmatrix} 66 & 1 & 0 \\ 50 & 1 & 1 \\ 38 & 0 & 1 \end{vmatrix} = 54,\ D_y = \begin{vmatrix} 1 & 66 & 0 \\ 0 & 50 & 1 \\ 1 & 38 & 1 \end{vmatrix} =$

78, and $D_z = \begin{vmatrix} 1 & 1 & 66 \\ 0 & 1 & 50 \\ 1 & 0 & 38 \end{vmatrix} = 22.$

So Jackie is $x = \dfrac{D_x}{D} = 54/2 = 27$ years old,

Rochelle is $y = \dfrac{D_y}{D} = 78/2 = 39$ years old,

and Alisha is $z = \dfrac{D_z}{D} = 22/2 = 11$ years old.

43. Let $x, y,$ and z be the scores in the first test, second test, and final exam, respectively. Then

$$
\begin{array}{rcl}
x + y + z &=& 180 \\
0.2x + 0.2y + 0.6z &=& 76 \\
0.1x + 0.2y + 0.7z &=& 83.
\end{array}
$$

One finds $D = \begin{vmatrix} 1 & 1 & 1 \\ 0.2 & 0.2 & 0.6 \\ 0.1 & 0.2 & 0.7 \end{vmatrix} = -0.04,$

$D_x = \begin{vmatrix} 180 & 1 & 1 \\ 76 & 0.2 & 0.6 \\ 83 & 0.2 & 0.7 \end{vmatrix} = -1.2,$

$D_y = \begin{vmatrix} 1 & 180 & 1 \\ 0.2 & 76 & 0.6 \\ 0.1 & 83 & 0.7 \end{vmatrix} = -2,$

and $D_z = \begin{vmatrix} 1 & 1 & 180 \\ 0.2 & 0.2 & 76 \\ 0.1 & 0.2 & 83 \end{vmatrix} = -4.$

Test scores are $x = \dfrac{D_x}{D} = 1.2/(0.04) = 30,$

$y = \dfrac{D_y}{D} = 2/(0.04) = 50,$ and

the final exam is $z = \dfrac{D_z}{D} = 4/(0.04) = 100.$

45. One finds $D = \begin{vmatrix} 0.2 & -0.3 & 1.2 \\ 0.25 & 0.35 & -0.9 \\ 2.4 & -1 & 1.25 \end{vmatrix} = -\dfrac{527}{800},$

$D_x = \begin{vmatrix} 13.11 & -0.3 & 1.2 \\ -1.575 & 0.35 & -0.9 \\ 42.02 & -1 & 1.25 \end{vmatrix} = -\dfrac{11,067}{1000},$

$D_y = \begin{vmatrix} 0.2 & 13.11 & 1.2 \\ 0.25 & -1.575 & -0.9 \\ 2.4 & 42.02 & 1.25 \end{vmatrix} = -\dfrac{64,821}{8000},$

and $D_z = \begin{vmatrix} 0.2 & -0.3 & 13.11 \\ 0.25 & 0.35 & -1.575 \\ 2.4 & -1 & 42.02 \end{vmatrix} = -\dfrac{3689}{500}.$

Note, $\dfrac{D_x}{D} = 16.8$, $\dfrac{D_y}{D} = 12.3$, and $\dfrac{D_z}{D} = 11.2$.

The solution set is $\{(16.8, 12.3, 11.2)\}$.

47. Let x, y, and z be the prices per gallon of regular, plus, and supreme gasoline, respectively.

One finds

$D = \begin{vmatrix} 1270 & 980 & 890 \\ 1450 & 1280 & 1050 \\ 1340 & 1190 & 1060 \end{vmatrix} = 18,038,000,$

$D_x = \begin{vmatrix} 4668.86 & 980 & 890 \\ 5626.32 & 1280 & 1050 \\ 5353.41 & 1190 & 1060 \end{vmatrix} = 25,235,200,$

$D_y = \begin{vmatrix} 1270 & 4668.86 & 890 \\ 1450 & 5626.32 & 1050 \\ 1340 & 5353.41 & 1060 \end{vmatrix} = 27,039,000,$

$D_z = \begin{vmatrix} 1270 & 980 & 4668.86 \\ 1450 & 1280 & 5626.32 \\ 1340 & 1190 & 5353.41 \end{vmatrix} = 28,842,800.$

Regular gas costs $\dfrac{D_x}{D} \approx \$1.399,$

plus costs $\dfrac{D_y}{D} \approx \$1.499,$ and

supreme costs $\dfrac{D_z}{D} \approx \$1.599$ per gallon.

49.

One finds $D = \begin{vmatrix} 1 & 1 & 1 & 1 \\ 2 & -1 & 1 & 3 \\ 1 & 2 & -1 & 2 \\ 1 & -1 & -1 & 4 \end{vmatrix} = 11,$

$D_w = \begin{vmatrix} 4 & 1 & 1 & 1 \\ 13 & -1 & 1 & 3 \\ -2 & 2 & -1 & 2 \\ 8 & -1 & -1 & 4 \end{vmatrix} = 11,$

$D_x = \begin{vmatrix} 1 & 4 & 1 & 1 \\ 2 & 13 & 1 & 3 \\ 1 & -2 & -1 & 2 \\ 1 & 8 & -1 & 4 \end{vmatrix} = -22,$

$D_y = \begin{vmatrix} 1 & 1 & 4 & 1 \\ 2 & -1 & 13 & 3 \\ 1 & 2 & -2 & 2 \\ 1 & -1 & 8 & 4 \end{vmatrix} = 33,$

and $D_z = \begin{vmatrix} 1 & 1 & 1 & 4 \\ 2 & -1 & 1 & 13 \\ 1 & 2 & -1 & -2 \\ 1 & -1 & -1 & 8 \end{vmatrix} = 22.$

Note, $\dfrac{D_w}{D} = 1$, $\dfrac{D_x}{D} = -2$, $\dfrac{D_y}{D} = 3$, and

$\dfrac{D_z}{D} = 2$. The solution set is $\{(1, -2, 3, 2)\}$.

51.

Note, $\begin{vmatrix} x & y & 1 \\ 3 & -5 & 1 \\ -2 & 6 & 1 \end{vmatrix} = 8 - 11x - 5y = 0.$

Since both points $(3, -5)$ and $(-2, 6)$ satisfies $8 - 11x - 5y = 0$, this is an equation of the line through the two points.

53. We will show it for one particular case and the other cases can proved similarly.

Let us suppose $A = \begin{bmatrix} a & b & c \\ 0 & 0 & 0 \\ g & h & i \end{bmatrix}.$

Then $|A| = a \cdot 0 \cdot i + b \cdot 0 \cdot g + c \cdot 0 \cdot h - g \cdot 0 \cdot c - h \cdot 0 \cdot a - i \cdot 0 \cdot b = 0$

55. We will prove it for one particular case and the other cases can be shown similarly.

Let us suppose $A = \begin{bmatrix} a & b & c \\ d & e & f \\ g & h & i \end{bmatrix}$ and

$B = \begin{bmatrix} a & b & c \\ d & e & f \\ kg & kh & ki \end{bmatrix}.$ Then $|B| =$

$aeki + bfkg + cdkh - kgec - khfa - kidb = k(aei + bfg + cdh - gec - hfa - idb) = k|A|.$

Review Exercises

1.

$\begin{bmatrix} 2+3 & -3+7 \\ -2+1 & 4+2 \end{bmatrix} = \begin{bmatrix} 5 & 4 \\ -1 & 6 \end{bmatrix}$

3.

$\begin{bmatrix} 4 & -6 \\ -4 & 8 \end{bmatrix} - \begin{bmatrix} 3 & 7 \\ 1 & 2 \end{bmatrix} =$

$\begin{bmatrix} 1 & -13 \\ -5 & 6 \end{bmatrix}$

5.

$$AB = \begin{bmatrix} 6-3 & 14-6 \\ -6+4 & -14+8 \end{bmatrix} = \begin{bmatrix} 3 & 8 \\ -2 & -6 \end{bmatrix}$$

7. $D + E$ is undefined

9.

$$AC = \begin{bmatrix} -2-9 \\ 2+12 \end{bmatrix} = \begin{bmatrix} -11 \\ 14 \end{bmatrix}$$

11.

$$EF = \begin{bmatrix} 3 & 2 & -1 \\ -12 & -8 & 4 \\ 9 & 6 & -3 \end{bmatrix}$$

13. $FG = [-3+2+2 \quad 2-3 \quad -1] =$
$[1 \quad -1 \quad -1]$

15. GF is undefined

17.

On $\begin{bmatrix} 2 & -3 & | & 1 & 0 \\ -2 & 4 & | & 0 & 1 \end{bmatrix}$, use $R_1 + R_2 \to R_2$

to get

$\begin{bmatrix} 2 & -3 & | & 1 & 0 \\ 0 & 1 & | & 1 & 1 \end{bmatrix}$, use $3R_2 + R_1 \to R_1$ to get

$\begin{bmatrix} 2 & 0 & | & 4 & 3 \\ 0 & 1 & | & 1 & 1 \end{bmatrix}$, use $\frac{1}{2}R_1 \to R_1$ to get

$\begin{bmatrix} 1 & 0 & | & 2 & 1.5 \\ 0 & 1 & | & 1 & 1 \end{bmatrix}$. So $A^{-1} = \begin{bmatrix} 2 & 1.5 \\ 1 & 1 \end{bmatrix}$.

19.

On $\begin{bmatrix} -1 & 0 & 0 & | & 1 & 0 & 0 \\ 1 & 1 & 0 & | & 0 & 1 & 0 \\ -2 & 3 & 1 & | & 0 & 0 & 1 \end{bmatrix}$,

use $R_1 + R_2 \to R_2$, $2R_2 + R_3 \to R_3$, and

$-1R_1 \to R_1$ to get

$\begin{bmatrix} 1 & 0 & 0 & | & -1 & 0 & 0 \\ 0 & 1 & 0 & | & 1 & 1 & 0 \\ 0 & 5 & 1 & | & 0 & 2 & 1 \end{bmatrix}$,

use $-5R_2 + R_3 \to R_3$ to get

$\begin{bmatrix} 1 & 0 & 0 & | & -1 & 0 & 0 \\ 0 & 1 & 0 & | & 1 & 1 & 0 \\ 0 & 0 & 1 & | & -5 & -3 & 1 \end{bmatrix}$.

So $G^{-1} = \begin{bmatrix} -1 & 0 & 0 \\ 1 & 1 & 0 \\ -5 & -3 & 1 \end{bmatrix}$.

21.

From Exercise 5, $(AB)^{-1} = \begin{bmatrix} 3 & 8 \\ -2 & -6 \end{bmatrix}^{-1}$.

On $\begin{bmatrix} 3 & 8 & | & 1 & 0 \\ -2 & -6 & | & 0 & 1 \end{bmatrix}$, use $-\frac{1}{2}R_2 \to R_2$

to get

$\begin{bmatrix} 3 & 8 & | & 1 & 0 \\ 1 & 3 & | & 0 & -0.5 \end{bmatrix}$, use $-3R_2 + R_1 \to R_1$

to get

$\begin{bmatrix} 0 & -1 & | & 1 & 1.5 \\ 1 & 3 & | & 0 & -0.5 \end{bmatrix}$, use $3R_1 + R_2 \to R_2$

to get

$\begin{bmatrix} 0 & -1 & | & 1 & 1.5 \\ 1 & 0 & | & 3 & 4 \end{bmatrix}$, use $R_1 \leftrightarrow R_2$ to get

$\begin{bmatrix} 1 & 0 & | & 3 & 4 \\ 0 & -1 & | & 1 & 1.5 \end{bmatrix}$, use $-1R_2 \to R_2$ to get

$\begin{bmatrix} 1 & 0 & | & 3 & 4 \\ 0 & 1 & | & -1 & -1.5 \end{bmatrix}$.

So $(AB)^{-1} = \begin{bmatrix} 3 & 4 \\ -1 & -1.5 \end{bmatrix}$.

23.

$$AA^{-1} = I = \begin{bmatrix} 1 & 0 \\ 0 & 1 \end{bmatrix}$$

25. $|A| = 8 - 6 = 2$

27. Expanding about the third column,

$$|G| = 1 \cdot \begin{vmatrix} -1 & 0 \\ 1 & 1 \end{vmatrix} = 1(-1) = -1.$$

29. Solution set is $\{(10/3, 17/3)\}$.

First solution is by Gaussian elimination.

On $\begin{bmatrix} 1 & 1 & | & 9 \\ 2 & -1 & | & 1 \end{bmatrix}$, use $-2R_1 + R_2 \to R_2$ to

get

$\begin{bmatrix} 1 & 1 & | & 9 \\ 0 & -3 & | & -17 \end{bmatrix}$, use $-\frac{1}{3}R_2 \to R_2$ to get

$\begin{bmatrix} 1 & 1 & | & 9 \\ 0 & 1 & | & 17/3 \end{bmatrix}$, use $-1R_2 + R_1 \to R_1$ to get

$$\begin{bmatrix} 1 & 0 & | & 10/3 \\ 0 & 1 & | & 17/3 \end{bmatrix}.$$

Secondly, by matrix inversion note $A^{-1} =$

$$\begin{bmatrix} 1/3 & 1/3 \\ 2/3 & -1/3 \end{bmatrix} \text{ and } A^{-1}\begin{bmatrix} 9 \\ 1 \end{bmatrix} = \begin{bmatrix} 10/3 \\ 17/3 \end{bmatrix}.$$

Thirdly, by Cramer's Rule note

$$D = \begin{vmatrix} 1 & 1 \\ 2 & -1 \end{vmatrix} = -3, D_x = \begin{vmatrix} 9 & 1 \\ 1 & -1 \end{vmatrix} = -10,$$

and $D_y = \begin{vmatrix} 1 & 9 \\ 2 & 1 \end{vmatrix} = -17.$

So $\dfrac{D_x}{D} = 10/3, \dfrac{D_y}{D} = 17/3.$

31. Solution set is $\{(-2,3)\}$.

First solution is by Gaussian elimination. On

$$\begin{bmatrix} 2 & 1 & | & -1 \\ 3 & 2 & | & 0 \end{bmatrix}, \text{ use } -\frac{3}{2}R_1 + R_2 \rightarrow R_2 \text{ to get}$$

$$\begin{bmatrix} 2 & 1 & | & -1 \\ 0 & 1/2 & | & 3/2 \end{bmatrix}, \text{ use } -2R_2 + R_1 \rightarrow R_1$$

and $2R_2 \rightarrow R_2$ to get

$$\begin{bmatrix} 2 & 0 & | & -4 \\ 0 & 1 & | & 3 \end{bmatrix}, \text{ use } \frac{1}{2}R_2 \rightarrow R_2 \text{ to get}$$

$$\begin{bmatrix} 1 & 0 & | & -2 \\ 0 & 1 & | & 3 \end{bmatrix}.$$

Secondly, by matrix inversion note $A^{-1} =$

$$\begin{bmatrix} 2 & -1 \\ -3 & 2 \end{bmatrix} \text{ and } A^{-1}\begin{bmatrix} -1 \\ 0 \end{bmatrix} = \begin{bmatrix} -2 \\ 3 \end{bmatrix}.$$

Thirdly, by Cramer's Rule note

$$D = \begin{vmatrix} 2 & 1 \\ 3 & 2 \end{vmatrix} = 1, D_x = \begin{vmatrix} -1 & 1 \\ 0 & 2 \end{vmatrix} = -2,$$

and $D_y = \begin{vmatrix} 2 & -1 \\ 3 & 0 \end{vmatrix} = 3.$

So $\dfrac{D_x}{D} = -2, \dfrac{D_y}{D} = 3.$

33. Solution set is $\{(x,y)|x - 5y = 9\}$.

Solution is by Gaussian elimination.

On $\begin{bmatrix} 1 & -5 & | & 9 \\ -2 & 10 & | & -18 \end{bmatrix}$, use $2R_1 + R_2 \rightarrow R_2$

to get $\begin{bmatrix} 1 & -5 & | & 9 \\ 0 & 0 & | & 0 \end{bmatrix}$. Dependent system.

This cannot be solved by matrix inversion

since $A^{-1} = \begin{bmatrix} 1 & -5 \\ -2 & 10 \end{bmatrix}^{-1}$ does not exist.

Nor can it be solved by Cramer's rule

since $|D| = \begin{vmatrix} 1 & -5 \\ -2 & 10 \end{vmatrix} = 0.$

35. Solution set is $\emptyset$ as shown by Gaussian

elimination. On $\begin{bmatrix} 0.05 & 0.1 & | & 1 \\ 10 & 20 & | & 20 \end{bmatrix}$, use

$-200R_1 + R_2 \rightarrow R_2$ to get

$$\begin{bmatrix} 0.05 & 0.1 & | & 1 \\ 0 & 0 & | & -180 \end{bmatrix}. \text{ Inconsistent.}$$

This cannot be solved by matrix inversion since

$$A^{-1} = \begin{bmatrix} 0.05 & 0.1 \\ 10 & 20 \end{bmatrix}^{-1} \text{ does not exist.}$$

Nor can it be solved by Cramer's rule

since $|D| = \begin{vmatrix} 0.05 & 0.1 \\ 10 & 20 \end{vmatrix} = 0.$

37. Solution set is $\{(1,2,3)\}$.

First solution is by Gaussian elimination.

On $\begin{bmatrix} 1 & 1 & -2 & | & -3 \\ -1 & 2 & -1 & | & 0 \\ -1 & -1 & 3 & | & 6 \end{bmatrix}$, use

$R_1 + R_2 \rightarrow R_2$ to get

$$\begin{bmatrix} 1 & 1 & -2 & | & -3 \\ 0 & 3 & -3 & | & -3 \\ 0 & 0 & 1 & | & 3 \end{bmatrix}, \text{ use}$$

$\frac{1}{3}R_2 \rightarrow R_2$ to get

$$\begin{bmatrix} 1 & 1 & -2 & | & -3 \\ 0 & 1 & -1 & | & -1 \\ 0 & 0 & 1 & | & 3 \end{bmatrix}, \text{ use}$$

$R_3 + R_2 \rightarrow R_2$ and $2R_3 + R_1 \rightarrow R_1$ to get

$$\begin{bmatrix} 1 & 1 & 0 & | & 3 \\ 0 & 1 & 0 & | & 2 \\ 0 & 0 & 1 & | & 3 \end{bmatrix}, \text{ use } -1R_2 + R_1 \rightarrow R_1$$

to get $\begin{bmatrix} 1 & 0 & 0 & | & 1 \\ 0 & 1 & 0 & | & 2 \\ 0 & 0 & 1 & | & 3 \end{bmatrix}.$

Secondly, by matrix inversion note $A^{-1} =$

$$\begin{bmatrix} 5/3 & -1/3 & 1 \\ 4/3 & 1/3 & 1 \\ 1 & 0 & 1 \end{bmatrix} \text{ and } A^{-1}\begin{bmatrix} -3 \\ 0 \\ 6 \end{bmatrix} = \begin{bmatrix} 1 \\ 2 \\ 3 \end{bmatrix}.$$

Thirdly, by Cramer's Rule note

$$D = \begin{vmatrix} 1 & 1 & -2 \\ -1 & 2 & -1 \\ -1 & -1 & 3 \end{vmatrix} = 3,$$

$$D_x = \begin{vmatrix} -3 & 1 & -2 \\ 0 & 2 & -1 \\ 6 & -1 & 3 \end{vmatrix} = 3,$$

$$D_y = \begin{vmatrix} 1 & -3 & -2 \\ -1 & 0 & -1 \\ -1 & 6 & 3 \end{vmatrix} = 6,$$

$$\text{and } D_z = \begin{vmatrix} 1 & 1 & -3 \\ -1 & 2 & 0 \\ -1 & -1 & 6 \end{vmatrix} = 9.$$

Then $\dfrac{D_x}{D} = 1$, $\dfrac{D_y}{D} = 2$, and $\dfrac{D_z}{D} = 3$.

39. Solution set is $\{(-3, 4, 1)\}$.

First solution is by Gaussian elimination.

$$\text{On } \begin{bmatrix} 0 & 1 & -3 & | & 1 \\ 1 & 2 & 0 & | & 5 \\ 1 & 0 & 4 & | & 1 \end{bmatrix}, \text{ use}$$

$-1R_3 + R_2 \to R_2$ and $R_1 \leftrightarrow R_3$ to get

$$\begin{bmatrix} 1 & 0 & 4 & | & 1 \\ 0 & 2 & -4 & | & 4 \\ 0 & 1 & -3 & | & 1 \end{bmatrix}, \text{ use } \frac{1}{2}R_2 \to R_2 \text{ to get}$$

$$\begin{bmatrix} 1 & 0 & 4 & | & 1 \\ 0 & 1 & -2 & | & 2 \\ 0 & 1 & -3 & | & 1 \end{bmatrix}, \text{ use } -1R_2 + R_3 \to R_3$$

to get

$$\begin{bmatrix} 1 & 0 & 4 & | & 1 \\ 0 & 1 & -2 & | & 2 \\ 0 & 0 & -1 & | & -1 \end{bmatrix}, \text{ use}$$

$-2R_3 + R_2 \to R_2$, $4R_3 + R_1 \to R_1$, and

$-1R_3 \to R_3$ to get

$$\begin{bmatrix} 1 & 0 & 0 & | & -3 \\ 0 & 1 & 0 & | & 4 \\ 0 & 0 & 1 & | & 1 \end{bmatrix}.$$

Secondly, by matrix inversion

$$A^{-1} = \begin{bmatrix} 4 & -2 & 3 \\ -2 & 3/2 & -3/2 \\ -1 & 1/2 & -1/2 \end{bmatrix}$$

and $A^{-1}\begin{bmatrix} 1 \\ 5 \\ 1 \end{bmatrix} = \begin{bmatrix} -3 \\ 4 \\ 1 \end{bmatrix}.$

Thirdly, by Cramer's Rule note

$$D = \begin{vmatrix} 0 & 1 & -3 \\ 1 & 2 & 0 \\ 1 & 0 & 4 \end{vmatrix} = 2,$$

$$D_x = \begin{vmatrix} 1 & 1 & -3 \\ 5 & 2 & 0 \\ 1 & 0 & 4 \end{vmatrix} = -6,$$

$$D_y = \begin{vmatrix} 0 & 1 & -3 \\ 1 & 5 & 0 \\ 1 & 1 & 4 \end{vmatrix} = 8,$$

$$\text{and } D_z = \begin{vmatrix} 0 & 1 & 1 \\ 1 & 2 & 5 \\ 1 & 0 & 1 \end{vmatrix} = 2.$$

Then $\dfrac{D_x}{D} = -3$, $\dfrac{D_y}{D} = 4$, and $\dfrac{D_z}{D} = 1$.

41. Solution set is

$$\left\{ \left(\frac{3y+3}{2}, y, \frac{1-y}{2} \right) \mid y \text{ is any real number} \right\}.$$

Solution is by Gaussian elimination.

$$\text{On } \begin{bmatrix} 1 & -1 & 1 & | & 2 \\ 1 & -2 & -1 & | & 1 \\ 2 & -3 & 0 & | & 3 \end{bmatrix}, \text{ use}$$

$-1R_1 + R_2 \to R_2$ and $-2R_1 + R_3 \to R_3$ to get

$$\begin{bmatrix} 1 & -1 & 1 & | & 2 \\ 0 & -1 & -2 & | & -1 \\ 0 & -1 & -2 & | & -1 \end{bmatrix}, \text{ use}$$

$-1R_2 + R_3 \to R_3$, $-1R_2 + R_1 \to R_1$,

and $-1R_2 \to R_2$ to get

$$\begin{bmatrix} 1 & 0 & 3 & | & 3 \\ 0 & 1 & 2 & | & 1 \\ 0 & 0 & 0 & | & 0 \end{bmatrix}. \text{ Since } z = \frac{1-y}{2}, \text{ we get}$$

$$x = 3 - 3z = 3 - 3\left(\frac{1-y}{2} \right) = \frac{3y+3}{2}.$$

This cannot be solved by matrix inversion

since $A^{-1} = \begin{bmatrix} 1 & -1 & 1 \\ 1 & -2 & -1 \\ 2 & -3 & 0 \end{bmatrix}^{-1}$ does not exist.

Nor can it be solved by Cramer's rule

since $|D| = \begin{vmatrix} 1 & -1 & 1 \\ 1 & -2 & -1 \\ 2 & -3 & 0 \end{vmatrix} = 0.$

43. Solution set is $\emptyset$ as shown by Gaussian

elimination. On $\begin{bmatrix} 1 & -3 & -1 & | & 2 \\ 1 & -3 & -1 & | & 1 \\ 1 & -3 & -1 & | & 0 \end{bmatrix}$, use

$-1R_1 + R_2 \to R_2$ and $-1R_1 + R_3 \to R_3$ to get

$\begin{bmatrix} 1 & -3 & -1 & | & 2 \\ 0 & 0 & 0 & | & -1 \\ 0 & 0 & 0 & | & -2 \end{bmatrix}$. Inconsistent.

This cannot be solved by matrix inversion

since $A^{-1} = \begin{bmatrix} 1 & -3 & -1 \\ 1 & -3 & -1 \\ 1 & -3 & -1 \end{bmatrix}^{-1}$ does not exist.

Nor can it be solved by Cramer's rule

since $|D| = \begin{vmatrix} 1 & -3 & -1 \\ 1 & -3 & -1 \\ 1 & -3 & -1 \end{vmatrix} = 0.$

45. Using $x = 9$ in $x + y = -3$, we find $9 + y = -3$ and $y = -12$. Solution set is $\{(9, -12)\}$.

47.

Note $\begin{bmatrix} x \\ y \end{bmatrix} = \begin{bmatrix} 1 & 1 \\ 2 & 1 \end{bmatrix}^{-1} \begin{bmatrix} 6 \\ 8 \end{bmatrix} =$

$\begin{bmatrix} -1 & 1 \\ 2 & -1 \end{bmatrix} \begin{bmatrix} 6 \\ 8 \end{bmatrix} = \begin{bmatrix} 2 \\ 4 \end{bmatrix}.$

The solution set is $\{(2, 4)\}$.

49. System of equations can be written as

$$x + y = -3$$
$$-x = 0.$$

Using $x = 0$ in $x + y = -3$, we get $y = -3$. Solution set is $\{(0, -3)\}$.

51. System of equations can be written as

$$\begin{bmatrix} 1 & 1 & 0 \\ 0 & 1 & 1 \\ 1 & 0 & 1 \end{bmatrix} \begin{bmatrix} x \\ y \\ z \end{bmatrix} = \begin{bmatrix} 1 \\ 1 \\ 1 \end{bmatrix}.$$

By using matrix inversion, we obtain

$\begin{bmatrix} x \\ y \\ z \end{bmatrix} = \begin{bmatrix} 1/2 & -1/2 & 1/2 \\ 1/2 & 1/2 & -1/2 \\ -1/2 & 1/2 & 1/2 \end{bmatrix} \begin{bmatrix} 1 \\ 1 \\ 1 \end{bmatrix} =$

$\begin{bmatrix} 1/2 \\ 1/2 \\ 1/2 \end{bmatrix}$. The solution set is $\{(1/2, 1/2, 1/2)\}$.

53. By using the inverse of the coefficient matrix, we get

$\begin{bmatrix} x \\ y \\ z \end{bmatrix} = \begin{bmatrix} 3/5 & -3/5 & 2/5 \\ 2/5 & 3/5 & -2/5 \\ -1/5 & 1/5 & 1/5 \end{bmatrix} \begin{bmatrix} -1 \\ 7 \\ 17 \end{bmatrix} =$

$\begin{bmatrix} 2 \\ -3 \\ 5 \end{bmatrix}$. Solution set is $\{(2, -3, 5)\}$.

55. Let x and y be the number of gallons of pollutant A and pollutant B, respectively. Then

$$10x + 6y = 4060$$
$$\frac{3}{4} = \frac{x}{y}.$$

Substitute $x = \frac{3}{4}y$ in $10x + 6y = 4060$. Solving for x, one finds the quantities discharged are $x \approx 225.56$ gallons of pollutant A and $y \approx 300.74$ gallons of pollutant B.

57. Let x, y, and z be the expenses including tax for water, gas, and electricity, respectively.

$$x + y + z = 189.83$$
$$\frac{x}{1.04} + \frac{y}{1.05} + \frac{z}{1.06} = 180$$
$$z = 2y$$

Solving the system, one finds the expenses including taxes are $x = \$22.88$ for water, $y = \$55.65$ for gas, and $z = \$111.30$ for electricity.

Chapter 6 Test

1.

On $\begin{bmatrix} 2 & -3 & | & 1 \\ 1 & 9 & | & 4 \end{bmatrix}$, use $-2R_2 + R_1 \rightarrow R_2$

to get

$\begin{bmatrix} 2 & -3 & | & 1 \\ 0 & -21 & | & -7 \end{bmatrix}$, use $\frac{1}{2}R_1 \rightarrow R_1$ and

$-\frac{1}{21}R_2 \rightarrow R_2$ to get

$\begin{bmatrix} 1 & -3/2 & | & 1/2 \\ 0 & 1 & | & 1/3 \end{bmatrix}$, use $\frac{3}{2}R_2 + R_1 \rightarrow R_1$

to get

$\begin{bmatrix} 1 & 0 & | & 1 \\ 0 & 1 & | & 1/3 \end{bmatrix}$. Solution set is $\{(1, 1/3)\}$.

2.

On $\begin{bmatrix} 2 & -1 & 1 & | & 5 \\ 1 & -2 & -1 & | & -2 \\ 3 & -1 & -1 & | & 6 \end{bmatrix}$, use

$-\frac{1}{2}R_1 + R_2 \rightarrow R_2$ and

$-\frac{3}{2}R_1 + R_3 \rightarrow R_3$ to get

$\begin{bmatrix} 2 & -1 & 1 & | & 5 \\ 0 & -3/2 & -3/2 & | & -9/2 \\ 0 & 1/2 & -5/2 & | & -3/2 \end{bmatrix}$, use

$-\frac{2}{3}R_2 \rightarrow R_2$ and $3R_3 + R_2 \rightarrow R_3$ to get

$\begin{bmatrix} 2 & -1 & 1 & | & 5 \\ 0 & 1 & 1 & | & 3 \\ 0 & 0 & -9 & | & -9 \end{bmatrix}$,

use $-\frac{1}{9}R_3 \rightarrow R_3$ and $R_2 + R_1 \rightarrow R_1$ to get

$\begin{bmatrix} 2 & 0 & 2 & | & 8 \\ 0 & 1 & 1 & | & 3 \\ 0 & 0 & 1 & | & 1 \end{bmatrix}$, use

$-1R_3 + R_2 \rightarrow R_2$ and $\frac{1}{2}R_1 \rightarrow R_1$ to get

$\begin{bmatrix} 1 & 0 & 1 & | & 4 \\ 0 & 1 & 0 & | & 2 \\ 0 & 0 & 1 & | & 1 \end{bmatrix}$, use $-1R_3 + R_1 \rightarrow R_1$

to get $\begin{bmatrix} 1 & 0 & 0 & | & 3 \\ 0 & 1 & 0 & | & 2 \\ 0 & 0 & 1 & | & 1 \end{bmatrix}$.

The solution set is $\{(3, 2, 1)\}$.

3.

On $\begin{bmatrix} 1 & -1 & -1 & | & 1 \\ 2 & 1 & -1 & | & 0 \\ 5 & -2 & -4 & | & 3 \end{bmatrix}$, use

$-2R_1 + R_2 \rightarrow R_2$ and

$-5R_1 + R_3 \rightarrow R_3$ to get

$\begin{bmatrix} 1 & -1 & -1 & | & 1 \\ 0 & 3 & 1 & | & -2 \\ 0 & 3 & 1 & | & -2 \end{bmatrix}$, use

$-1R_3 + R_2 \rightarrow R_3$ and $\frac{1}{3}R_2 \rightarrow R_2$ to get

$\begin{bmatrix} 1 & -1 & -1 & | & 1 \\ 0 & 1 & 1/3 & | & -2/3 \\ 0 & 0 & 0 & | & 0 \end{bmatrix}$, use

$R_2 + R_1 \rightarrow R_1$ to get

$\begin{bmatrix} 1 & 0 & -2/3 & | & 1/3 \\ 0 & 1 & 1/3 & | & -2/3 \\ 0 & 0 & 0 & | & 0 \end{bmatrix}$.

Since $x = \frac{2}{3}z + \frac{1}{3}$, we find $3x = 2z + 1$ and

$z = \frac{3x - 1}{2}$. Since $y = -\frac{1}{3}z - \frac{2}{3}$, we get

$$y = -\frac{1}{3}\left(\frac{3x - 1}{2}\right) - \frac{2}{3}$$
$$y = \frac{-3x - 3}{6}$$
$$y = \frac{-x - 1}{2}$$

The solution set is

$$\left\{\left(x, \frac{-x - 1}{2}, \frac{3x - 1}{2}\right) \mid x \text{ is any real number}\right\}.$$

4.

$A + B = \begin{bmatrix} 3 & -4 \\ -6 & 10 \end{bmatrix}$

5.

$2A - B = \begin{bmatrix} 2 & -2 \\ -4 & 8 \end{bmatrix} - \begin{bmatrix} 2 & -3 \\ -4 & 6 \end{bmatrix} =$

$\begin{bmatrix} 0 & 1 \\ 0 & 2 \end{bmatrix}$

6.

$AB = \begin{bmatrix} 2 + 4 & -3 - 6 \\ -4 - 16 & 6 + 24 \end{bmatrix} = \begin{bmatrix} 6 & -9 \\ -20 & 30 \end{bmatrix}$

7.
$$AC = \begin{bmatrix} -2-1 \\ 4+4 \end{bmatrix} = \begin{bmatrix} -3 \\ 8 \end{bmatrix}$$

8. CB is undefined

9. $FG = [-2 \quad 3-2 \quad 1+1] = [-2 \quad 1 \quad 2]$

10.
$$EF = \begin{bmatrix} 2(1) & 2(0) & 2(-1) \\ 3(1) & 3(0) & 3(-1) \\ -1(1) & -1(0) & -1(-1) \end{bmatrix}$$
$$= \begin{bmatrix} 2 & 0 & -2 \\ 3 & 0 & -3 \\ -1 & 0 & 1 \end{bmatrix}$$

11.

On $\begin{bmatrix} 1 & -1 & | & 1 & 0 \\ -2 & 4 & | & 0 & 1 \end{bmatrix}$, use

$2R_1 + R_2 \to R_2$ to get

$\begin{bmatrix} 1 & -1 & | & 1 & 0 \\ 0 & 2 & | & 2 & 1 \end{bmatrix}$, use

$\frac{1}{2}R_2 + R_1 \to R_1$ and $\frac{1}{2}R_2 \to R_2$

to get $\begin{bmatrix} 1 & 0 & | & 2 & 1/2 \\ 0 & 1 & | & 1 & 1/2 \end{bmatrix}$.

Then $A^{-1} = \begin{bmatrix} 2 & 1/2 \\ 1 & 1/2 \end{bmatrix}$.

12.

On $\begin{bmatrix} -2 & 3 & 1 & | & 1 & 0 & 0 \\ -3 & 1 & 3 & | & 0 & 1 & 0 \\ 0 & 2 & -1 & | & 0 & 0 & 1 \end{bmatrix}$,

use $-\frac{1}{2}R_1 \to R_1$ and $-\frac{3}{2}R_1 + R_2 \to R_2$ to get

$\begin{bmatrix} 1 & -3/2 & -1/2 & | & -1/2 & 0 & 0 \\ 0 & -7/2 & 3/2 & | & -3/2 & 1 & 0 \\ 0 & 2 & -1 & | & 0 & 0 & 1 \end{bmatrix}$, use

$\frac{4}{7}R_2 + R_3 \to R_3$, $-\frac{3}{7}R_2 + R_1 \to R_1$,

and $-\frac{2}{7}R_2 \to R_2$ to get

$\begin{bmatrix} 1 & 0 & -8/7 & | & 1/7 & -3/7 & 0 \\ 0 & 1 & -3/7 & | & 3/7 & -2/7 & 0 \\ 0 & 0 & -1/7 & | & -6/7 & 4/7 & 1 \end{bmatrix}$, use

$-3R_3 + R_2 \to R_2$, $-8R_3 + R_1 \to R_1$,

and $-7R_3 \to R_3$ to get

$\begin{bmatrix} 1 & 0 & 0 & | & 7 & -5 & -8 \\ 0 & 1 & 0 & | & 3 & -2 & -3 \\ 0 & 0 & 1 & | & 6 & -4 & -7 \end{bmatrix}$.

Then $G^{-1} = \begin{bmatrix} 7 & -5 & -8 \\ 3 & -2 & -3 \\ 6 & -4 & -7 \end{bmatrix}$.

13. $|A| = 4 - 2 = 2$

14. $|B| = 12 - 12 = 0$

15. Expanding about the third row, we find

$$|G| = -2\begin{vmatrix} -2 & 1 \\ -3 & 3 \end{vmatrix} + (-1)\begin{vmatrix} -2 & 3 \\ -3 & 1 \end{vmatrix} =$$
$$-2(-3) + (-1)(7) = -1$$

16.

One finds $D = \begin{vmatrix} 1 & -1 \\ -2 & 4 \end{vmatrix} = 2$,

$D_x = \begin{vmatrix} 2 & -1 \\ 2 & 4 \end{vmatrix} = 10$, $D_y = \begin{vmatrix} 1 & 2 \\ -2 & 2 \end{vmatrix} = 6$.

Then $\frac{D_x}{D} = 5$ and $\frac{D_y}{D} = 3$.

The solution set is $\{(5,3)\}$.

17. Cramer's Rule is not applicable since

$$D = \begin{vmatrix} 2 & -3 \\ -4 & 6 \end{vmatrix} = 0. \text{ Rather, multiply first}$$

equation by 2 and add to the second one.

$$\begin{array}{rcl} 4x - 6y & = & 12 \\ -4x + 6y & = & 1 \\ \hline 0 & = & 13 \end{array}$$

Inconsistent and the solution set is $\emptyset$.

18.

One finds $D = \begin{vmatrix} -2 & 3 & 1 \\ -3 & 1 & 3 \\ 0 & 2 & -1 \end{vmatrix} = -1$,

$D_x = \begin{vmatrix} -2 & 3 & 1 \\ -4 & 1 & 3 \\ 0 & 2 & -1 \end{vmatrix} = -6$,

$D_y = \begin{vmatrix} -2 & -2 & 1 \\ -3 & -4 & 3 \\ 0 & 0 & -1 \end{vmatrix} = -2$, and

$$D_z = \begin{vmatrix} -2 & 3 & -2 \\ -3 & 1 & -4 \\ 0 & 2 & 0 \end{vmatrix} = -4.$$

Then $\dfrac{D_x}{D} = 6$, $\dfrac{D_y}{D} = 2$, and $\dfrac{D_z}{D} = 4$.

Solution set is $\{(6, 2, 4)\}$.

19. The inverse of coefficient matrix is given

by $A^{-1} = \begin{bmatrix} 2 & 1/2 \\ 1 & 1/2 \end{bmatrix}$. Since

$A^{-1} \begin{bmatrix} 1 \\ -8 \end{bmatrix} = \begin{bmatrix} -2 \\ -3 \end{bmatrix}$, the solution

set is $\{(-2, -3)\}$.

20. The inverse of coefficient matrix was found

in Exercise 12: $A^{-1} = \begin{bmatrix} 7 & -5 & -8 \\ 3 & -2 & -3 \\ 6 & -4 & -7 \end{bmatrix}$.

Since $A^{-1} \begin{bmatrix} 1 \\ 0 \\ -1 \end{bmatrix} = \begin{bmatrix} 15 \\ 6 \\ 13 \end{bmatrix}$, the

solution set is $\{(15, 6, 13)\}$.

21. Corresponding system of equations is

$$\begin{aligned} x - y &= 12 \\ 35y - 10x &= 730. \end{aligned}$$

Solving this system, one finds $x = 46$ copies
were bought and $y = 34$ copies were sold.

22. Substitute $(0, 3)$, $(1, -1/2)$, and $(4, 3)$ into
$y = ax^2 + b\sqrt{x} + c$. Then we obtain

$$\begin{aligned} c &= 3 \\ a + b + c &= -\frac{1}{2} \\ 16a + 2b + c &= 3. \end{aligned}$$

Solving this system, one finds $a = 0.5$,
$b = -4$, $c = 3$, and the graph is
given by $y = 0.5x^2 - 4\sqrt{x} + 3$.

Tying It All Together

1. Simplify the left-hand side.

$$\begin{aligned} 2x + 6 - 5x &= 7 \\ -3x &= 1 \end{aligned}$$

Solution set is $\{-1/3\}$.

2. Multiply each side of the equation by 30.

$$\begin{aligned} \frac{1}{2}x - \frac{1}{6} &= \frac{4}{5} \\ 15x - 5 &= 24 \\ 15x &= 29 \end{aligned}$$

Solution set is $\{29/15\}$.

3. Multiply $\dfrac{1}{2}$ to $(2x - 2)$.

$$\begin{aligned} (x - 1)(6x - 8) &= 4 \\ 6x^2 - 14x + 8 &= 4 \\ 6x^2 - 14x + 4 &= 0 \\ 2(3x^2 - 7x + 2) &= 0 \\ 2(3x - 1)(x - 2) &= 0 \end{aligned}$$

Solution set is $\{1/3, 2\}$.

4. Simplify left-hand side.

$$\begin{aligned} 1 - (4x - 2) &= 9 \\ -4x + 3 &= 9 \\ -4x &= 6 \end{aligned}$$

Solution set is $\{-3/2\}$.

5. The solution set is $\{(4, -2)\}$ as can be seen from
the point of intersection.

6. Using $y = 1 - 2x$ in $2x + 6y = 2$, we get

$$\begin{aligned} 2x + 6(1 - 2x) &= 2 \\ 2x + 6 - 12x &= 2 \\ -10x &= -4. \end{aligned}$$

Using $x = \dfrac{2}{5}$ in $y = 1 - 2x$, we have

$y = 1 - \dfrac{4}{5} = \dfrac{1}{5}$. Solution set is $\left\{ \left(\dfrac{2}{5}, \dfrac{1}{5} \right) \right\}$.

7. Multiply second equation by 6 and add to the first one.

$$\begin{aligned} 2x - 0.06y &= 20 \\ \underline{18x + 0.06y} &= \underline{120} \\ 20x &= 140 \end{aligned}$$

Using $x = 7$ in $2x - 0.06y = 20$, we find

$$\begin{aligned} 14 - 0.06y &= 20 \\ -0.06y &= 6 \\ y &= -100. \end{aligned}$$

The solution set is $\{(7, -100)\}$.

8.

On $\begin{bmatrix} 2 & -1 & -1 \\ 1 & 3 & -11 \end{bmatrix}$, use $-2R_2 + R_1 \to R_2$

to get

$\begin{bmatrix} 2 & -1 & -1 \\ 0 & -7 & 21 \end{bmatrix}$, use $-\dfrac{1}{7}R_2 \to R_2$

and $\dfrac{1}{2}R_1 \to R_1$ to get

$\begin{bmatrix} 1 & -1/2 & -1/2 \\ 0 & 1 & -3 \end{bmatrix}$, use $\dfrac{1}{2}R_2 + R_1 \to R_1$

to get $\begin{bmatrix} 1 & 0 & -2 \\ 0 & 1 & -3 \end{bmatrix}$.

Solution set is $\{(-2, -3)\}$.

9. Inverse of the coefficient matrix A is

$$A^{-1} = \begin{bmatrix} -1/2 & -5/2 \\ -1/2 & -3/2 \end{bmatrix}.$$

Since $A^{-1} \begin{bmatrix} -7 \\ 1 \end{bmatrix} = \begin{bmatrix} 1 \\ 2 \end{bmatrix}$,

the solution set is $\{(1, 2)\}$.

10.

One finds $D = \begin{vmatrix} 4 & -3 \\ 3 & -5 \end{vmatrix} = -11$,

$D_x = \begin{vmatrix} 5 & -3 \\ 1 & -5 \end{vmatrix} = -22$, and

$D_y = \begin{vmatrix} 4 & 5 \\ 3 & 1 \end{vmatrix} = -11$.

Then $\dfrac{D_x}{D} = 2$ and $\dfrac{D_y}{D} = 1$.

The solution set is $\{(2, 1)\}$.

11. Note, $6 = 3x - 5y$ and $1 = 3x - 5y$. So, we get $6 = 1$, a false statement. Thus, the solution set is $\emptyset$.

12. Multiply the first equation by 2 and add this to the second equation.

$$\begin{aligned} 2x - 4y &= 6 \\ \underline{-2x + 4y} &= \underline{-6} \\ 0 &= 0 \end{aligned}$$

A dependent system and the solution set is

$$\{(x, y) \mid x - 2y = 3\}.$$

13. Substitute $x = y - 1$ in $x^2 + y^2 = 25$. Then

$$\begin{aligned} (y^2 - 2y + 1) + y^2 &= 25 \\ 2y^2 - 2y - 24 &= 0 \\ 2(y^2 - y - 12) &= 0 \\ 2(y - 4)(y + 3) &= 0. \end{aligned}$$

If $y = 4$, then $x = y - 1 = 4 - 1 = 3$.
If $y = -3$, then $x = (-3) - 1 = -4$.
The solution set is $\{(3, 4), (-4, -3)\}$.

14. Substituting $y = x^2 - 1$ in $x + y = 1$, we get

$$\begin{aligned} x + (x^2 - 1) &= 1 \\ x^2 + x - 2 &= 0 \\ (x + 2)(x - 1) &= 0. \end{aligned}$$

If $x = -2$, then $y = x^2 - 1 = (-2)^2 - 1 = 3$.
If $x = 1$, then $y = 1^2 - 1 = 0$.
Solution set is $\{(-2, 3), (1, 0)\}$.

15. Adding the given equations, we obtain

$$\begin{aligned} x^2 - y^2 &= 1 \\ \underline{x^2 + y^2} &= \underline{3} \\ 2x^2 &= 4 \\ x^2 &= 2 \\ x &= \pm\sqrt{2}. \end{aligned}$$

Substitute $x = \pm\sqrt{2}$ into $x^2 + y^2 = 3$.
Then $2 + y^2 = 3$ and so $y = \pm 1$. The solution set is $\{(\sqrt{2}, \pm 1), (-\sqrt{2}, \pm 1)\}$.

16. Multiply the first equation by -2 and add it to the second equation.

$$
\begin{array}{rcr}
-2x^2 + 2y^2 &=& -2 \\
2x^2 + 3y^2 &=& 2 \\
\hline
5y^2 &=& 0 \\
y &=& 0
\end{array}
$$

Substitute $y = 0$ into $x^2 - y^2 = 1$. Then $x^2 = 1$ and so $x = \pm 1$. The solution set is $\{(\pm 1, 0)\}$.

For Thought

1. False, vertex is $(0, -1/2)$.   2.   True

3. True, since $p = 3/2$ and the focus is $(4, 5)$, vertex is $(4 - 3/2, 5) = (5/2, 5)$.

4. False, focus is at $(0, 1/4)$ since parabola opens upward.   5.   True, $p = 1/4$.

6. False. Since $p = 4$ and the vertex is $(2, -1)$, equation of parabola is $y = \frac{1}{16}(x - 2)^2 - 1$ and x-intercepts are $(6, 0)$, $(-2, 0)$.

7. False, if $x = 0$ then $y = 0$ and y-intercept is $(0, 0)$.   8.   True

9. False. Since $p = 1/4$ and the vertex is $(5, 4)$, the focus is $(5, 4 + 1/4) = (-5, 17/4)$.

10. False, it opens to the left.

7.1 Exercises

1. Note $p = 1$.

   Vertex $(0, 0)$, focus $(0, 1)$, and directrix $y = -1$

3. Note $p = -1/2$. Vertex $(1, 2)$, focus $(1, 3/2)$, and directrix $y = 5/2$

5. Note $p = 3/4$. Vertex $(3, 1)$, focus $(15/4, 1)$, and directrix $x = 9/4$

7. Since the vertex is equidistant from $y = -2$ and $(0, 2)$, the vertex is $(0, 0)$ and $p = 2$. Then
$$a = \frac{1}{4p} = \frac{1}{8} \text{ and an equation is } y = \frac{1}{8}x^2.$$

9. Since the vertex is equidistant from $y = 3$ and $(0, -3)$, the vertex is $(0, 0)$ and $p = -3$. Then
$$a = \frac{1}{4p} = -\frac{1}{12} \text{ and an equation is } y = -\frac{1}{12}x^2.$$

11. One finds $p = \frac{3}{2}$. So $a = \frac{1}{4p} = \frac{1}{6}$ and

   vertex is $\left(3, 5 - \frac{3}{2}\right) = \left(3, \frac{7}{2}\right)$. Parabola

   is given by $y = \frac{1}{6}(x - 3)^2 + \frac{7}{2}$.

13. One finds $p = -\frac{5}{2}$. So $a = \frac{1}{4p} = -\frac{1}{10}$ and

   vertex is $\left(1, -3 + \frac{5}{2}\right) = \left(1, -\frac{1}{2}\right)$. Parabola

   is given by $y = -\frac{1}{10}(x - 1)^2 - \frac{1}{2}$.

15. One finds $p = 0.2$. So $a = \frac{1}{4p} = 1.25$ and

   vertex is $(-2, 1.2 - 0.2) = (-2, 1)$. Parabola

   is given by $y = 1.25(x + 2)^2 + 1$.

17. Since $p = 1$, we get $a = \frac{1}{4p} = \frac{1}{4}$.

   An equation is $y = \frac{1}{4}x^2$.

19. Since $p = -\frac{1}{4}$, we get $a = \frac{1}{4p} = -1$.

   An equation is $y = -x^2$

21. Note, vertex is $(1, 0)$ and since $a = \frac{1}{4p} = 1$,

   we find $p = \frac{1}{4}$.   The focus is $(1, 0 + p) =$

   $\left(1, \frac{1}{4}\right)$ and the directrix is $y = 0 - p$ or

   $y = -\frac{1}{4}$.

23. Note, vertex is $(3, 0)$ and since $a = \frac{1}{4p} = \frac{1}{4}$,

   we find $p = 1$. The focus is $(3, 0 + p) = (3, 1)$

   and the directrix is $y = 0 - p$ or $y = -1$.

25. Note, vertex is $(3, 4)$ and since $a = \frac{1}{4p} = -2$,

   we find $p = -\frac{1}{8}$.   The focus is $(3, 4 + p) =$

   $\left(3, \frac{31}{8}\right)$ and the directrix is $y = 4 - p$ or

   $y = \frac{33}{8}$.

27. Completing the square, we obtain
$$y = (x^2 - 8x + 16) - 16 + 3$$
$$y = (x - 4)^2 - 13.$$

   Since $\frac{1}{4p} = 1$, $p = 0.25$. Since vertex is

   $(4, -13)$, focus is $(4, -13 + 0.25) = (4, -12.75)$, and directrix is $y = -13 - p = -13.25$.

29. Completing the square, we obtain

$$y = 2(x^2 + 6x + 9) + 5 - 18$$
$$y = 2(x + 3)^2 - 13.$$

Since $\dfrac{1}{4p} = 2$, $p = 1/8$.

Since vertex is $(-3, -13)$, the focus is $(-3, -13 + 1/8) = (-3, -103/8)$, and directrix is $y = -13 - 1/8 = -105/8$.

31. Completing the square, we get

$$y = -2(x^2 - 3x + 9/4) + 1 + 9/2$$
$$y = -2(x - 3/2)^2 + 11/2.$$

Since $\dfrac{1}{4p} = -2$, $p = -0.125$.

Since vertex is $(1.5, 5.5)$, the focus is $(1.5, 5.5 - 0.125) = (1.5, 5.375)$, and directrix is $y = 5.5 + 0.125 = 5.625$.

33. Completing the square,

$$y = 5(x^2 + 6x + 9) - 45$$
$$y = 5(x + 3)^2 - 45.$$

Since $\dfrac{1}{4p} = 5$, we have $p = 0.05$.

Since vertex is $(-3, -45)$, the focus is $(-3, -45 + 0.05) = (-3, -44.95)$, and directrix is $y = -45 - .05 = -45.05$.

35. Completing the square, we get

$$y = \frac{1}{8}(x^2 - 4x + 4) + \frac{9}{2} - \frac{1}{2}$$
$$y = \frac{1}{8}(x - 2)^2 + 4.$$

Since $\dfrac{1}{4p} = 1/8$, we find $p = 2$.

Since vertex is $(2, 4)$, the focus is $(2, 4 + 2) = (2, 6)$, and directrix is $y = 4 - 2$ or $y = 2$.

37. Note, $a = 1$, $b = -4$. So $x = \dfrac{-b}{2a} = \dfrac{4}{2} = 2$ and since $\dfrac{1}{4p} = a = 1$, $p = 1/4$.

Substituting $x = 2$, $y = 2^2 - 4(2) + 3 = -1$.

Vertex is $(2, -1)$ and focus is $(2, -1 + p) = (2, -3/4)$, directrix is $y = -1 - p = -5/4$, and parabola opens up since $a > 0$.

39. Note, $a = -1$, $b = 2$. So $x = \dfrac{-b}{2a} = \dfrac{-2}{-2} = 1$ and since $\dfrac{1}{4p} = a = -1$, $p = -1/4$.

Substituting $x = 1$, $y = -(1)^2 + 2(1) - 5 = -4$.

Vertex is $(1, -4)$ and focus is $(1, -4 + p) = (1, -17/4)$, directrix is $y = -4 - p = -15/4$, and parabola opens down since $a < 0$.

41. Note, $a = 3$, $b = -6$. So $x = \dfrac{-b}{2a} = \dfrac{6}{6} = 1$ and since $\dfrac{1}{4p} = a = 3$, $p = 1/12$. Substituting $x = 1$, $y = 3(1)^2 - 6(1) + 1 = -2$.

Vertex is $(1, -2)$ and focus is $(1, -2 + p) = (1, -23/12)$, directrix is $y = -2 - p = -25/12$, and parabola opens up since $a > 0$.

43. Note, $a = -1/2$, $b = -3$. So $x = \dfrac{-b}{2a} = \dfrac{3}{-1} = -3$ and since $\dfrac{1}{4p} = a = -1/2$, $p = -1/2$. Substituting $x = -3$, we have $y = -\dfrac{1}{2}(-3)^2 - 3(-3) + 2 = \dfrac{13}{2}$. Vertex is $\left(-3, \dfrac{13}{2}\right)$ and focus is $\left(-3, \dfrac{13}{2} + p\right) = (-3, 6)$, directrix is $y = \dfrac{13}{2} - p = 7$, and parabola opens down since $a < 0$.

45. Note, $y = \frac{1}{4}x^2 + 5$ is of the form

$y = a(x-h)^2 + k$. So $h = 0$, $k = 5$, and

$\frac{1}{4p} = a = \frac{1}{4}$ from which we have $p = 1$. The

vertex is $(h, k) = (0, 5)$, focus is $(0, 5+p) = (0, 6)$, directrix is $y = 5 - p = 4$ and parabola opens up since $a > 0$.

47. From the given focus and directrix, one

finds $p = 1/4$. So $a = \frac{1}{4p} = 1$, vertex is

$(h, k) = (1/2, -2 - p) = (1/2, -9/4)$,

axis of symmetry is $x = 1/2$, and parabola

is given by $y = a(x-h)^2 + k = \left(x - \frac{1}{2}\right)^2 - \frac{9}{4}$.

If $y = 0$ then $x - \frac{1}{2} = \pm\frac{3}{2}$ or $x = 2, -1$.

The x-intercepts are $(2, 0), (-1, 0)$.

If $x = 0$ then $y = \left(0 - \frac{1}{2}\right)^2 - \frac{9}{4} = -2$ and

y-intercept is $(0, -2)$.

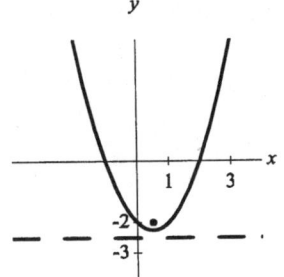

49. From the given focus and directrix one

finds $p = -1/4$. So $a = \frac{1}{4p} = -1$, vertex is

$(h, k) = (-1/2, 6 - p) = (-1/2, 25/4)$,

axis of symmetry is $x = -1/2$, and parabola is given by

$$y = a(x-h)^2 + k = -\left(x + \frac{1}{2}\right)^2 + \frac{25}{4}.$$

If $y = 0$ then $x + \frac{1}{2} = \pm\frac{5}{2}$ or $x = -3, 2$.

The x-intercepts are $(-3, 0), (2, 0)$.

If $x = 0$ then $y = -\left(0 + \frac{1}{2}\right)^2 + \frac{25}{4} = 6$

and y-intercept is $(0, 6)$.

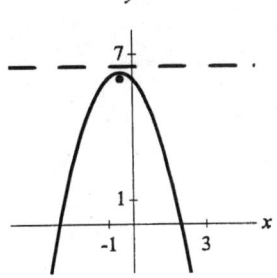

51. Since $\frac{1}{2}(x+2)^2 + 2$ is of the form $a(x-h)^2 + k$,

vertex is $(h, k) = (-2, 2)$ and axis of symmetry

is $x = -2$. If $y = 0$ then $0 = \frac{1}{2}(x+2)^2 + 2$; this

has no solution since left-hand side is always positive. No x-intercept. If $x = 0$ then

$y = \frac{1}{2}(0 + 2)^2 + 2 = 4$. y-intercept is $(0, 4)$.

Since $\frac{1}{4p} = a = \frac{1}{2}$, $p = \frac{1}{2}$, focus is $(h, k+p) = (-2, 5/2)$, and directrix is $y = k - p = 3/2$.

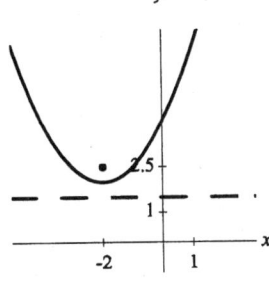

53. Since $-\frac{1}{4}(x+4)^2+2$ is of the form $a(x-h)^2+k$, vertex is $(h,k)=(-4,2)$ and axis of symmetry is $x=-4$. If $y=0$ then

$$\frac{1}{4}(x+4)^2 = 2$$
$$x+4 = \pm\sqrt{8}.$$

x-intercepts are $(-4\pm2\sqrt{2},0)$. If $x=0$, then $y=-\frac{1}{4}(0+4)^2+2=-2$. y-intercept is $(0,-2)$. Since $\frac{1}{4p}=a=-\frac{1}{4}$, $p=-1$, focus is $(h,k+p)=(-4,1)$, and directrix is $y=k-p=3$.

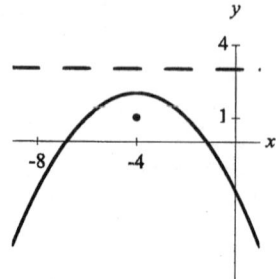

55. Since $\frac{1}{2}x^2-2$ is of the form $a(x-h)^2+k$, vertex is $(h,k)=(0,-2)$ and axis of symmetry is $x=0$. If $y=0$, then

$$\frac{1}{2}x^2 = 2$$
$$x^2 = 4.$$

x-intercepts are $(\pm2,0)$. If $x=0$ then $y=\frac{1}{2}(0)^2-2=-2$. y-intercept is $(0,-2)$.

Since $\frac{1}{4p}=a=\frac{1}{2}$, $p=1/2$, focus is $(h,k+p)=(0,-3/2)$, and directrix is $y=k-p=-5/2$.

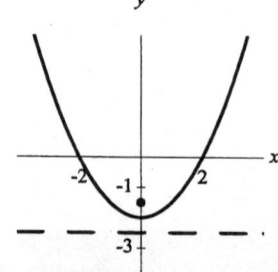

57. Since $y=(x-2)^2$ is of the form $a(x-h)^2+k$, vertex is $(h,k)=(2,0)$, and axis of symmetry is $x=2$. If $y=0$ then $(x-2)^2=0$ and x-intercept is $(2,0)$. If $x=0$ then $y=(0-2)^2=4$ and y-intercept is $(0,4)$. Since $\frac{1}{4p}=a=1$, $p=1/4$, focus is $(h,k+p)=(2,1/4)$, and directrix is $y=k-p=-1/4$.

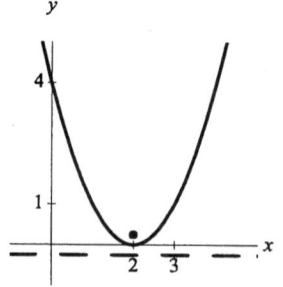

59. By completing the square, we obtain

$$y=\frac{1}{3}(x-3/2)^2-3/4.$$

Vertex is $(h,k)=(3/2,-3/4)$ and axis of symmetry is $x=3/2$. If $y=0$, then

$$\frac{1}{3}\left(x-\frac{3}{2}\right)^2 = \frac{3}{4}$$
$$\left(x-\frac{3}{2}\right)^2 = \frac{9}{4}$$
$$x = \frac{3}{2}\pm\frac{3}{2}.$$

x-intercepts are $(3,0),(0,0)$. If $x=0$, then $y=\frac{1}{3}\left(0-\frac{3}{2}\right)^2-\frac{3}{4}=0$ and y-intercept is $(0,0)$. Since $\frac{1}{4p}=a=1/3$, $p=3/4$, focus is $(h,k+p)=(3/2,0)$, and directrix is $y=k-p=-3/2$ or $y=-3/2$.

61. Since $x = -y^2$ is of the form $x = a(y-h)^2 + k$, vertex is $(k,h) = (0,0)$ and axis of symmetry is $y = 0$. If $y = 0$ then $x = -0^2 = 0$ and x-intercept is $(0,0)$. If $x = 0$ then $0 = -y^2$ and y-intercept is $(0,0)$. Since $\dfrac{1}{4p} = a = -1$, $p = -1/4$, focus is $(k+p,h) = (-1/4,0)$, and directrix is $x = k - p = 1/4$.

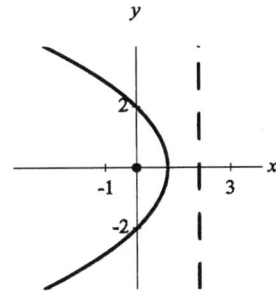

63. Since $x = -\dfrac{1}{4}y^2 + 1$ is of the form

$$x = a(y-h)^2 + k,$$

vertex is $(k,h) = (1,0)$ and axis of symmetry is $y = 0$. If $y = 0$ then $x = 1$ and x-intercept is $(1,0)$. If $x = 0$ then $\dfrac{1}{4}y^2 = 1$, $y^2 = 4$, and y-intercepts are $(0,\pm 2)$. Since $\dfrac{1}{4p} = a = -\dfrac{1}{4}$, $p = -1$, focus is $(k+p,h) = (0,0)$, and directrix is $x = k - p = 2$ or $x = 2$.

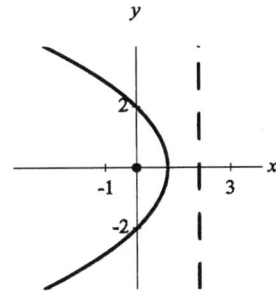

65. By completing the square, we obtain

$$x = (y + 1/2)^2 - 25/4.$$

Vertex is $(k,h) = (-25/4, -1/2)$ and axis of symmetry is $y = -1/2$. If $y = 0$, then $x = 1/4 - 25/4 = -6$. x-intercept is $(-6,0)$. If $x = 0$, then

$$\left(y + \frac{1}{2}\right)^2 = \frac{25}{4}$$

$$y = -\frac{1}{2} \pm \frac{5}{2}$$
$$y = 2, -3.$$

y-intercepts are $(0,2), (0,-3)$. Since $\dfrac{1}{4p} = a = 1$, $p = 1/4$, focus is $(k+p,h) = (-6,-1/2)$, and directrix is $x = k - p = -13/2$.

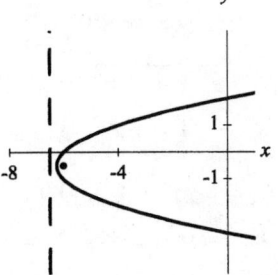

67. By completing the square, we get

$$x = -\frac{1}{2}(y+1)^2 - 7/2.$$

Vertex is $(k,h) = (-7/2,-1)$ and axis of symmetry is $y = -1$. If $y = 0$, then $x = -1/2 - 7/2 = -4$. x-intercept is $(-4,0)$. If $x = 0$, then

$$0 = -\frac{1}{2}(y+1)^2 - 7/2 < 0$$

which is inconsistent and so there is no y-intercept. Since $\dfrac{1}{4p} = a = -1/2$, $p = -1/2$, focus is $(k+p,h) = (-4,-1)$, and directrix is $x = k - p = -3$.

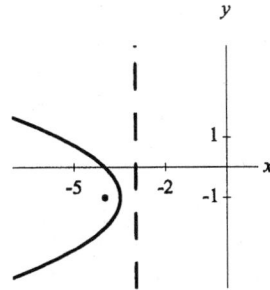

69. Since $x = 2(y-1)^2 + 3$ is of the form $a(y-h)^2 + k$, we find that the vertex is $(k,h) = (3,1)$ and axis of symmetry is $y = 1$. If $y = 0$ then $x = 2(-1)^2 + 3 = 5$ and x-intercept is $(5,0)$. If $x = 0$, we obtain $2(y-1)^2 + 3 = 0$ which is inconsistent since the left-hand side is always positive. No y-intercept.

Since $\dfrac{1}{4p} = a = 2$, $p = 1/8$, focus is $(k+p, h) =$ $(25/8, 1)$, and directrix is $x = k - p = 23/8$.

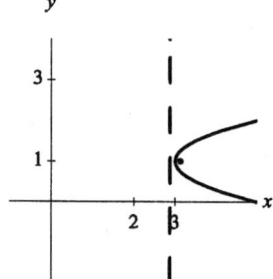

71. Since $x = -\dfrac{1}{2}(y + 2)^2 + 1$ is of the form $a(y - h)^2 + k$, vertex is $(k, h) = (1, -2)$ and axis of symmetry is $y = -2$. If $y = 0$, then $x = -\dfrac{1}{2}(2)^2 + 1 = -1$ and x-intercept is $(-1, 0)$. If $x = 0$, then

$$\frac{1}{2}(y + 2)^2 = 1$$
$$(y + 2)^2 = 2$$
$$y = -2 \pm \sqrt{2}.$$

y-intercepts are $(0, -2 \pm \sqrt{2})$. Since $\dfrac{1}{4p} = a = -\dfrac{1}{2}$, $p = -\dfrac{1}{2}$, focus is $(k + p, h) = (1/2, -2)$, and directrix is $x = k - p = 3/2$.

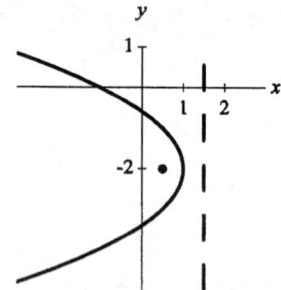

73. Since focus is 1 unit above the vertex $(1, 4)$, we obtain $p = 1$. Then $a = \dfrac{1}{4p} = \dfrac{1}{4}$ and parabola is given by $y = \dfrac{1}{4}(x - 1)^2 + 4$.

75. Since vertex $(0, 0)$ is 2 units to the right of the directrix, we find $p = 2$ and parabola opens to the right. Then $a = \dfrac{1}{4p} = \dfrac{1}{8}$ and the parabola is given by $x = \dfrac{1}{8}y^2$.

77. Since the parabola opens up, $p = 55(12)$ inches, and the vertex is $(0, 0)$. Note, $\dfrac{1}{4p} = \dfrac{1}{2640}$. Thus, the parabola is given by $y = \dfrac{1}{2640}x^2$. Thickness at the outside edge is $23 + \dfrac{1}{2640}(100)^2 \approx 26.8$ in.

79. $y = x^2$ has vertex $(0, 0)$ and opens up. The second parabola can be written as

$$y = 2(x - 1)^2 + 3$$

and its vertex is $(1, 3)$. In the given viewing window these graphs look alike.

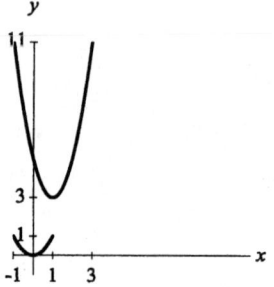

81. Two functions are $f_1(x) = \sqrt{-x}$ and $f_2(x) = -\sqrt{-x}$ where $x \leq 0$.

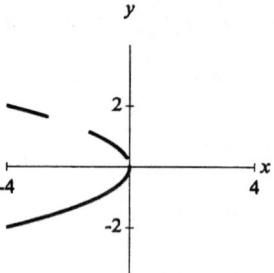

For Thought

1. False, y-intercepts are $(\pm 3, 0)$.

2. True, since it can be written as $\dfrac{x^2}{1/2} + y^2 = 1$.

3. True, length of the major axis is $2a = 2(5) = 10$.

4. True, if $y = 0$ then $x^2 = \dfrac{1}{0.5} = 2$ and $x = \pm\sqrt{2}$.

5. True, if $x = 0$ then $y^2 = 3$ and $y = \pm\sqrt{3}$.

6. False, the center is not a point on the circle.

7. True   **8.**   False, $(3, -1)$ satisfies equation.

9. False. No point satisfies the equation since the left-hand side is always positive.

10. False, since the circle can be written as $(x-2)^2 + (y+1/2)^2 = 53/4$ and so the radius is $\sqrt{53}/2$.

7.2 Exercises

1. Foci $(\pm\sqrt{5}, 0)$, vertices $(\pm 3, 0)$, center $(0, 0)$

3. Foci $(2, 1 \pm \sqrt{5})$, vertices $(2, 4)$ and $(2, -2)$, center $(2, 1)$

5. Since $c = 2$ and $b = 3$, we get
$$a^2 = b^2 + c^2 = 9 + 4 = 13 \text{ and } a = \sqrt{13}.$$
Ellipse is given by $\dfrac{x^2}{13} + \dfrac{y^2}{9} = 1$.

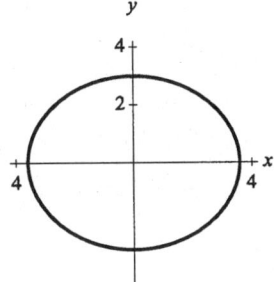

7. Since $c = 4$ and $a = 5$, we find
$$b^2 = a^2 - c^2 = 25 - 16 = 9 \text{ and } b = 3.$$
Ellipse is given by $\dfrac{x^2}{25} + \dfrac{y^2}{9} = 1$.

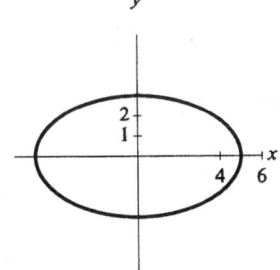

9. Since $c = 2$ and $b = 2$, we obtain
$$a^2 = b^2 + c^2 = 4 + 4 = 8 \text{ and } a = \sqrt{8}.$$
Ellipse is given by $\dfrac{x^2}{4} + \dfrac{y^2}{8} = 1$.

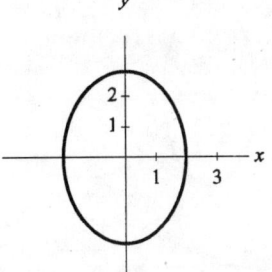

11. Since $c = 4$ and $a = 7$, we obtain
$$b^2 = a^2 - c^2 = 49 - 16 = 33 \text{ and } b = \sqrt{33}.$$
Ellipse is given by $\dfrac{x^2}{33} + \dfrac{y^2}{49} = 1$.

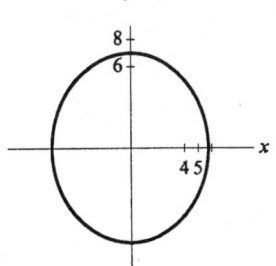

13. Since $c = \sqrt{a^2 - b^2} = \sqrt{16 - 4} = 2\sqrt{3}$, the foci are $(\pm 2\sqrt{3}, 0)$

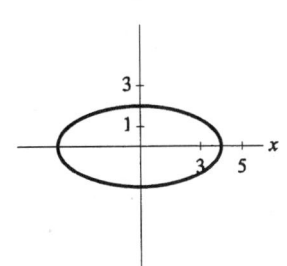

15. Since $c = \sqrt{a^2 - b^2} = \sqrt{36 - 9} = \sqrt{27}$, the foci are $(0, \pm 3\sqrt{3})$

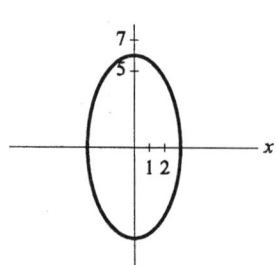

17. Since $c = \sqrt{a^2 - b^2} = \sqrt{25 - 1} = \sqrt{24}$, the foci are $(\pm 2\sqrt{6}, 0)$

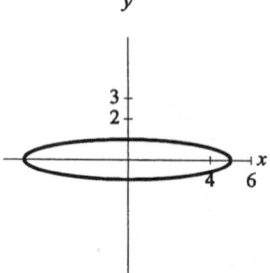

19. Since $c = \sqrt{a^2 - b^2} = \sqrt{25 - 9} = \sqrt{16}$, the foci are $(0, \pm 4)$

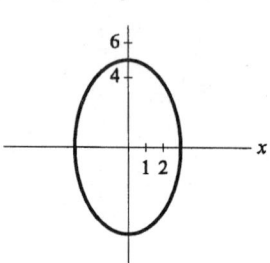

21. From $x^2 + \dfrac{y^2}{9} = 1$, one finds $c = \sqrt{a^2 - b^2}$ $= \sqrt{9 - 1} = \sqrt{8}$ and foci are $(0, \pm 2\sqrt{2})$.

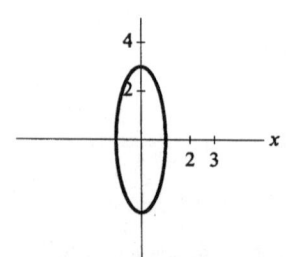

23. From $\dfrac{x^2}{9} + \dfrac{y^2}{4} = 1$, one finds $c = \sqrt{a^2 - b^2}$ $= \sqrt{9 - 4} = \sqrt{5}$ and foci are $(\pm\sqrt{5}, 0)$.

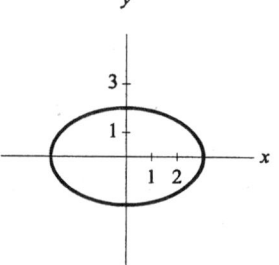

25. Since $c = \sqrt{a^2 - b^2} = \sqrt{16 - 9} = \sqrt{7}$, the foci are $(1 \pm c, -3) = (1 \pm \sqrt{7}, -3)$.

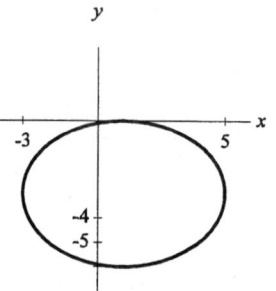

27. Since $c = \sqrt{a^2 - b^2} = \sqrt{25 - 9} = \sqrt{16} = 4$, the foci are $(3, -2 \pm c)$, or $(3, 2)$ and $(3, -6)$.

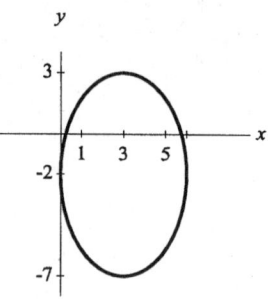

29. Since $\dfrac{(x + 4)^2}{36} + (y + 3)^2 = 1$, $c = \sqrt{a^2 - b^2}$ $= \sqrt{36 - 1} = \sqrt{35}$ and foci are $(-4 \pm \sqrt{35}, -3)$.

31. If one applies the method of completing the square, one obtains

$$9(x^2 - 2x + 1) + 4(y^2 + 4y + 4) = 11 + 9 + 16$$
$$9(x-1)^2 + 4(y+2)^2 = 36$$
$$\frac{(x-1)^2}{4} + \frac{(y+2)^2}{9} = 1.$$

From $a^2 = b^2 + c^2$ with $a = 3$ and $b = 2$, one finds $c = \sqrt{5}$. The foci are $(1, -2 \pm \sqrt{5})$ and a sketch of the ellipse is given.

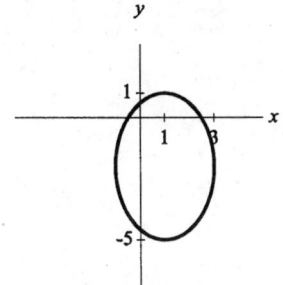

33. Since $\dfrac{x^2}{16} + \dfrac{y^2}{4} = 1$, we get $c = \sqrt{a^2 - b^2}$
$= \sqrt{16 - 4} = \sqrt{12}$, and the foci are $(\pm 2\sqrt{3}, 0)$.

35. Since $\dfrac{(x+1)^2}{4} + \dfrac{(y+2)^2}{16} = 1$, we get
$c = \sqrt{a^2 - b^2} = \sqrt{16 - 4} = \sqrt{12}$, and the foci
are $(-1, -2 \pm 2\sqrt{3})$.

37. $x^2 + y^2 = 4$

39. Since $r = \sqrt{(4-0)^2 + (5-0)^2} = \sqrt{41}$,
the circle is given by $x^2 + y^2 = 41$.

41. Since $r = \sqrt{(4-2)^2 + (1+3)^2} = \sqrt{20}$,
the circle is given by $(x-2)^2 + (y+3)^2 = 20$.

43. Since center is $\left(\dfrac{3-1}{2}, \dfrac{4+2}{2}\right) = (1, 3)$ and
$r = \sqrt{(3-1)^2 + (4-3)^2} = \sqrt{5}$, the circle
is given by $(x-1)^2 + (y-3)^2 = 5$.

45. center $(0, 0)$, radius 10

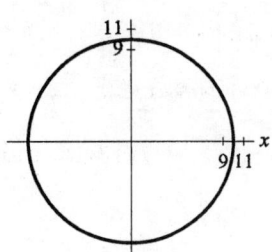

47. center $(1, 2)$, radius 2

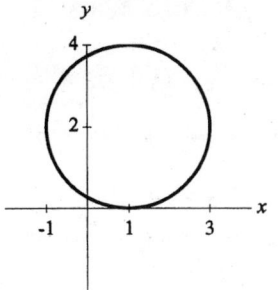

49. center $(-2, -2)$, radius $\sqrt{8}$ or $2\sqrt{2}$

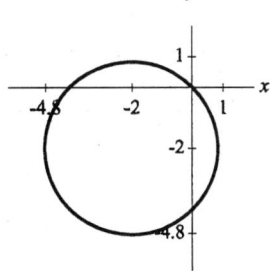

51. Completing the square, we obtain

$$x^2 + (y^2 + 2y + 1) = 8 + 1$$
$$x^2 + (y+1)^2 = 9.$$

The center is $(0, -1)$ with radius 3.

53. Completing the square, we find

$$x^2 + 8x + 16 + y^2 - 10y + 25 = 16 + 25$$
$$(x+4)^2 + (y-5)^2 = 41.$$

The center is $(-4, 5)$ with radius $\sqrt{41}$.

55. Completing the square, we find

$$(x^2 + 4x + 4) + y^2 = 5 + 4$$
$$(x+2)^2 + y^2 = 9.$$

The center is $(-2, 0)$ with radius 3.

57. Completing the square, we have

$$x^2 - x + \frac{1}{4} + y^2 + y + \frac{1}{4} = \frac{1}{2} + \frac{1}{4} + \frac{1}{4}$$

$$(x - 1/2)^2 + (y + 1/2)^2 = 1.$$

The center is $(1/2, -1/2)$ with radius 1.

59. Completing the square, we get

$$x^2 + \frac{2}{3}x + \frac{1}{9} + y^2 + \frac{1}{3}y + \frac{1}{36} = \frac{1}{9} + \frac{1}{9} + \frac{1}{36}$$

$$(x + 1/3)^2 + (y + 1/6)^2 = 1/4.$$

The center is $(-1/3, -1/6)$ with radius $1/2$.

61. Divide equation by 2 and complete the square.

$$x^2 + 2x + y^2 = 1/2$$

$$(x^2 + 2x + 1) + y^2 = 1/2 + 1$$

$$(x + 1)^2 + y^2 = 3/2$$

The center is $(-1, 0)$ with radius $\sqrt{\frac{3}{2}}$ or $\frac{\sqrt{6}}{2}$.

63. Completing the square, we get

$$y^2 - y + x^2 = 0$$

$$y^2 - y + \frac{1}{4} + x^2 = \frac{1}{4}$$

$$(y - 1/2)^2 + x^2 = \frac{1}{4},$$

and one identifies a circle.

65. Divide equation by 4.

$$x^2 + 3y^2 = 1$$

$$x^2 + \frac{y^2}{1/3} = 1$$

This is an ellipse.

67. Solve for y and complete the square.

$$y = -2x^2 - 4x + 4$$

$$y = -2(x^2 + 2x) + 4$$

$$y = -2(x^2 + 2x + 1) + 4 + 2$$

$$y = -2(x + 1)^2 + 6$$

One identifies a parabola.

69. Note, $(y - 2)^2 = (2 - y)^2$. Then we solve for x.

$$2 - x = (y - 2)^2$$

$$x = -(y - 2)^2 + 2$$

This is a parabola.

71. Simplify and note $(x - 4)^2 = (4 - x)^2$.

$$2(x - 4)^2 = 4 - y^2$$

$$2(x - 4)^2 + y^2 = 4$$

$$\frac{(x - 4)^2}{2} + \frac{y^2}{4} = 1$$

One identifies an ellipse.

73. Divide equation by 9 to get the circle given by $x^2 + y^2 = \frac{1}{9}$.

75. From the foci, we obtain $c = 2$ and $a^2 = b^2 + c^2 = b^2 + 4$. Equation of the ellipse is of the form $\frac{x^2}{b^2 + 4} + \frac{y^2}{b^2} = 1$. Substitute $x = 2$ and $y = 3$.

$$\frac{4}{b^2 + 4} + \frac{9}{b^2} = 1$$

$$4b^2 + 9(b^2 + 4) = b^2(b^2 + 4)$$

$$0 = b^4 - 9b^2 - 36$$

$$0 = (b^2 - 12)(b^2 + 3)$$

So $b^2 = 12$ and $a^2 = 12 + 4 = 16$.

Ellipse is given by $\frac{x^2}{16} + \frac{y^2}{12} = 1$.

77. If c is the distance between the center and focus $(0, 0)$ then the other focus is $(2c, 0)$. Since the distance between the x-intercepts is $6 + 2c$, which is also the length of the major axis, then $6 + 2c = 2a$. Since $2a$ is the sum of the distances of $(0, 5)$ from the foci,

$$5 + \sqrt{25 + 4c^2} = 2a$$

$$5 + \sqrt{25 + 4c^2} = 6 + 2c$$

$$\sqrt{25 + 4c^2} = 1 + 2c$$

$$25 + 4c^2 = 1 + 4c + 4c^2$$

$$24 = 4c$$

$$6 = c.$$

The other focus is $(2c, 0) = (12, 0)$.

79. Since the sun is a focus of the elliptical orbit, the length of the major axis is $2a = 521$ (the sum of the shortest distance, $P = 1$ AU, and longest distance, $A = 520$ AU, between the orbit and the sun, respectively). In addition, $c = 259.5$ AU (which is the distance from the center to a focus). The eccentricity is given by $e = \dfrac{c}{a} = \dfrac{259.5}{260.5} \approx .996$. Orbit's equation is

$$\frac{x^2}{260.5^2} + \frac{y^2}{520} = 1.$$

81. If $2a$ is the sum of the distances from Haley's comet to the two foci and c is the distance from the sun to the center of the ellipse then, $c = a - 8 \times 10^7$.

Since $0.97 = c/a$, we get

$$0.97 = \frac{a - 8 \times 10^7}{a}$$

and the solution of this equation is $a \approx 2.667 \times 10^9$. So $c = a(0.97) \approx 2.587 \times 10^9$. The maximum distance from the sun is $c + a \approx 5.25 \times 10^9$ km.

83. Solving for y, one finds

$$\begin{aligned} y^2 &= 6360^2 - x^2 \\ y &= \pm\sqrt{6360^2 - x^2}. \end{aligned}$$

A sketch of the circle is given.

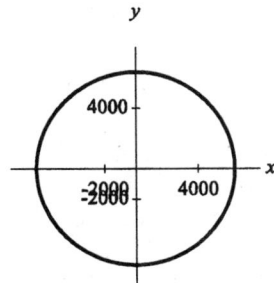

85.

a) As derived below, an equation of the tangent line is given by

$$\frac{x_1 x}{a^2} + \frac{y_1 y}{b^2} = 1$$

$$\frac{-4x}{25} + \frac{\frac{9}{5}y}{9} = 1$$

$$\frac{1}{5}y = 1 + \frac{4x}{25}$$

$$y = \frac{4x}{5} + 5.$$

b) The tangent line and the ellipse intersect at the point $\left(-4, \dfrac{9}{5}\right)$ as shown.

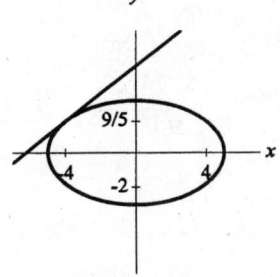

87. A parabolic reflector is preferable since otherwise one would have to place the moving quarterback on one focus of the ellipse.

89. If (x, y) is any point on the ellipse and the sum of the distances from the two foci $(0, \pm c)$ is $2a$, then

$$\sqrt{x^2 + (y - c)^2} + \sqrt{x^2 + (y + c)^2} = 2a.$$

Simplify the inside of the radical and transpose a term to the right-hand side.

$$\sqrt{x^2 + y^2 - 2yc + c^2} = 2a - \sqrt{x^2 + y^2 + 2cy + c^2}$$

Squaring both sides, we find $x^2 + y^2 - 2yc + c^2 = 4a^2 - 4a\sqrt{x^2 + y^2 + 2cy + c^2} + x^2 + y^2 + 2cy + c^2$. Cancel like terms and simplify.

$$\begin{aligned} -4yc - 4a^2 &= -4a\sqrt{x^2 + y^2 + 2cy + c^2} \\ yc + a^2 &= a\sqrt{x^2 + y^2 + 2cy + c^2} \\ y^2c^2 + 2yca^2 + a^4 &= a^2(x^2 + y^2 + 2cy + c^2) \\ y^2c^2 + a^4 &= a^2(x^2 + y^2 + c^2) \\ a^4 - a^2c^2 &= a^2x^2 + y^2(a^2 - c^2) \end{aligned}$$

Let $b^2 = a^2 - c^2$. So $a^2b^2 = a^2x^2 + b^2y^2$ and $\dfrac{x^2}{b^2} + \dfrac{y^2}{a^2} = 1$ is an ellipse with foci $(0, \pm c)$ and one finds that the x-intercepts are $(\pm b, 0)$.

For Thought

1. False, it is a parabola.

2. False, there is no y-intercept.

3. True   **4.**   True

5. False, $y = \dfrac{b}{a}x$ is an asymptote.

6. True   **7.**   True, $c = \sqrt{16 + 9} = 5$.

8. True, $c = \sqrt{3 + 5} = \sqrt{8}$.

9. False, $y = \dfrac{2}{3}x$ is an asymptote.

10. False, it is a circle centered at $(0,0)$.

7.3 Exercises

1. Vertices $(\pm 1, 0)$, foci $(\pm \sqrt{2}, 0)$,
asymptotes $y = \pm x$

3. Vertices $(1, \pm 3)$, foci $(1, \pm \sqrt{10})$

Since the slope of the asymptotes are ± 3, the equations of the asymptotes are of the form $y = \pm 3x + b$. If we substitute $(1, 0)$ into $y = \pm 3x + b$, then $0 = \pm 3(1) + b$ or $b = \mp 3$. Thus, the asymptotes are $y = 3x - 3$ and $y = -3x + 3$.

5. Note, $c = \sqrt{a^2 + b^2} = \sqrt{2^2 + 3^2} = \sqrt{13}$.

Foci $(\pm \sqrt{13}, 0)$, asymptotes $y = \pm \dfrac{b}{a}x = \pm \dfrac{3}{2}x$

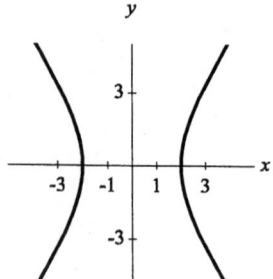

7. Note, $c = \sqrt{a^2 + b^2} = \sqrt{2^2 + 5^2} = \sqrt{29}$.

Foci $(0, \pm \sqrt{29})$, asymptotes $y = \pm \dfrac{a}{b}x = \pm \dfrac{2}{5}x$

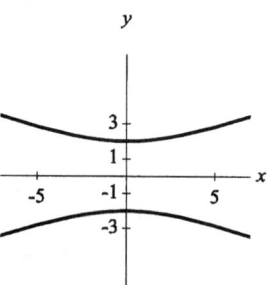

9. Note, $c = \sqrt{a^2 + b^2} = \sqrt{2^2 + 1^2} = \sqrt{5}$.

Foci $(\pm \sqrt{5}, 0)$, asymptotes $y = \pm \dfrac{b}{a}x = \pm \dfrac{1}{2}x$

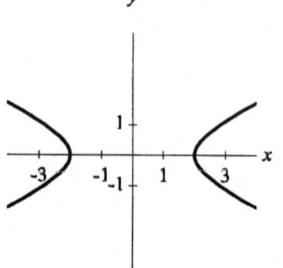

11. Note, $c = \sqrt{a^2 + b^2} = \sqrt{1^2 + 3^2} = \sqrt{10}$.

Foci $(\pm \sqrt{10}, 0)$, asymptotes $y = \pm \dfrac{b}{a}x = \pm 3x$

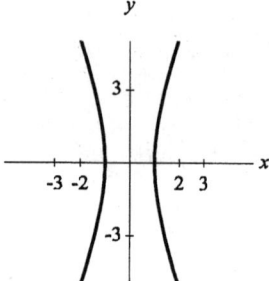

13. Dividing by 144, we get $\dfrac{x^2}{9} - \dfrac{y^2}{16} = 1$.

Note, $c = \sqrt{a^2 + b^2} = \sqrt{3^2 + 4^2} = 5$.

Foci $(\pm 5, 0)$, asymptotes $y = \pm \dfrac{b}{a}x = \pm \dfrac{4}{3}x$

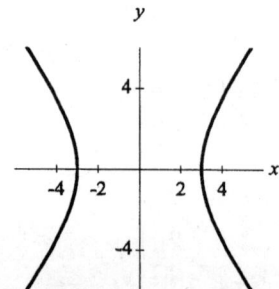

15. Note, $c = \sqrt{a^2 + b^2} = \sqrt{1^2 + 1^2} = \sqrt{2}$.

Foci $(\pm\sqrt{2}, 0)$, asymptotes $y = \pm\dfrac{b}{a}x = \pm x$

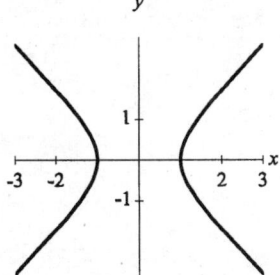

17. Note, $c = \sqrt{a^2 + b^2} = \sqrt{2^2 + 3^2} = \sqrt{13}$.
Since the center is $(-1, 2)$, we find that the foci are $(-1 \pm \sqrt{13}, 2)$. Solving for y in $y - 2 = \pm\dfrac{3}{2}(x + 1)$, we obtain that the asymptotes are $y = \dfrac{3}{2}x + \dfrac{7}{2}$ and $y = -\dfrac{3}{2}x + \dfrac{1}{2}$.

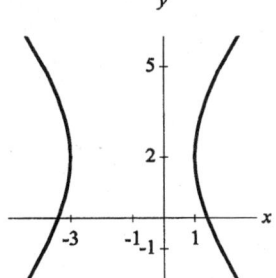

19. Note, $c = \sqrt{a^2 + b^2} = \sqrt{2^2 + 1^2} = \sqrt{5}$.
Since the center is $(-2, 1)$, the foci are $(-2, 1 \pm \sqrt{5})$. Solving for y in $y - 1 = \pm 2(x + 2)$, we obtain that the asymptotes are $y = 2x + 5$ and $y = -2x - 3$.

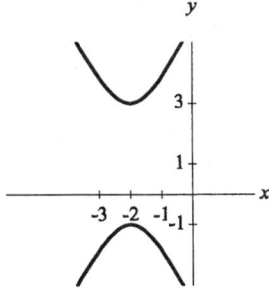

21. Note, $c = \sqrt{a^2 + b^2} = \sqrt{4^2 + 3^2} = 5$.
Since the center is $(-2, 3)$, the foci are $(3, 3)$ and $(-7, 3)$. Solving for y in

$y - 3 = \pm\dfrac{3}{4}(x + 2)$, we find that the asymptotes are $y = \dfrac{3}{4}x + \dfrac{9}{2}$ and $y = -\dfrac{3}{4}x + \dfrac{3}{2}$.

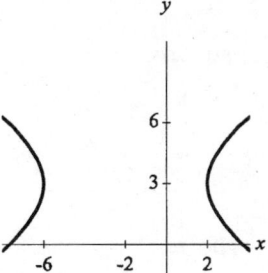

23. Note, $c = \sqrt{a^2 + b^2} = \sqrt{1^2 + 1^2} = \sqrt{2}$.
Since the center is $(3, 3)$, the foci are $(3, 3 \pm \sqrt{2})$. Solving for y in $y - 3 = \pm(x - 3)$, we obtain that the asymptotes are $y = x$ and $y = -x + 6$.

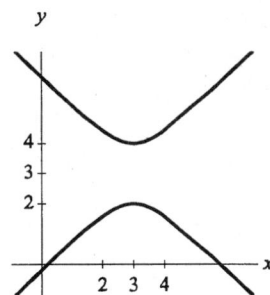

25. Since the x-intercepts are $(\pm 6, 0)$, the hyperbola is given by $\dfrac{x^2}{6^2} - \dfrac{y^2}{b^2} = 1$.
From the asymptotes one gets $\dfrac{1}{2} = \dfrac{b}{6}$ and $b = 3$. Equation is $\dfrac{x^2}{36} - \dfrac{y^2}{9} = 1$.

27. Since the x-intercepts are $(\pm 3, 0)$, the hyperbola is given by $\dfrac{x^2}{3^2} - \dfrac{y^2}{b^2} = 1$.
From the foci, $c = 5$ and $b^2 = c^2 - a^2 = 5^2 - 3^2 = 16$. Equation is $\dfrac{x^2}{9} - \dfrac{y^2}{16} = 1$.

29. By using the vertices of the fundamental rectangle and since it opens sideways, one gets $a = 3$, $b = 5$, and the center is at the origin. Equation is $\dfrac{x^2}{9} - \dfrac{y^2}{25} = 1$.

31. $\dfrac{x^2}{9} - \dfrac{y^2}{16} = 1$

33. $\dfrac{y^2}{9} - \dfrac{(x-1)^2}{9} = 1$

35. Completing the square,

$$
\begin{aligned}
y^2 - (x^2 - 2x) &= 2 \\
y^2 - (x^2 - 2x + 1) &= 2 - 1 \\
y^2 - (x-1)^2 &= 1
\end{aligned}
$$

one gets a hyperbola.

37. $y = x^2 + 2x$ is a parabola.

39. Simplifying,

$$
\begin{aligned}
25x^2 + 25y^2 &= 2500 \\
x^2 + y^2 &= 100
\end{aligned}
$$

one gets a circle.

41. Simplifying,

$$
\begin{aligned}
25x &= -100y^2 + 2500 \\
x &= -4y^2 + 100
\end{aligned}
$$

one gets a parabola.

43. Completing the square,

$$
\begin{aligned}
2(x^2 - 2x) + 2(y^2 - 4y) &= -9 \\
2(x^2 - 2x + 1) + 2(y^2 - 4y + 4) &= -9 + 2 + 8 \\
2(x-1)^2 + 2(y-2)^2 &= 1 \\
(x-1)^2 + (y-2)^2 &= \frac{1}{2}
\end{aligned}
$$

one finds a circle.

45. Completing the square,

$$
\begin{aligned}
2(x^2 + 2x) + y^2 + 6y &= -7 \\
2(x^2 + 2x + 1) + y^2 + 6y + 9 &= -7 + 2 + 9 \\
2(x+1)^2 + (y+3)^2 &= 4 \\
\frac{(x+1)^2}{2} + \frac{(y+3)^2}{4} &= 1
\end{aligned}
$$

one gets an ellipse.

47. From the center $(0,0)$ and vertex $(0,8)$ one gets $a = 8$. From the foci $(0, \pm 10)$, $c = 10$. So $b^2 = c^2 - a^2 = 10^2 - 8^2 = 36$.

Hyperbola is given by $\dfrac{y^2}{64} - \dfrac{x^2}{36} = 1$.

49. Multiply $16y^2 - x^2 = 16$ by 9 and add to $9x^2 - 4y^2 = 36$.

$$
\begin{aligned}
-9x^2 + 144y^2 &= 144 \\
9x^2 - 4y^2 &= 36 \\
\hline
140y^2 &= 180 \\
y^2 &= \frac{9}{7} \\
y &= \frac{3\sqrt{7}}{7}
\end{aligned}
$$

Using $y^2 = \dfrac{9}{7}$ in $x^2 = 16(y^2 - 1)$, we get

$x^2 = 16\left(\dfrac{2}{7}\right)$ or $x = \dfrac{4\sqrt{14}}{7}$.

Exact location is $\left(\dfrac{4\sqrt{14}}{7}, \dfrac{3\sqrt{7}}{7}\right)$.

51. Since $c^2 = a^2 + b^2 = 1^2 + 1^2 = 2$, the foci of $x^2 - y^2 = 1$ are $A(\sqrt{2}, 0)$ and $B(-\sqrt{2}, 0)$. Note, $y^2 = x^2 - 1$. Suppose (x, y) is a point on the hyperbola whose distance from B is twice the distance between (x, y) and A. Then

$$
\begin{aligned}
2\sqrt{(x - \sqrt{2})^2 + y^2} &= \sqrt{(x + \sqrt{2})^2 + y^2} \\
4((x - \sqrt{2})^2 + y^2) &= (x + \sqrt{2})^2 + y^2
\end{aligned}
$$

$$
\begin{aligned}
4(x - \sqrt{2})^2 - (x + \sqrt{2})^2 + 3y^2 &= 0 \\
3x^2 - 10\sqrt{2}x + 6 + 3y^2 &= 0 \\
3x^2 - 10\sqrt{2}x + 6 + 3(x^2 - 1) &= 0 \\
6x^2 - 10\sqrt{2}x + 3 &= 0.
\end{aligned}
$$

Solving for x, one finds $x = \dfrac{3\sqrt{2}}{2}$ and $x = \dfrac{\sqrt{2}}{6}$; the second value must be excluded since it is out of the domain. Substituting $x = \dfrac{3\sqrt{2}}{2}$ into $y^2 = x^2 - 1$, one obtains $y = \pm\dfrac{\sqrt{14}}{2}$.

By the symmetry of the hyperbola, there are four points that are twice as far from one focus as they are from the other focus. Namely, these points are

$$
\left(\pm\frac{3\sqrt{2}}{2}, \pm\frac{\sqrt{14}}{2}\right).
$$

53. Note, the asymptotes are $y = \pm x$.
The difference is $50 - \sqrt{50^2 - 1} \approx 0.01$

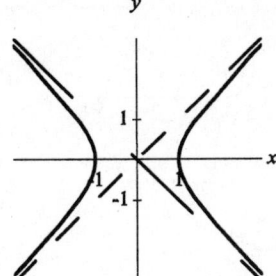

57. Since the asymptotes are $y = \pm x$, the lines parallel to the asymptotes and passing through (x_1, y_1) are $f(x) = x + (y_1 - x_1)$ and $g(x) = -x + (x_1 + y_1)$. Similarly, the lines through (x_2, y_2) and parallel to $y = \pm x$ are $s(x) = x + (y_2 - x_2)$ and $t(x) = -x + (x_2 + y_2)$. To find the point of intersection between lines f and t, we solve a system of equations. Then

$$x + (y_1 - x_1) = -x + (x_2 + y_2)$$
$$x = \frac{(x_2 + y_2) - (y_1 - x_1)}{2}.$$

Substituting into f, one finds

$$y = \frac{(x_2 + y_2) - (y_1 - x_1)}{2} + (y_1 - x_1)$$
$$y = \frac{(x_2 + y_2) + (y_1 - x_1)}{2}.$$

The point of intersection between f and t is
$A =$
$$\left(\frac{(x_2 + y_2) - (y_1 - x_1)}{2}, \frac{(x_2 + y_2) + (y_1 - x_1)}{2} \right).$$

Similarly, one can show that the point of intersection between g and s is $B =$
$$\left(\frac{(x_1 + y_1) - (y_2 - x_2)}{2}, \frac{(y_2 - x_2) + (x_1 + y_1)}{2} \right).$$

Let $A = (A_1, A_2)$ and $B = (B_1, B_2)$. To show that the line joining A and B passes through the origin it will be sufficient to show $\frac{A_2}{A_1} = \frac{B_2}{B_1}$. Rewriting $\frac{A_2}{A_1}$ and noting that $x_2^2 - y_2^2 = 1$, one obtains

$$\frac{A_2}{A_1} = \frac{(x_2 + y_2) + (y_1 - x_1)}{(x_2 + y_2) - (y_1 - x_1)}$$

$$= \frac{(x_2 + y_2) + (y_1 - x_1)}{(x_2 + y_2) - (y_1 - x_1)} \cdot \frac{x_2 - y_2}{x_2 - y_2}$$

$$= \frac{(x_2 + y_2)(x_2 - y_2) + (y_1 - x_1)(x_2 - y_2)}{(x_2 + y_2)(x_2 - y_2) - (y_1 - x_1)(x_2 - y_2)}$$

$$= \frac{1 + (y_1 - x_1)(x_2 - y_2)}{1 - (y_1 - x_1)(x_2 - y_2)}$$

$$= \frac{1 + (x_1 - y_1)(y_2 - x_2)}{1 - (x_1 - y_1)(y_2 - x_2)}$$

$$= \frac{(x_1 - y_1)(y_2 - x_2) + 1}{1 - (x_1 - y_1)(y_2 - x_2)}$$

$$= \frac{(y_2 - x_2) + (x_1 + y_1)}{(x_1 + y_1) - (y_2 - x_2)} \cdot \frac{x_1 - y_1}{x_1 - y_1}$$

$$= \frac{B_2}{B_1}.$$

Hence, the line through A and B passes through the origin.

Review Exercises

1. If $y = 0$, then by factoring we get $0 = (x + 6)(x - 2)$ and the x-intercepts are $(2, 0)$ & $(-6, 0)$.

If $x = 0$, then $y = -12$ and y-intercept is $(0, -12)$.

By completing the square, we obtain $y = (x + 2)^2 - 16$, vertex $(h, k) = (-2, -16)$, and the axis of symmetry is $x = -2$.

Since $p = \frac{1}{4a} = \frac{1}{4}$, the focus is $(h, k + p) = (-2, -63/4)$ and directrix is $y = k - p$ or $y = -65/4$.

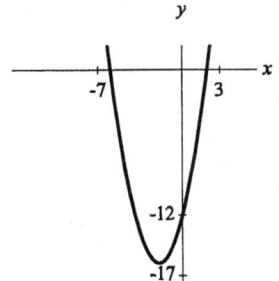

3. If $y = 0$, then by factoring we find $0 = x(6 - 2x)$ and the x-intercepts are $(0, 0)$ & $(3, 0)$.

If $x = 0$, then $y = 0$ and y-intercept is $(0, 0)$.

By completing the square, one gets $y = -2(x - 3/2)^2 + 9/2$,

vertex $(h, k) = (3/2, 9/2)$,
and axis of symmetry is $x = 3/2$.

Since $p = \dfrac{1}{4a} = -\dfrac{1}{8}$, the focus is

$(h, k + p) = (3/2, 35/8)$ and
the directrix is $y = k - p$ or $y = 37/8$.

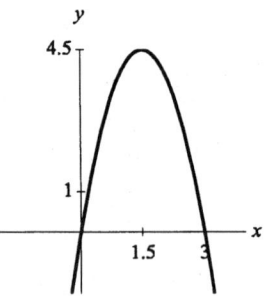

5. By completing the square, we have
$x = (y+2)^2 - 10$. If $x = 0$, then $y + 2 = \pm\sqrt{10}$
and the y-intercepts are $(0, -2 \pm \sqrt{10})$.

If $y = 0$, then $x = -6$ and x-intercept is
$(-6, 0)$. Since $x = (y + 2)^2 - 10$ is of the
form $x = a(y - h)^2 + k$, the vertex is $(k, h) =$
$(-10, -2)$, axis of symmetry is $y = -2$.

Since $p = \dfrac{1}{4a} = \dfrac{1}{4}$, the focus is $(k + p, h) =$
$(-39/4, -2)$ and directrix is $x = k - p$ or
$x = -41/4$.

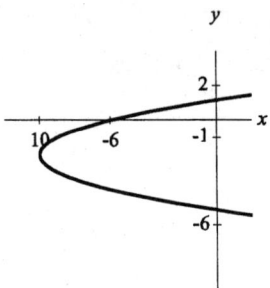

7. Since $c = \sqrt{a^2 - b^2} = \sqrt{36 - 16} = 2\sqrt{5}$,
the foci are $(0, \pm 2\sqrt{5})$

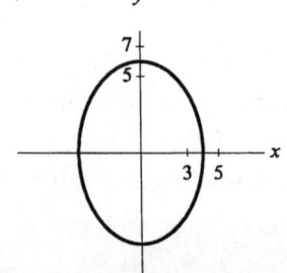

9. Since $c = \sqrt{a^2 - b^2} = \sqrt{24 - 8} = 4$, the
foci are $(1, 1 \pm 4)$ or $(1, 5)$ & $(1, -3)$.

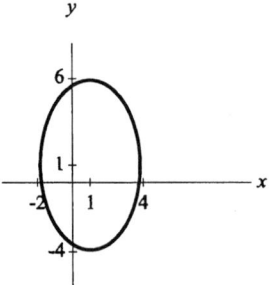

11. Since $c = \sqrt{a^2 - b^2} = \sqrt{10 - 8} = \sqrt{2}$,
the foci are $(1, -3 \pm \sqrt{2})$.

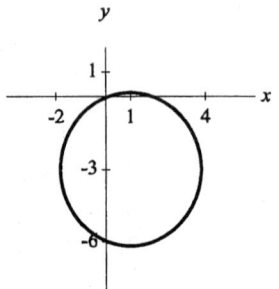

13. center $(0, 0)$, radius 9

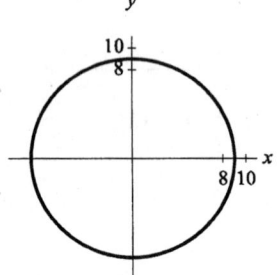

15. center $(-1, 0)$, radius 2

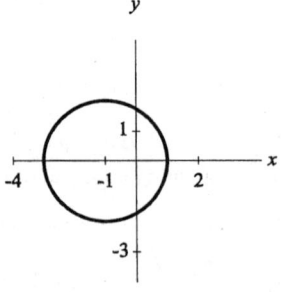

17. Completing the square, we have

$$x^2 + 5x + \frac{25}{4} + y^2 = -\frac{1}{4} + \frac{25}{4}$$

$$\left(x + \frac{5}{2}\right)^2 + y^2 = 6,$$

and so center is $(-5/2, 0)$ and radius is $\sqrt{6}$.

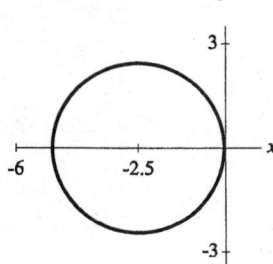

19. $x^2 + (y + 4)^2 = 9$

21. $(x + 2)^2 + (y + 7)^2 = 6$

23. Since $c = \sqrt{a^2 + b^2} = \sqrt{8^2 + 6^2} = 10$,

foci $(\pm 10, 0)$, asymptotes $y = \pm\frac{b}{a} = \pm\frac{3}{4}x$

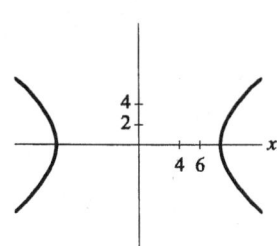

25. Since $c = \sqrt{a^2 + b^2} = \sqrt{8^2 + 4^2} = 4\sqrt{5}$,

the foci are $(4, 2 \pm 4\sqrt{5})$. Solving

for y in $y - 2 = \pm\frac{8}{4}(x - 4)$,

asymptotes are $y = 2x - 6$ and $y = -2x + 10$.

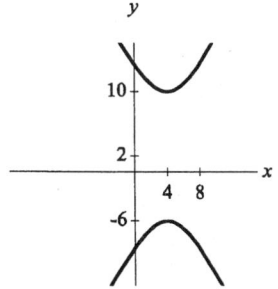

27. Completing the square, we have

$$x^2 - 4x - 4(y^2 - 8y) = 64$$

$$x^2 - 4x + 4 - 4(y^2 - 8y + 16) = 64 + 4 - 64$$

$$(x - 2)^2 - 4(y - 4)^2 = 4$$

$$\frac{(x - 2)^2}{4} - (y - 4)^2 = 1,$$

and so $c = \sqrt{a^2 + b^2} = \sqrt{2^2 + 1^2} = \sqrt{5}$, and the foci are $(2 \pm \sqrt{5}, 4)$. Solving for y in $y - 4 = \pm\frac{1}{2}(x - 2)$, we get that the asymptotes are $y = \frac{1}{2}x + 3$ and $y = -\frac{1}{2}x + 5$.

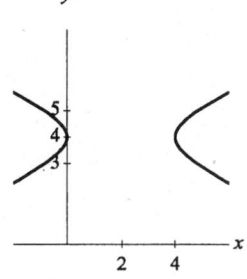

29. Hyperbola   **31.** Ellipse

33. Parabola   **35.** Hyperbola

37. $x^2 + y^2 = 4$ is a circle

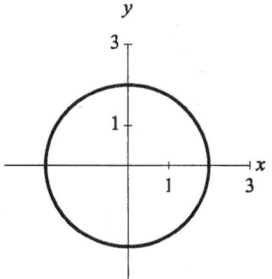

39. Since $4y = x^2 - 4$, $y = \frac{1}{4}x^2 - 1$.

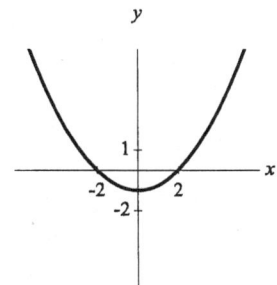

41. Since $x^2 + 4y^2 = 4$, $\dfrac{x^2}{4} + y^2 = 1$.

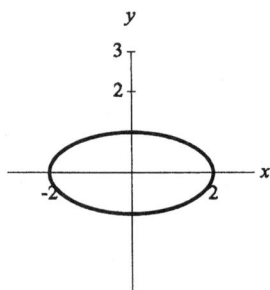

43. Since $x^2 - 4x + 4 + 4y^2 = 4$, we find

$(x - 2)^2 + 4y^2 = 4$ and $\dfrac{(x-2)^2}{4} + y^2 = 1$.

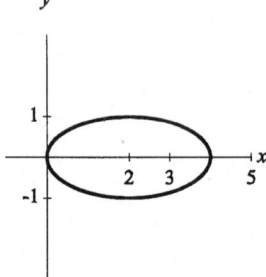

45. Since the vertex is midway between the

focus $(1, 3)$ and directrix $x = \dfrac{1}{2}$, the

vertex is $\left(\dfrac{3}{4}, 3\right)$ and $p = \dfrac{1}{4}$. Since

$a = \dfrac{1}{4p} = 1$, parabola is given by

$x = (y - 3)^2 + \dfrac{3}{4}$.

47. From the foci and vertices one gets
$c = 4$ and $a = 6$, respectively. Since
$b^2 = a^2 - c^2 = 36 - 16 = 20$, the

ellipse is given by $\dfrac{x^2}{36} + \dfrac{y^2}{20} = 1$.

49. Radius is $\sqrt{(-1-1)^2 + (-1-3)^2} = \sqrt{20}$.
Equation is $(x - 1)^2 + (y - 3)^2 = 20$.

51. From the foci and x-intercepts one gets
$c = 3$ and $a = 2$, respectively. Since
$b^2 = c^2 - a^2 = 9 - 4 = 5$, the

hyperbola is given by $\dfrac{x^2}{4} - \dfrac{y^2}{5} = 1$.

53. Since the center of the circle is $(-2, 3)$ and
the raidus is 3, we have $(x+2)^2 + (y-3)^2 = 9$.

55. Note, the center of the ellipse is $(-2, 1)$. By
using the lengths of the major axis, we get
$a = 3$ and $b = 1$.

Thus, an equation is $\dfrac{(x+2)^2}{9} + (y-1)^2 = 1$.

57. Note, the center of the hyperbola is $(2, 1)$. By
using the fundamental rectangle, we find $a = 3$
and $b = 2$.

Thus, an equation is $\dfrac{(y-1)^2}{9} - \dfrac{(x-2)^2}{4} = 1$

59. The equation is of the form $\dfrac{x^2}{100^2} - \dfrac{y^2}{b^2} = 1$.

Since the graph passes through $(120, 24\sqrt{11})$,
we get

$$\begin{aligned}
\frac{120^2}{100^2} - \frac{(24\sqrt{11})^2}{b^2} &= 1 \\
1.44 - \frac{6336}{b^2} &= 1 \\
b^2 &= \frac{6336}{0.44} \\
b^2 &= 120^2.
\end{aligned}$$

Equation is $\dfrac{x^2}{100^2} - \dfrac{y^2}{120^2} = 1$.

61. Note $c = 30$ and $a = 34$. Then an equation we

can use is of the form $\dfrac{x^2}{34^2} + \dfrac{y^2}{b^2} = 1$.

Since $b^2 = a^2 - c^2 = 34^2 - 30^2 = 16^2$, the

equation is $\dfrac{x^2}{34^2} + \dfrac{y^2}{16^2} = 1$.

To find h, let $x = 32$. Substituting, one finds

$$\begin{aligned}
\frac{32^2}{34^2} + \frac{y^2}{16^2} &= 1 \\
y^2 &= \left(1 - \frac{32^2}{34^2}\right)16^2 \\
y &\approx 5.407.
\end{aligned}$$

Thus, $h = 2y \approx 10.81$ feet.

Chapter 7 Test

1. Circle $x^2 + y^2 = 8$

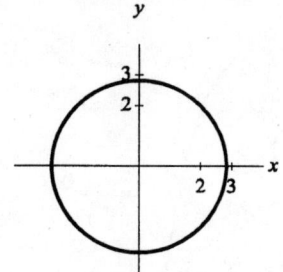

2. Ellipse $\dfrac{x^2}{9} + \dfrac{y^2}{100} = 1$

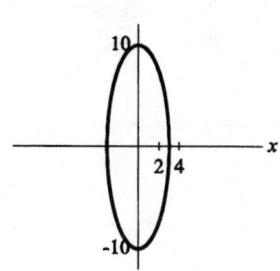

3. Parabola $y = x^2 + 6x + 8$

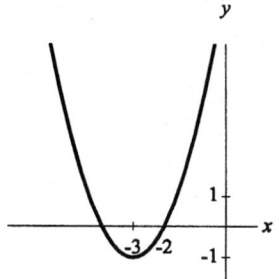

4. Hyperbola $\dfrac{y^2}{25} - \dfrac{x^2}{9} = 1$

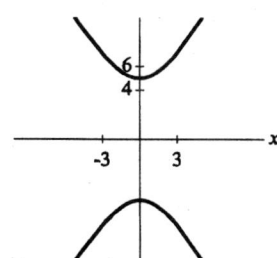

5. Circle $(x + 3)^2 + (y - 1)^2 = 10$

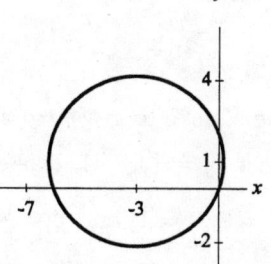

6. Hyperbola $\dfrac{(x - 2)^2}{9} - \dfrac{(y + 3)^2}{4} = 1$

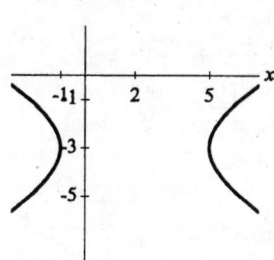

7. By completing the square, the equation can be written as $(x - 4)^2 - 16 = y^2$. This is a hyperbola.

8. By completing the square, we can rewrite the equation as $(x - 4)^2 - 16 = y$. This is a parabola.

9. By completing the square, we can rewrite the equation as $-(x - 4)^2 + 16 = y^2$. This is a circle.

10. By completing the square, the equation can be written as $-(x - 4)^2 + 16 = 8y^2$. This is an ellipse.

11. Since $(2\sqrt{3})^2 = 12$, the circle is given by $(x + 3)^2 + (y - 4)^2 = 12$.

12. Midway between the focus $(2, 0)$ and directrix $x = -2$ is the vertex $(0, 0)$. Since $p = 2$, $a = \dfrac{1}{4p} = \dfrac{1}{8}$. Equation is $x = \dfrac{1}{8}y^2$.

13. Equation is of the form $\dfrac{x^2}{2^2} + \dfrac{y^2}{a^2} = 1$.

Since $a^2 = b^2 + c^2 = 2^2 + \sqrt{6}^2 = 10$, equation is $\dfrac{x^2}{4} + \dfrac{y^2}{10} = 1$.

14. From the foci and vertices one gets $c = 8$ and $a = 6$. Since $b^2 = c^2 - a^2 = 8^2 - 6^2 = 28$, equation is $\dfrac{x^2}{36} - \dfrac{y^2}{28} = 1$.

15. Complete the square to get $y = (x-2)^2 - 4$. So vertex is $(h, k) = (2, -4)$, axis of symmetry $x = 2$, $a = 1$, and $p = \dfrac{1}{4a} = \dfrac{1}{4}$. Focus is $(h, k+p) = (2, -15/4)$ and directrix is $y = k - p = -17/4$.

16. Since $\dfrac{x^2}{16} + \dfrac{y^2}{4} = 1$, we get
$c = \sqrt{a^2 - b^2} = \sqrt{4^2 - 2^2} = 2\sqrt{3}$.
Foci are $(\pm 2\sqrt{3}, 0)$, length of major axis is $2a = 8$, length of minor axis is $2b = 4$.

17. Since $y^2 - \dfrac{x^2}{16} = 1$, we obtain
$c = \sqrt{a^2 + b^2} = \sqrt{1^2 + 4^2} = \sqrt{17}$.
Foci are $(0, \pm\sqrt{17})$, vertices $(0, \pm 1)$, asymptotes $y = \pm\dfrac{1}{4}x$, length of transverse axis is $2a = 2$, length of conjugate axis is $2b = 8$.

18. Completing the square, we find
$$x^2 + x + \frac{1}{4} + y^2 - 3y + \frac{9}{4} = -\frac{1}{4} + \frac{1}{4} + \frac{9}{4}$$
$$\left(x + \frac{1}{2}\right)^2 + \left(y - \frac{3}{2}\right)^2 = \frac{9}{4}.$$
The center is $\left(-\dfrac{1}{2}, \dfrac{3}{2}\right)$ and radius is $\dfrac{3}{2}$.

19. Since $\dfrac{x^2}{225} + \dfrac{y^2}{81} = 1$, we find $a = 15$ and $b = 9$.
Note, $c = \sqrt{a^2 - c^2} = \sqrt{15^2 - 9^2} = 12$.
Since the foci are $(\pm 12, 0)$, the distance from the point of generation of the waves to the kidney stones is $2c = 24$ cm.

Tying It All Together

1. Parabola $y = 6x - x^2$

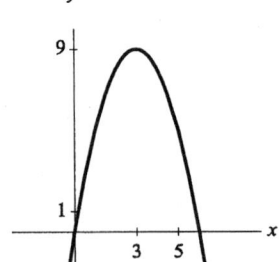

2. Line $y = 6x$

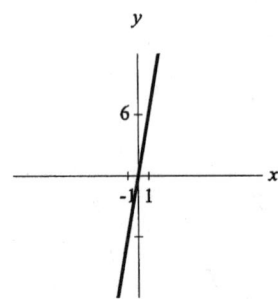

3. Parabola $y = 6 - x^2$

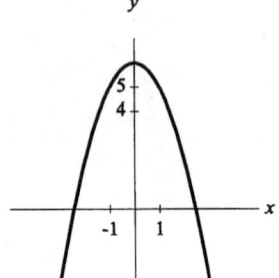

4. Circle $x^2 + y^2 = 6$

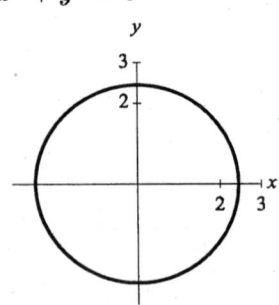

5. Line $y = x + 6$

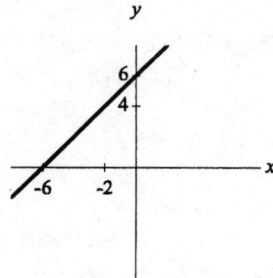

6. Parabola $x = 6 - y^2$

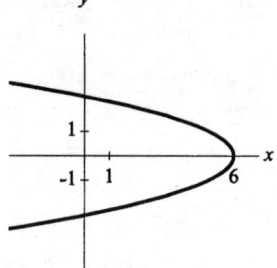

7. Parabola $y = (6 - x)^2$

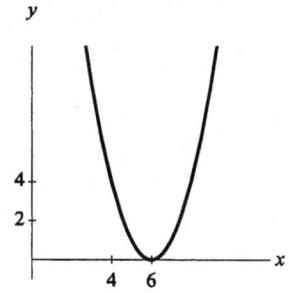

8. $y = |x + 6|$ goes through $(-6, 0)$, $(-5, 1)$

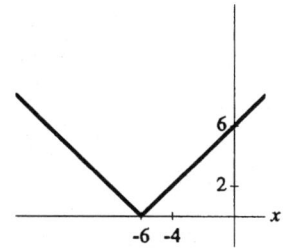

9. $y = 6^x$ goes through $(0, 1)$, $(1, 6)$

10. $y = \log_6(x)$ goes through $(6, 1)$, $(1/6, -1)$

11. $y = \dfrac{1}{x^2 - 6}$ has asymptotes $x = \pm\sqrt{6}$

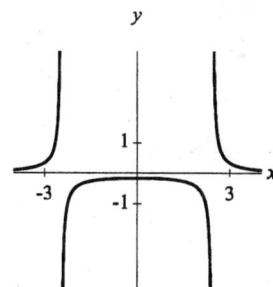

12. Ellipse $\dfrac{x^2}{9} + \dfrac{y^2}{4} = 1$

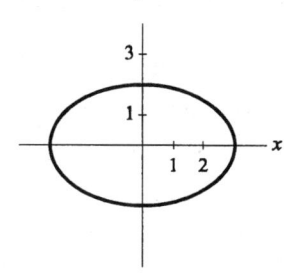

13. Hyperbola $\dfrac{x^2}{9} - \dfrac{y^2}{4} = 1$

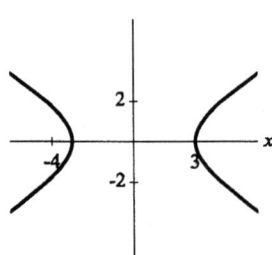

14. $y = x(6 - x^2)$ goes through $(0,0)$, $(\pm\sqrt{6}, 0)$

15. Line $2x + 3y = 6$

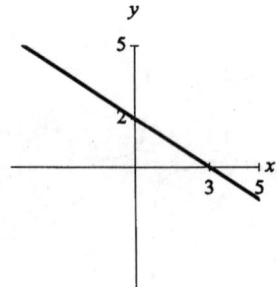

16. By completing the square, the equation can be written as $\dfrac{(x-3)^2}{3} - \dfrac{y^2}{3} = 1$. This is a hyperbola.

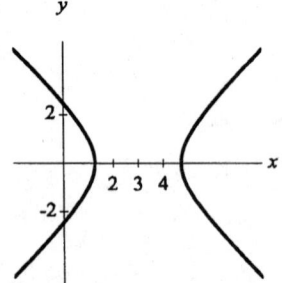

17. Solving for x, we get

$$\begin{aligned} 3x - 9 + 5 &= -9 \\ 3x &= -5. \end{aligned}$$

The solution set is $\{-5/3\}$.

18. Solving for x, we find

$$\begin{aligned} 5x - 8x + 12 &= 17 \\ -3x &= 5. \end{aligned}$$

The solution set is $\{-5/3\}$.

19. Solving for x, we obtain

$$\begin{aligned} 2x - \frac{2}{3} - \frac{3}{2} + 3x &= \frac{3}{2} \\ 5x - \frac{13}{6} &= \frac{3}{2} \\ 5x &= \frac{11}{3}. \end{aligned}$$

The solution set is $\{11/15\}$.

20. Solving for x, we have

$$\begin{aligned} -4x + 2 &= 3x + 2 \\ -7x &= 0. \end{aligned}$$

The solution set is $\{0\}$.

21. Solving for x, we find

$$\begin{aligned} \frac{1}{2}x - \frac{1}{4}x &= \frac{3}{2} + \frac{1}{3} \\ \frac{1}{4}x &= \frac{11}{6} \\ x &= \frac{44}{6}. \end{aligned}$$

Solution set is $\{22/3\}$.

22. Multiplying both sides by 100, we obtain

$$\begin{aligned} 5(x - 20) + 2(x + 10) &= 270 \\ 5x - 100 + 2x + 20 &= 270 \\ 7x &= 350. \end{aligned}$$

The solution set is $\{50\}$.

23. By using the quadratic formula to solve $2x^2 + 31x - 51 = 0$, one obtains

$$x = \frac{-31 \pm \sqrt{31^2 - 4(2)(-51)}}{4}$$

$$= \frac{-31 \pm \sqrt{1369}}{4}$$

$$= \frac{-31 \pm 37}{4}$$

$$= \frac{3}{2}, -17.$$

The solution set is $\left\{\frac{3}{2}, -17\right\}$.

24. Since $x(2x + 31) = 0$, the solution set is $\left\{0, -\frac{31}{2}\right\}$.

25. By using the quadratic formula to solve $x^2 - 34x + 286 = 0$, one obtains

$$x = \frac{34 \pm \sqrt{34^2 - 4(1)(286)}}{2}$$

$$x = \frac{34 \pm \sqrt{12}}{2}$$

$$x = \frac{34 \pm 2\sqrt{3}}{2}$$

$$x = 17 \pm \sqrt{3}.$$

Solution set is $\left\{17 \pm \sqrt{3}\right\}$.

26. In solving $x^2 - 34x + 290 = 0$, one can use the quadratic formula. Then

$$x = \frac{34 \pm \sqrt{34^2 - 4(1)(290)}}{2}$$

$$x = \frac{34 \pm \sqrt{-4}}{2}$$

$$x = \frac{34 \pm 2i}{2}$$

$$x = 17 \pm i.$$

The solution set is $\{17 \pm i\}$.

For Thought

1. True

2. False, the domain of a finite sequence is $\{1, 2, ..., n\}$ for some positive integer n.

3. True

4. False, n is the independent variable.

5. False, the first four terms are $1, -8, 27, -64$.

6. False, $a_5 = -3 + 4 \cdot 6 = 21$.

7. False, the common difference is $d = 4 - 7 = -3$.

8. False, there is no common difference.

9. False, since if $14 = 4 + 2d$ then $d = 5$ and $a_4 = 4 + 3 \cdot 5 = 19$.   **10.** True

8.1 Exercises

1. $a_1 = 1^2 = 1$, $a_2 = 2^2 = 4$, $a_3 = 3^2 = 9$, $a_4 = 4^2 = 16$, $a_5 = 5^2 = 25$, $a_6 = 6^2 = 36$, $a_7 = 7^2 = 49$.

First seven terms are $1, 4, 9, 16, 25, 36, 49$.

3. $b_1 = \dfrac{(-1)^2}{2} = \dfrac{1}{2}$, $b_2 = \dfrac{(-1)^3}{3} = -\dfrac{1}{3}$,

$b_3 = \dfrac{(-1)^4}{4} = \dfrac{1}{4}$, $b_4 = \dfrac{(-1)^5}{5} = -\dfrac{1}{5}$,

$b_5 = \dfrac{(-1)^6}{6} = \dfrac{1}{6}$, $b_6 = \dfrac{(-1)^7}{7} = -\dfrac{1}{7}$,

$b_7 = \dfrac{(-1)^8}{8} = \dfrac{1}{8}$, $b_8 = \dfrac{(-1)^9}{9} = -\dfrac{1}{9}$.

First eight terms are $\dfrac{1}{2}, -\dfrac{1}{3}, \dfrac{1}{4}, -\dfrac{1}{5}, \dfrac{1}{6}, -\dfrac{1}{7}, \dfrac{1}{8}, -\dfrac{1}{9}$.

5. $c_1 = (-2)^0 = 1$, $c_2 = (-2)^1 = -2$, $c_3 = (-2)^2 = 4$, $c_4 = (-2)^3 = -8$, $c_5 = (-2)^4 = 16$, $c_6 = (-2)^5 = -32$.
First six terms are $1, -2, 4, -8, 16, -32$

7. $a_1 = 2^1 = 2$, $a_2 = 2^0 = 1$, $a_3 = 2^{-1} = \dfrac{1}{2}$, $a_4 = 2^{-2} = \dfrac{1}{4}$, $a_5 = 2^{-3} = \dfrac{1}{8}$.
First five terms are $2, 1, \dfrac{1}{2}, \dfrac{1}{4}, \dfrac{1}{8}$.

9. $a_1 = -6 + 0 = -6$, $a_2 = -6 - 4 = -10$, $a_3 = -6 - 8 = -14$, $a_4 = -6 - 12 = -18$, $a_5 = -6 - 16 = -22$. First five terms are $-6, -10, -14, -18, -22$.

11. $b_1 = 5 + 0 = 5$, $b_2 = 5 + 0.5 = 5.5$, $b_3 = 5 + 1 = 6$, $b_4 = 5 + 1.5 = 6.5$, $b_5 = 5 + 2 = 7$, $b_6 = 5 + 2.5 = 7.5$, $b_7 = 5 + 3 = 8$. First seven terms are $5, 5.5, 6, 6.5, 7, 7.5, 8$.

13. $a_1 = -0.1 + 9 = 8.9$, $a_2 = -0.2 + 9 = 8.8$, $a_3 = -0.3 + 9 = 8.7$, $a_4 = -0.4 + 9 = 8.6$, $a_{10} = -1 + 9 = 8$. First four terms are $8.9, 8.8, 8.7, 8.6$ and and 10th term is 8.

15. $a_1 = 8 + 0 = 8$, $a_2 = 8 - 3 = 5$, $a_3 = 5 - 3 = 2$, $a_4 = 2 - 3 = -1$, and $a_{10} = 8 - 27 = -19$. First four terms are $8, 5, 2, -1$ and 10th term is -19.

17. One gets $a_1 = \dfrac{4}{2+1} = \dfrac{4}{3}$, $a_2 = \dfrac{4}{4+1} = \dfrac{4}{5}$, $a_3 = \dfrac{4}{6+1} = \dfrac{4}{7}$, $a_4 = \dfrac{4}{8+1} = \dfrac{4}{9}$, $a_{10} = \dfrac{4}{20+1} = \dfrac{4}{21}$
The first four terms are $\dfrac{4}{3}, \dfrac{4}{5}, \dfrac{4}{7}, \dfrac{4}{9}$ and 10th term is $\dfrac{4}{21}$.

19. One gets $a_1 = \dfrac{(-1)^1}{2 \cdot 3} = -\dfrac{1}{6}$, $a_2 = \dfrac{(-1)^2}{3 \cdot 4} = \dfrac{1}{12}$, $a_3 = \dfrac{(-1)^3}{4 \cdot 5} = -\dfrac{1}{20}$, $a_4 = \dfrac{(-1)^4}{5 \cdot 6} = \dfrac{1}{30}$, $a_{10} = \dfrac{(-1)^{10}}{11 \cdot 12} = \dfrac{1}{132}$.
The first four terms are $-\dfrac{1}{6}, \dfrac{1}{12}, -\dfrac{1}{20}, \dfrac{1}{30}$ and 10th term is $\dfrac{1}{132}$.

21. $a_1 = 2! = 2$, $a_2 = 4! = 24$, $a_3 = 6! = 720$, $a_4 = 8! = 40,320$, and $a_5 = 10! = 3,628,800$

23. $a_1 = \dfrac{2^1}{1!} = 2$, $a_2 = \dfrac{2^2}{2!} = 2$, $a_3 = \dfrac{2^3}{3!} = \dfrac{4}{3}$,

$a_4 = \dfrac{2^4}{4!} = \dfrac{2}{3}$, and $a_5 = \dfrac{2^5}{5!} = \dfrac{4}{15}$

25. $b_1 = \dfrac{1!}{0!} = 1$, $b_2 = \dfrac{2!}{1!} = 2$, $b_3 = \dfrac{3!}{2!} = 3$,

$b_4 = \dfrac{4!}{3!} = 4$, and $b_5 = \dfrac{5!}{4!} = \dfrac{5 \cdot 4!}{4!} = 5$

27. $c_1 = \dfrac{(-1)^1}{2!} = -\dfrac{1}{2}$, $c_2 = \dfrac{(-1)^2}{3!} = \dfrac{1}{6}$,

$c_3 = \dfrac{(-1)^3}{4!} = -\dfrac{1}{24}$, $c_4 = \dfrac{(-1)^4}{5!} = \dfrac{1}{120}$,

and $c_5 = \dfrac{(-1)^5}{6!} = -\dfrac{1}{720}$.

29. Note $t_n = \dfrac{1}{(n+2)!e^n}$. Then we obtain

$t_1 = \dfrac{1}{3!e} = \dfrac{1}{6e}$, $t_2 = \dfrac{1}{4!e^2} = \dfrac{1}{24e^2}$,

$t_3 = \dfrac{1}{5!e^3} = \dfrac{1}{120e^3}$, $t_4 = \dfrac{1}{6!e^4} = \dfrac{1}{720e^4}$,

and $t_5 = \dfrac{1}{7!e^5} = \dfrac{1}{5040e^5}$.

31. $a_n = 2n$ **33.** $a_n = 2n + 7$

35. $a_n = (-1)^{n+1}$ **37.** $a_n = n^3$

39. $a_n = e^n$ **41.** $a_n = \dfrac{1}{2^{n-1}}$

43. $a_2 = 3a_1 + 2 = 3(-4) + 2 = -10$,
$a_3 = 3a_2 + 2 = 3(-10) + 2 = -28$,
$a_4 = 3a_3 + 2 = 3(-28) + 2 = -82$,
$a_5 = 3a_4 + 2 = 3(-82) + 2 = -244$,
$a_6 = 3a_5 + 2 = 3(-244) + 2 = -730$,
$a_7 = 3a_6 + 2 = 3(-730) + 2 = -2188$,
$a_8 = 3a_7 + 2 = 3(-2188) + 2 = -6562$.
First four terms $-4, -10, -28, -82$ and
8th term is -6562.

45. One finds $a_2 = a_1^2 - 3 = 2^2 - 3 = 1$,

$a_3 = a_2^2 - 3 = 1^2 - 3 = -2$,

$a_4 = a_3^2 - 3 = (-2)^2 - 3 = 1$.

By a repeating pattern, $a_8 = 1$. First

four terms are $2, 1, -2, 1$ and 8th term is 1.

47. $a_2 = a_1 + 7 = (-15) + 7 = -8$,

$a_3 = a_2 + 7 = (-8) + 7 = -1$,

$a_4 = a_3 + 7 = (-1) + 7 = 6$.

There is a common difference of 7.
So $a_8 = 34$. First four terms are $-15, -8, -1, 6$
and 8th term is 34.

49. First four terms are $6, 3, 0, -3$ and
10th term is $a_{10} = 6 + 9(-3) = -21$.

51. First four terms are $1, 0.9, 0.8, 0.7$ and
10th term is $c_{10} = 1 + 9(-0.1) = 0.1$.

53. One finds $w_1 = -\dfrac{1}{3}(1) + 5 = \dfrac{14}{3}$,

$w_2 = -\dfrac{1}{3}(2) + 5 = \dfrac{13}{3}$,

$w_3 = -\dfrac{1}{3}(3) + 5 = \dfrac{12}{3}$,

$w_4 = -\dfrac{1}{3}(4) + 5 = \dfrac{11}{3}$, and

$w_{10} = -\dfrac{1}{3}(10) + 5 = \dfrac{5}{3}$.

First four terms are $\dfrac{14}{3}, \dfrac{13}{3}, \dfrac{12}{3}, \dfrac{11}{3}$

and 10th term is $\dfrac{5}{3}$.

55. Yes, $d = 1$ is the common difference.

57. No, there is no common difference.

59. No, there is no common difference.

61. Yes, $d = \dfrac{\pi}{4}$ is the common difference.

63. Since $d = 5$ and $a_1 = 1$, $a_n = a_1 + (n-1)d$
$= 1 + (n-1)5$. Then $a_n = 5n - 4$.

65. Since $d = 2$ and $a_1 = 0$, $a_n = a_1 + (n-1)d$
$= 0 + (n-1)2$. So $a_n = 2n - 2$.

67. Since $d = -4$ and $a_1 = 5$, $a_n = a_1 + (n-1)d$
$= 5 + (n-1)(-4)$. So $a_n = -4n + 9$.

69. Since $d = 0.1$ and $a_1 = 1$, $a_n = a_1 + (n-1)d$
$= 1 + (n-1)(0.1)$. So $a_n = 0.1n + 0.9$.

71. Since $d = \dfrac{\pi}{6}$ and $a_1 = \dfrac{\pi}{6}$, $a_n = a_1 + (n-1)d$
$= \dfrac{\pi}{6} + (n-1)\dfrac{\pi}{6}$. So $a_n = \dfrac{\pi}{6}n$.

73. Since $d = 15$ and $a_1 = 20$, we have

$$a_n = a_1 + (n-1)d = 20 + (n-1)15.$$

Then $a_n = 15n + 5$.

75. Since $a_n = a_1 + (n-1)d$, we get
$a_8 = (-3) + (7)5 = 32.$

77. Since $a_n = a_1 + (n-1)d$, we find

$$\begin{aligned} a_1 + 2d &= 6 \\ a_1 + 6d &= 18 \end{aligned}$$

Subtracting the first equation from the second, we get $4d = 12$ and $d = 3$. From the first equation, $a_1 + 6 = 6$ and $a_1 = 0$. Then $a_{10} = a_1 + 9d = 0 + 9 \cdot 3 = 27$. So $a_{10} = 27$.

79. Since $a_n = a_1 + (n-1)d$, we get
$a_{21} = 12 + 20d = 96$. So $20d = 84$ and
the common difference is $d = 4.2$.

81. Since $a_n = a_1 + (n-1)d$, we find

$$\begin{aligned} a_1 + 2d &= 10 \\ a_1 + 6d &= 20. \end{aligned}$$

Subtracting the first equation from the second, we obtain $4d = 10$ and $d = 2.5$. From the first equation, we get $a_1 + 5 = 10$ and $a_1 = 5$. Then $a_n = 5 + (n-1)(2.5) = 2.5n + 2.5$. So $a_n = 2.5n + 2.5$.

83. $a_1 = 3$, $a_n = a_{n-1} + 9$

85. $a_1 = \dfrac{1}{3}$, $a_n = 3a_{n-1}$

87. $a_1 = 16$, $a_n = \sqrt{a_{n-1}}$

89. A formula for the MSRP is $a_n = 35,265(1.06)^n$ where n is the number of years since 2002. For the years 2003-2007, the MSRP are $a_1 = \$37,381$, $a_2 = \$39,624$, $a_3 = \$42,001$, $a_4 = \$44,521$, and $a_5 = \$47,193$.

91. The number of pages read on the nth day of November is $a_n = 5 + 3(n-1)$ or $a_n = 3n + 2$. On November 30, they will read $a_{30} = 92$ pages.

93. In year n, the annual cost is

$$a_n = 43,000 + 1800(n - 2001).$$

In 2010, the cost is projected to be $a_{2010} = 43,000 + 1800(2010 - 2001) = \$59,200$.

95. Since there are four corners, $C_n = 4$.
If the corner tiles are in place, the number of edge tiles needed for each side is $2(n-1)$. Since there are four sides, $E_n = 8(n-1)$.

If the corner tiles and edge tiles are in place, the interior tiles will occupy an $(n-1)$-by-$(n-1)$ square.

Note, the area of a tile is $\dfrac{1}{4}$ ft^2. Then the number of interior tiles is $I_n = \dfrac{(n-1)^2}{\frac{1}{4}}$ or equivalently $I_n = 4n^2 - 8n + 4$ for $n = 1, 2, 3, \ldots$

For Thought

1. True, $\displaystyle\sum_{i=1}^{3}(-2)^i = (-2)^1 + (-2)^2 + (-2)^3 = -6$

2. True, $\displaystyle\sum_{i=1}^{6}(0 \cdot i + 5) = (0 \cdot 1 + 5) + (0 \cdot 2 + 5) + (0 \cdot 3 + 5) + (0 \cdot 4 + 5) + (0 \cdot 5 + 5) + (0 \cdot 6 + 5) = 30.$

3. True, since $\displaystyle\sum_{i=1}^{k}(5i) = (5 \cdot 1) + \ldots + (5 \cdot k) = $

$5(1 + 2 + \ldots + k) = 5\displaystyle\sum_{i=1}^{k}(i).$

4. True, $\displaystyle\sum_{i=1}^{k}(i^2 + 1) = (1^2 + 1) + \ldots + (k^2 + 1) = $

$(1^2 + 2^2 + \ldots + k^2) + \underbrace{(1 + \ldots + 1)}_{k\ times} = \displaystyle\sum_{i=1}^{k}i^2 + k.$

5. False, there are ten terms.

6. True, if $j = i - 1$ and $i = 2$ then $j = 1$.

7. True

8. True, $1 + 2 + \ldots + n = \dfrac{n}{2}(1 + n)$ represents an arithemetic series.

9. False, for if $a_1 = 8$, $a_n = 68$, $d = 2$ then $68 = 8 + (n-1)2$ and $n = 31$. The sum of the arithmetic series is $S_{31} = \dfrac{31}{2}(8 + 68)$.

10. False, $\sum_{i=1}^{10} i^2$ is not an arithmetic series.

8.2 Exercises

1. $1^2 + 2^2 + 3^2 + 4^2 + 5^2 = 55$

3. $2^0 + 2^1 + 2^2 + 2^3 = 15$

5. $1 + 1 + 2 + 6 + 24 = 34$

7. $\underbrace{3 + ... + 3}_{10 \ times} = 30$

9. $(2+5) + (4+5) + (6+5) + (8+5) + (10+5)$
$+(12+5) + (14+5) = 91$

11. $[(-1)^7 + (-1)^8] + ... + [(-1)^{43} + (-1)^{44}] =$
$0 + ... + 0 = 0$

13. $0 + 0 + 0 + 3(2)(1) + 4(3)(2) + 5(4)(3) = 90$

15. $\sum_{i=1}^{6} i$

17. $\sum_{i=1}^{5} (-1)^i (2i - 1)$

19. $\sum_{i=1}^{5} i^2$

21. $\sum_{i=1}^{\infty} \left(-\frac{1}{2}\right)^{i-1}$

23. $\sum_{i=1}^{\infty} \ln(x_i)$

25. $\sum_{i=1}^{11} ar^{i-1}$

27. Let $j = i - 1$. New series is $\sum_{j=0}^{31} (-1)^{j+1}$

29. Let $j = i - 3$. New series is $\sum_{j=1}^{10} (2j + 7)$

31. Let $j = x - 2$. New series is $\sum_{j=0}^{8} \frac{10!}{(j+2)!(8-j)!}$

33. Let $j = n + 3$. New series is $\sum_{j=5}^{\infty} \frac{5^{j-3} \cdot e^{-5}}{(j-3)!}$

35. $0.5 + 0.5r + 0.5r^2 + 0.5r^3 + 0.5r^4 + 0.5r^5$

37. $a^4 + a^3 b + a^2 b^2 + ab^3 + b^4$

39. $a^2 + 2ab + b^2$

41. $\dfrac{6 + 23 + 45}{3} = \dfrac{74}{3}$

43. $\dfrac{-6 + 0 + 3 + 4 + 3 + 92}{6} = 16$

45. $\dfrac{\sqrt{2} + \pi + 33.6 - 19.4 + 52}{5} \approx 14.151$

47. Since $a_1 = -3$, $a_{12} = 6(12) - 9 = 63$, and
$n = 12$, the sum is $S_{12} = \dfrac{12}{2}(-3 + 63) = 360$.

49. Since $a_3 = 0.7$, $a_{15} = -0.5$, and there are
$n = 13$ terms in the series, the sum is
$S_{13} = \dfrac{13}{2}(0.7 - 0.5) = 1.3$.

51. Note $a_1 = 1$, $a_n = 47$, $n = 47$. Then the
sum is $S_{47} = \dfrac{47}{2}(1 + 47) = 1128$.

53. Since $-16 = 8 + (n-1)(-3)$, we get $n = 9$.
Thus, the sum is $S_9 = \dfrac{9}{2}(8 + (-16)) = -36$.

55. Since $55 = 3 + (n-1)4$, we obtain $n = 14$.
The sum is $S_{14} = \dfrac{14}{2}(3 + 55) = 406$.

57. Since $5 = \dfrac{1}{2} + (n-1)\dfrac{1}{4}$, we get $n = 19$.
Thus, the sum is $S_{19} = \dfrac{19}{2}(1/2 + 5) = 52.25$.

59. Note, the mean of an arithmetic sequence with
n terms is $\overline{a} = \dfrac{a_1 + a_n}{2}$. Thus, $\overline{a} = \dfrac{2 + 16}{2} = 9$.

61. Note, the mean of an arithmetic sequence with
n terms is $\overline{a} = \dfrac{a_1 + a_n}{2}$. Then $\overline{a} = \dfrac{2 + 142}{2} = 72$.

63. Since $a_1 = 30,000$, $d = 1,000$, and
$a_{30} = 30,000 + 29(1,000) = 59,000$, we find
his total salary for 30 years of work is
$S_{30} = \dfrac{30}{2}(30,000 + 59,000) = \$1,335,000$.

The mean annual salary for 30 years is
$\dfrac{S_{30}}{30} = \$44,500$.

65. In the nth level, the number of cans is $a_n = 12(10-n)$. In the mountain of cans there are $\sum\limits_{i=1}^{9}(120-12i) = \sum\limits_{j=1}^{9}(120-12(10-j)) =$

$$\sum_{j=1}^{9} 12j = \frac{9}{2}(12+108) = 540 \text{ cans.}$$

67. At 5% compounded annually, the future value of $\$1000$ after i years is $\$1000(1.05)^i$. Then the amount in the account on January 1, 2000 is $\sum\limits_{i=1}^{10} 1000(1.05)^i$.

69. With a calculator, one finds $200 + 200(.63) + 200(.63)^2 + 200(.63)^3 \approx 455$ mg.

71. Since $a_9 = 101$, $a_{60} = 356$, and there are $n = 52$ terms from a_9 to a_{60}, we get that the mean of the 9th through the 60th terms is $\dfrac{1}{52}\sum\limits_{i=9}^{60} a_i = \dfrac{1}{52}\dfrac{52}{2}(101+356) = 228.5$.

73. One finds $a_2 = a_1^2 - 3 = (-2)^2 - 3 = 1$ and $a_3 = a_2^2 - 3 = 1^2 - 3 = -2$. By a repeating pattern, one derives $a_7 = a_9 = -2$ and $a_8 = a_{10} = 1$. The mean from the 7th term through the 10th terms is

$$\frac{-2+1-2+1}{4} = -\frac{1}{2}.$$

For Thought

1. False, the ratios $\dfrac{6}{2}$ and $\dfrac{24}{6}$ are not equal

2. True, $a_n = 3 \cdot 2^3 \cdot 2^{-n} = 24\left(\dfrac{1}{2}\right)^n$.

3. True, $a_1 = 5(0.3)^1 = 1.5$.

4. False, since $a_n = \left(\dfrac{1}{5}\right)^n$, the common ratio is $\dfrac{1}{5}$.

5. True

6. False, and note that the ratio 2 does not satisfy $|r| < 1$ and one cannot use the formula $S = \dfrac{a}{1-r}$.

7. False, $\sum\limits_{i=1}^{9} 3(0.6)^i = \dfrac{1.8(1-0.6^9)}{1-0.6}$.

8. True, $\sum\limits_{i=0}^{4} 2(10)^i = \dfrac{2(1-10^5)}{1-10} = 22,222$.

9. True, $\sum\limits_{i=1}^{\infty} 3(0.1)^i = \dfrac{0.3}{1-0.1} = \dfrac{1}{3}$.

10. True, $\sum\limits_{i=1}^{\infty}\left(\dfrac{1}{2}\right)^i = \dfrac{1/2}{1-1/2} = 1$.

8.3 Exercises

1. $a_1 = 3 \cdot 2^0 = 3$, $a_2 = 3 \cdot 2^1 = 6$, $a_3 = 3 \cdot 2^2 = 12$, $a_4 = 3 \cdot 2^3 = 24$

First four terms are $3, 6, 12, 24$, and common ratio is 2.

3. $a_1 = 800 \cdot \left(\dfrac{1}{2}\right)^1 = 400$, $a_2 = 800 \cdot \left(\dfrac{1}{2}\right)^2 = 200$,

$a_3 = 800 \cdot \left(\dfrac{1}{2}\right)^3 = 100$, $a_4 = 800 \cdot \left(\dfrac{1}{2}\right)^4 = 50$

First four terms are $400, 200, 100, 50$, and common ratio is $\dfrac{1}{2}$.

5. $a_1 = \left(-\dfrac{2}{3}\right)^0 = 1$, $a_2 = \left(-\dfrac{2}{3}\right)^1 = -\dfrac{2}{3}$,

$a_3 = \left(-\dfrac{2}{3}\right)^2 = \dfrac{4}{9}$, $a_4 = \left(-\dfrac{2}{3}\right)^3 = -\dfrac{8}{27}$

First four terms are $1, -\dfrac{2}{3}, \dfrac{4}{9}, -\dfrac{8}{27}$, and common ratio is $-\dfrac{2}{3}$.

7. $\dfrac{1}{2}$ **9.**   10

11. -2 **13.** -1

15. Since $r = \dfrac{1/3}{1/6} = 2$, we get $a_n = \dfrac{1}{6}2^{n-1}$.

17. Since $r = \dfrac{0.09}{0.9} = 0.1$, we find $a_n = (0.9)(0.1)^{n-1}$.

19. Since $r = \dfrac{-12}{4} = -3$, we have $a_n = 4 \cdot (-3)^{n-1}$.

21. Arithmetic, since $\dfrac{1}{3} - \dfrac{1}{6} = \dfrac{1}{2} - \dfrac{1}{3} = \dots = \dfrac{1}{6}$.

23. Geometric, since $\dfrac{1/3}{1/6} = \dfrac{2/3}{1/3} = \dots = 2$.

25. Neither, since there is no common difference or ratio.

27. $a_1 = 2 \cdot 1 = 2$, $a_2 = 2 \cdot 2 = 4$, $a_3 = 2 \cdot 3 = 6$, $a_4 = 2 \cdot 4 = 8$. First four terms are $2, 4, 6, 8$. Arithmetic sequence.

29. One finds $a_1 = 1^2 = 1$, $a_2 = 2^2 = 4$, $a_3 = 3^2 = 9$, $a_4 = 4^2 = 16$. First four terms are $1, 4, 9, 16$. Neither geometric nor arithmetic.

31. $a_1 = \left(\dfrac{1}{2}\right)^1 = \dfrac{1}{2}$, $a_2 = \left(\dfrac{1}{2}\right)^2 = \dfrac{1}{4}$,

$a_3 = \left(\dfrac{1}{2}\right)^3 = \dfrac{1}{8}$, $a_4 = \left(\dfrac{1}{2}\right)^4 = \dfrac{1}{16}$. First four

terms are $\dfrac{1}{2}, \dfrac{1}{4}, \dfrac{1}{8}, \dfrac{1}{16}$. Geometric sequence.

33. $b_1 = 2^3 = 8$, $b_2 = 2^5 = 32$, $b_3 = 2^7 = 128$, $b_4 = 2^9 = 512$. First four terms are $8, 32, 128, 512$. Geometric sequence.

35. $c_2 = -3c_1 = -3 \cdot 3 = -9$, $c_3 = -3c_2 = -3 \cdot (-9) = 27$, $c_4 = -3c_3 = -3 \cdot 27 = -81$. First four terms are $3, -9, 27, -81$. Geometric sequence.

37. Since $a_n = a_1 r^{n-1}$, we obtain

$$\frac{3}{1024} = 3\left(\frac{1}{2}\right)^{n-1}$$
$$\frac{1}{1024} = \left(\frac{1}{2}\right)^{n-1}$$
$$\left(\frac{1}{2}\right)^{10} = \left(\frac{1}{2}\right)^{n-1}$$

So $n - 1 = 10$ and the number of terms is $n = 11$.

39. Since $a_n = a_1 r^{n-1}$, we have $\dfrac{1}{81} = a_1 \left(\dfrac{1}{3}\right)^5$.

Solving, one finds the first term is $a_1 = 3$.

41. Since $a_n = a_1 r^{n-1}$, we obtain $6 = \dfrac{2}{3} r^2$.

So $9 = r^2$ and the common ratio is $r = 3$ or $r = -3$.

43. Since $a_6 = a_3 r^3$, we obtain $96 = -12r^3$ and $r = -2$. Since $a_3 = a_1 r^2$, we get $-12 = a_1(-2)^2$. Then $a_1 = -3$ and $a_n = (-3)(-2)^{n-1}$.

45. Since $a = 1$, $r = 2$, and $n = 5$, we find that the sum is $S = \dfrac{1(1 - 2^5)}{1 - 2} = 31$.

47. Since $a = 6$, $r = \dfrac{1}{3}$, and $n = 5$, we find that

the sum is $S = \dfrac{6\left(1 - \left(\dfrac{1}{3}\right)^5\right)}{1 - \dfrac{1}{3}} = \dfrac{242}{27}$.

49. $\displaystyle\sum_{i=1}^{8} 1.5 \, (-2)^{i-1} = \dfrac{1.5(1 - (-2)^8)}{1 - (-2)} = -127.5$

51. $\dfrac{2(1 - 1.05^{12})}{1 - 1.05} \approx 31.8343$

53. $\dfrac{200(1 - 1.01^8)}{1 - 1.01} \approx 1657.13$

55. $\displaystyle\sum_{n=1}^{5} 3\left(-\dfrac{1}{3}\right)^{n-1}$

57. $\displaystyle\sum_{n=1}^{\infty} 0.6(0.1)^{n-1}$

59. $\displaystyle\sum_{n=1}^{\infty} (-4.5)\left(-\dfrac{1}{3}\right)^{n-1}$

61. $S = \dfrac{a_1}{1-r} = \dfrac{3}{1-(-1/3)} = \dfrac{3}{4/3} = \dfrac{9}{4}$

63. $S = \dfrac{a_1}{1-r} = \dfrac{0.9}{1-0.1} = \dfrac{0.9}{0.9} = 1$

65. $S = \dfrac{a_1}{1-r} = \dfrac{-9.9}{1-(-1/3)} = \dfrac{-9.9}{4/3} =$

$-\dfrac{99}{10} \cdot \dfrac{3}{4} = -\dfrac{297}{40}$

67. $S = \dfrac{a_1}{1-r} = \dfrac{0.34}{1-0.01} = \dfrac{0.34}{0.99} = \dfrac{34}{99}$

69. No sum, since $|r| = |-1.06| > 1$.

71. $S = \dfrac{a_1}{1-r} = \dfrac{0.6}{1-0.1} = \dfrac{0.6}{0.9} = \dfrac{2}{3}$

73. $S = \dfrac{a_1}{1-r} = \dfrac{34}{1-(-0.7)} = \dfrac{34}{1.7} = 20$

75. $\displaystyle\sum_{i=2}^{\infty} 4(0.1)^i = \dfrac{0.04}{1-0.1} = \dfrac{0.04}{0.9} = \dfrac{4}{90} = \dfrac{2}{45}$

77. $8.2 + 0.05454... = 8.2 + 0.1(0.5454...) =$

$8.2 + 0.1 \sum\limits_{i=0}^{\infty} 0.54(0.01)^i =$

$8.2 + (0.1)\dfrac{0.54}{1 - 0.01} = 8.2 + \dfrac{0.054}{0.99} =$

$8.2 + \dfrac{54}{990} = \dfrac{8118 + 54}{990} = \dfrac{8172}{990} = \dfrac{454}{55}$

79. Using a formula for S_n, the sum is

$$\sum\limits_{n=1}^{25} 100(.69)^{n-1} \quad = \quad \dfrac{100 \left(1 - (.69)^{25}\right)}{1 - .69}$$

$$\approx \quad 322.55 \ \text{mg.}$$

81. The total number of subscriptions in June is

$$\sum\limits_{i=1}^{30} 2^{i-1} = \dfrac{1 - 2^{30}}{1 - 2} = 1,073,741,823.$$

83. At the end of the nth quarter the amount is $a_n = 4000(1.02)^n$. At the end of the 37th quarter, it is $a_{37} = 4000(1.02)^{37} = \$8,322.74$.

85. At the end of the 12th month, the amount in the account is $\sum\limits_{i=1}^{12} 200(1.01)^i =$

$$\dfrac{200(1.01)(1 - 1.01^{12})}{1 - 1.01} = \$2561.87 \ .$$

87. The value of the annuity immediately after the last payment is $\sum\limits_{i=0}^{359} 100 \left(1 + \dfrac{0.09}{12}\right)^i =$

$$\dfrac{100(1 - 1.0075^{360})}{1 - 1.0075} = \$183,074.35 \ .$$

89. Assume the ball has a small radius. Approximately, the distance it travels before it comes to a rest is

$$9 + \dfrac{2}{3} \cdot 9 + \dfrac{2}{3} \cdot 9 + \left(\dfrac{2}{3}\right)^2 \cdot 9 + \left(\dfrac{2}{3}\right)^2 \cdot 9 + ... =$$

$$9 + 18\left[\dfrac{2}{3} + \left(\dfrac{2}{3}\right)^2 + \left(\dfrac{2}{3}\right)^3 + ...\right] =$$

$$9 + 18\dfrac{2/3}{1 - 2/3} = 45 \ \text{feet.}$$

91. The first amount spent in Hammond is $2 million, then 75% of $2 million is the next amount spent, and so on. So the total economic impact is

$\sum\limits_{i=0}^{\infty} 2(10^6)(0.75)^i = \dfrac{2,000,000}{1 - 0.75} =$ $\$8,000,000.$

93. Yes, since if $a = d = 0$ and r is any positive number, then the arithmetic sequence is also a geometric sequence. This sequence is $0, 0, ...$

95. If $0 < r < 1$, then $r^x \to 0$ as $x \to \infty$. If $r > 1$, then $|r^x| \to \infty$ as $x \to \infty$.

The graphs of $y = 2^x$ and $y = \left(\dfrac{1}{2}\right)^x$ are shown below.

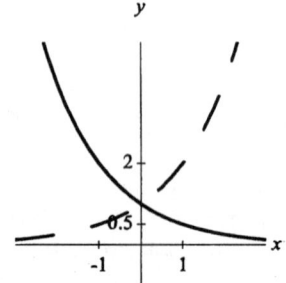

For Thought

1. False, there are $90 \cdot 26$ different codes.

2. False, there are 24^3 different possible fraternity names.

3. False, there are $3 \cdot 5 \cdot 3 \cdot 1 = 45$ different outfits.

4. True, $5! = 120$.

5. True, $P(10, 2) = \dfrac{10!}{8!} = 10 \cdot 9 = 90$.

6. False, the number of ways is 4^{20} **7.**   True

8. True, since $\dfrac{1000(999)998!}{998!} = 999,000.$

9. False, $P(10, 1) = \dfrac{10!}{9!} = 10$ and

$$P(10, 9) = \dfrac{10!}{1!} = 10!.$$

10. False, $P(29, 1) = \dfrac{29!}{28!} = \dfrac{29 \cdot 28!}{28!} = 29.$

8.4 Exercises

1. The number of possible outcomes is $2 \cdot 2$ or 4.

3. The number of possible outcomes is $2 \cdot 2 \cdot 2$ or 8.

5. Number of routes is $2 \cdot 4 = 8$.

7. There are $3! = 6$ different schedules.
   These schedules are given by:

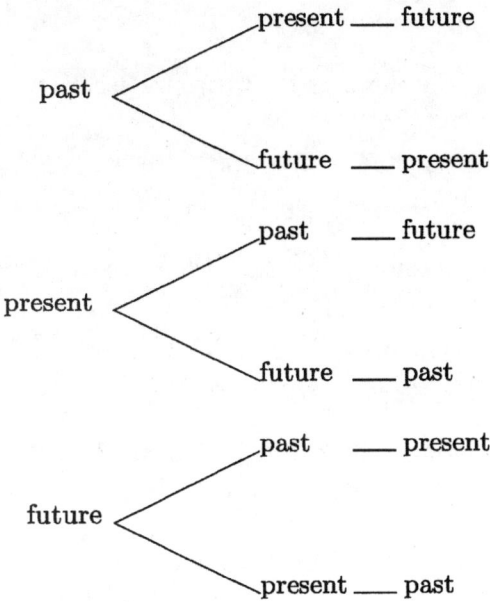

9. There are $4^5 = 1024$ different possible hands.

11. There are 8 optional items since $2^8 = 256$.

13. $\dfrac{7(6)(5)4!}{4!} = 7(6)(5) = 210$

15. $\dfrac{5!}{0!} = \dfrac{120}{1} = 120$

17. $\dfrac{78(77!)}{77!} = 78$

19. $\dfrac{4!}{2!} = 12$

21. $\dfrac{5!}{2!} = 60$

23. $\dfrac{7!}{4!} = 210$ **25.** $\dfrac{9!}{4!} = 15,120$

27. $\dfrac{5!}{0!} = 120$ **29.** $\dfrac{11!}{8!} = 990$

31. $\dfrac{99!}{99!} = 1$

33. $\dfrac{105!}{103!} = \dfrac{105(104)(103!)}{103!} = 105(104) = 10,920$

35. Hercules can perform 12 tasks in
   $12! = 479,001,600$ ways.

37. The number of ways a first, second, and a third
   restaurant can be chosen is
   $$P(15,3) = \frac{15!}{12!} = 2730 \text{ ways.}$$

39. Since there are 8 half-hour shows from
   6:00 p.m. to 10:00 p.m., the number of possible
   different schedules is $P(26,8) \approx 6.3 \times 10^{10}$.

41. Since each question has 4 possible answers, the
   number of ways to answer 6 questions is 4^6 or
   4096.

43. There are $4 \cdot 3 \cdot 6 = 72$ possible prizes.

45. There are $P(26,3) = 15,600$ possible
   passwords.

47. There are $3 \cdot 10^4 = 30,000$ possible phone
   numbers.

49. Superman has $4! = 24$ ways to arrange these
   rescues.

51. Since the true-false questions can be answered
   in 2^5 ways and the multiple choice questions in
   4^6 ways, the test can be answered in $2^5 \cdot 4^6 = 131,072$ ways.

53. On her list will be $P(7,3) = 210$ possible
   words.

For Thought

1. True, the number of ways is
   $C(5,3) = C(5,2) = 10$.

2. False, the number of ways is 3^5.

3. True, $P(5,2) = 20$.

4. False, it contains $n+1$ terms.   **5.** True

6. True, $C(n, r)$ is the number of subsets of size r, and a set with n elements has 2^n subsets.

7. False, $P(8, 3) = 336$ and $C(8, 3) = 56$.

8. True   **9.**   False, since $P(8, 3) = 336$ and $P(8, 5) = 6720$.   **10.**   True

8.5 Exercises

1. 10   **3.** 7   **5.** $\dfrac{5!}{4!1!} = 5$

7. $\dfrac{8!}{4!4!} = 70$ **9.** $\dfrac{7!}{3!4!} = 35$

11. $\dfrac{10!}{10!0!} = 1$ **13.** $\dfrac{12!}{0!12!} = 1$

15. There are $C(4, 2) = 6$ possible selections. They are Alice and Brenda, Alice and Carol, Alice and Dolores Brenda and Carol, Brenda and Dolores, Carol and Dolores.

17. There are $C(5, 2) = 10$ possible selections.

19. $C(5, 3) = 10$ ways for an in-depth interview

21. $C(49, 6) = 13,983,816$ ways to choose six numbers

23. $C(52, 5) = 2,598,960$ possible poker hands

25. These assignments can be done in
$\dfrac{10!}{5!3!2!} = 2520$ ways.

27. Since A occurs 4 times, the number of permutations is $\dfrac{7!}{4!} = 210$.

29. There are 3, 6, 10, and $C(n, 2)$ distinct chords, respectively,.

31. There are $\dfrac{10!}{3!2!5!} = 2520$ possible assignments.

33. There are $3^3 = 27$ ways to watch.

35. $C(8, 3) = 56$ ways to make the selection

37. Since 3 men can be selected in $C(9, 3)$ ways and 2 women in $C(6, 2)$ ways, the team can be selected in $C(9, 3) \cdot C(6, 2) = 1260$ ways.

39. $12! = 479,001,600$ ways to return the papers

41. $6 \cdot 6 = 36$ possible outcomes

43. Since there are 4! ways to arrange the bands in a line and 3! ways to line up the floats, the parade can be lead in $4!3! = 144$ ways.

45. Use the numbers 1,2,1 from Pascal's triangle. Then $(x+y)^2 = 1x^2 + 2xy + 1y^2 = x^2 + 2xy + y^2$.

47. Use the numbers 1,2,1 from Pascal's triangle. Then $(2a + (-3))^2 = 1(2a)^2 + 2(2a)(-3) + 1(-3)^2$
$= 4a^2 - 12a + 9$.

49. Use the numbers 1,3,3,1 from Pascal's triangle. Then $(a - 2)^3 =$
$1(a^3) + 3(a^2)(-2) + 3(a)(-2)^2 + 1(-2)^3 =$
$a^3 - 6a^2 + 12a - 8$.

51. Use the numbers 1,3,3,1 from Pascal's triangle. Thus, $(2a + b^2)^3 =$
$1(2a)^3 + 3(2a)^2(b^2) + 3(2a)(b^2)^2 + 1(b^2)^3 =$
$8a^3 + 12a^2b^2 + 6ab^4 + b^6$.

53. Use the numbers 1,4,6,4,1 from Pascal's triangle. Then we get $(x - 2y)^4 =$
$1(x)^4 + 4(x)^3(-2y) + 6(x)^2(-2y)^2 + 4(x)(-2y)^3$
$+1(-2y)^4 =$
$x^4 - 8x^3y + 24x^2y^2 - 32xy^3 + 16y^4$.

55. Use the numbers 1,4,6,4,1 from Pascal's triangle. Then we obtain $(x^2 + 1)^4 =$
$1(x^2)^4 + 4(x^2)^3(1) + 6(x^2)^2(1)^2 + 4(x^2)(1)^3$
$+1(1)^4 = x^8 + 4x^6 + 6x^4 + 4x^2 + 1$.

57. Use the Binomial Theorem. So $(x + y)^9 =$
$\begin{pmatrix} 9 \\ 0 \end{pmatrix} x^9 + \begin{pmatrix} 9 \\ 1 \end{pmatrix} x^8 y + \begin{pmatrix} 9 \\ 2 \end{pmatrix} x^7 y^2 + ... =$
$x^9 + 9x^8y + 36x^7y^2 + ...$

59. Use the Binomial Theorem. Then $(2x - y)^{12} =$
$\begin{pmatrix} 12 \\ 0 \end{pmatrix} (2x)^{12} + \begin{pmatrix} 12 \\ 1 \end{pmatrix} (2x)^{11}(-y) +$
$\begin{pmatrix} 12 \\ 2 \end{pmatrix} (2x)^{10}(-y)^2 + ... =$
$4096x^{12} - 24,576x^{11}y + 67,584x^{10}y^2 + ...$

61. Use the Binomial Theorem. So $(2s - 0.5t)^8 =$

$$\binom{8}{0}(2s)^8 + \binom{8}{1}(2s)^7(-0.5t) +$$

$$\binom{8}{2}(2s)^6(-0.5t)^2 + \ldots =$$

$$256s^8 - 512s^7t + 448s^6t^2 + \ldots$$

63. Use the Binomial Theorem.
Thus, $(m^2 - 2w^3)^9 =$

$$\binom{9}{0}(m^2)^9 + \binom{9}{1}(m^2)^8(-2w^3) +$$

$$\binom{9}{2}(m^2)^7(-2w^3)^2 + \ldots =$$

$$m^{18} - 18m^{16}w^3 + 144m^{14}w^6 + \ldots$$

65. $\binom{8}{5} = 56$

67. $\binom{13}{5}(-2)^5 = -41,184$

69. By looking at a particular term, one finds

$$(a + [b + c])^{12} = \ldots + \binom{12}{10}a^2[b + c]^{10} + \ldots =$$

$$\ldots + \binom{12}{10}a^2\left[\ldots + \binom{10}{6}b^4c^6 + \ldots\right] + \ldots$$

So the coefficient of $a^2b^4c^6$ is

$$\binom{12}{10}\binom{10}{6} = \frac{12!}{2!4!6!} = 13,860.$$

71. The coefficient of a^3b^7 in $(a + b + 2c)^{10}$ is the number of rearrangements of aaabbbbbbb, which is $\dfrac{10!}{3!7!} = 120$.

For Thought

1. False, $P(E) = \dfrac{n(E)}{n(S)}$.

2. False, there are $2^4 = 16$ outcomes.

3. True, since possible outcomes are (H, H), (H, T), (T, H), and (T, T) then

$$P(at\ least\ one\ tail) = \frac{3}{4} = 0.75\ .$$

4. True, since $\dfrac{5}{6}$ is the probability of not getting a four when a die is tossed then

$$P(at\ least\ one\ four) = 1 - P(no\ four) =$$

$$1 - \frac{5}{6} \cdot \frac{5}{6} = \frac{11}{36}.$$

5. False, since $\dfrac{1}{2}$ is the probability of getting a tail when a coin is tossed,

$$P(at\ least\ one\ head) = 1 - P(four\ tails)$$

$$= 1 - \left(\frac{1}{2}\right)^4 = \frac{15}{16}.$$

6. False, the complement is getting either no head or exactly one head.

7. True, the complement of getting exactly 3 tails in a toss of three coins is getting at least one head.

8. False, the odds in favor of snow is

$$\frac{P(snow)}{P(no\ snow)} = \frac{7/10}{3/10} = \frac{7}{3}\ \text{i.e. 7 to 3.}$$

9. False, $P(E) = 3/7$.

10. False, it is equivalent to 1 to 4.

8.6 Exercises

1. $\{(H, H), (H, T), (T, H), (T, T)\}$

3. $\{(H, 1), (H, 2), (H, 3), (H, 4), (H, 5), (H, 6),$
$(T, 1), (T, 2), (T, 3), (T, 4), (T, 5), (T, 6)\}$

5. Note, (H, T) and (T, H) are the only outcomes with exactly one head. Since the sample space has 4 elements, the probability of getting exactly one head is $\dfrac{n\{(H, T), (T, H)\}}{4} = \dfrac{1}{2}$.

7. Note, $(6, 6)$ is the only outcome with exactly two sixes. Since the sample space has 36 elements, the probability of getting exactly two sixes is $\dfrac{n\{(6, 6)\}}{36} = \dfrac{1}{36}$.

9. Note, $(H, 5)$ is the only outcome with heads and a five. Since the sample space has 12 elements, the probability of getting heads and a five is $\dfrac{n\{(H, 5)\}}{12} = \dfrac{1}{12}$.

11. Note, (T, T, T) is the only outcome with all tails. Since the sample space has 8 elements, the probability of getting all tails is

$$\frac{n\{(T, T, T)\}}{8} = \frac{1}{8}.$$

13. (a) $\dfrac{n(\{3, 4, 5, 6\})}{6} = \dfrac{4}{6} = \dfrac{2}{3}$,

(b) $\dfrac{n(\{1, 2, 3, 4, 5, 6\})}{6} = \dfrac{6}{6} = 1$,

(c) $\dfrac{n(\{1, 2, 3, 5, 6\})}{6} = \dfrac{5}{6}$,

(d) 0, (e) $\dfrac{n(\{1\})}{6} = \dfrac{1}{6}$

15. (a) $\dfrac{n\{(T, T)\}}{4} = 1/4$,

(b) $1 - P(2\ tails) = 1 - 1/4 = 3/4$,

(c) $\dfrac{n\{(H, H)\}}{4} = 1/4$,

(d) $\dfrac{n\{(T, T), (H, T), (T, H)\}}{4} = 3/4$

17. Note there are 36 possible outcomes.

(a) $\dfrac{n\{(3, 3)\}}{36} = 1/36$,

(b) Since there are 11 possible outcomes with at least one 3, $P(at\ least\ one\ 3) = 11/36$,

(c) Since there are 5 possible outcomes with a sum of 6, $P(sum\ is\ 6) = 5/36$,

(d) $1 - P(sum\ is\ 2) = 1 - 1/36 = 35/36$,

(e) $P(sum\ is\ 2) = 1/36$

19. (a) 1/3, (b) 2/3, (c) 0

21. (a) 3/13, (b) 9/13, (c) 9/13,

(d) $P(marble\ is\ yellow) = 4/13$

23. There are 72 possible outcomes.

(a) $\dfrac{n\{(1, 9)\}}{72} = 1/72$,

(b) $\dfrac{n\{(1, 3), (3, 1)\}}{72} = 2/72 = 1/36$,

(c) $\dfrac{n\{(1, 4), (4, 1), (2, 3), (3, 2)\}}{72} = 4/72 =$

$1/18$

25. (a) $\dfrac{1}{C(52, 5)} = \dfrac{1}{2,598,960}$

(b) By using the counting principle, the probability of one 3, one 4, one 5, one 6, and one 7 is $\dfrac{4^5}{C(52, 5)} = \dfrac{1024}{2,598,960}$.

27. $P(A \cup B) = P(A) + P(B) - P(A \cap B) = 0.8 + 0.6 - 0.5 = 0.9$

29. Since $P(E \cup F) = P(E) + P(F) - P(E \cap F)$, we get $0.8 = 0.2 + 0.7 - P(E \cap F)$. Then $P(E \cap F) = 0.2 + 0.7 - 0.8 = 0.1$

31. $P(A \cup B) = P(A) + P(B) - P(A \cap B) = 0.2 + 0.3 - 0 = 0.5$

33. Yes, since an outcome cannot have a sum that is both 4 and 5.

35. No, since the sum in the outcome $(1, 1)$ is 2 (which is less than 5) and is even.

37. Yes, since there is no outcome with a two and with a sum greater than nine. Note, in $(2, 6)$, the highest possible sum is 8.

39. $P(high{-}risk\ or\ a\ woman) = P(high{-}risk) + P(woman) - P(high - risk\ and\ a\ woman) = 38\% + 64\% - 24\% = 78\%$.

41. One finds the probabilities $P(3\ boys) = P(3\ girls) = 1/8$. By the addition rule for mutually exclusive events, we get $P(3\ boys\ or\ 3\ girls) = 1/8 + 1/8 = 1/4$.

43. One obtains the probabilities $P(heart) = 13/52$, $P(king) = 4/52$, $P(heart\ and\ king) = 1/52$. By the addition rule, we get $P(heart\ or\ king) = 13/52 + 4/52 - 1/52 = 4/13$.

45. (a) 34%, (b) $22\% + 18\% = 40\%$, (c) $34\% + 22\% + 18\% = 74\%$

47. $1 - P(surviving) = 1 - 0.001 = 0.999$

49. Note there are 36 possible outcomes.

(a) $\dfrac{n\{(4, 4)\}}{36} = 1/36$,

(b) $1 - P(pair\ of\ 4's) = 1 - 1/36 = 35/36$,

(c) 35/36, since this is the same event as (b)

51. Since $\dfrac{4/5}{1/5} = 4$, the odds in favor of rain is 4 to 1.

53. Odds in favor of the eye of the hurricane coming ashore is $\dfrac{0.8}{0.2} = 4$, i.e., 4 to 1

55. (a) since $\dfrac{1/4}{3/4} = \dfrac{1}{3}$, the odds in favor of stock market going up is 1 to 3

(b) odds against market going up is 3 to 1

57. If p is the probability of rain today, then

$$\frac{p}{1-p} = \frac{4}{1}$$
$$p = 4 - 4p$$
$$p = \frac{4}{5}.$$

The probability of rain today is $\dfrac{4}{5}$ or 80%.

59. (a) 1 to 9 (b) 9/10

61. Since $P(2\ heads) = \dfrac{C(4,2)}{2^4} = 3/8$ and $\dfrac{3/8}{5/8} = 3/5$, the odds in favor of getting 2 heads in four tosses is 3 to 5

63. Since $\dfrac{P(sum\ of\ 7)}{P(sum\ that\ is\ not\ 7)} = \dfrac{1/6}{5/6} = 1/5$, odds in favor of getting a sum of 7 is 1 to 5.

65. 1 to $1,999,999$

67. $\dfrac{1}{1+31} = 1/32$

69. A and B are mutually exclusive if $P(A \cap B) = 0$. They are complementary events if $P(A) + P(B) = 1$

For Thought

1. True, for when $n = 1$ we get $4 \cdot 1 - 2 = 2 \cdot 1^2$ which is a true staetment.

2. False, when $n = 3$ one gets $27 < 12 + 15$, which is inconsistent.

3. False, $\dfrac{100-1}{100+1} \approx 0.98$.

4. False, it can prove it for the positive integers.

5. True **6.** False, rather $\displaystyle\sum_{i=1}^{k+1} \dfrac{1}{i(i+1)} = \dfrac{k+1}{k+2}$.

7. False **8.** False, since $0 > 0$ is inconsistent.

9. True

To see this let S_n be the inequality $n^2 - n > 0$. Note S_2 is true since $2^2 - 2 > 0$ holds. Suppose S_k is true. One needs to show S_{k+1} is true. Then we have

$$\begin{aligned} (k+1)^2 - (k+1) &= (k^2 + 2k + 1) - k - 1 \\ &= (k^2 - k) + 2k \\ &> 0 + 2k \ \text{ since } S_k \text{ is true} \\ &= 2k \\ &> 0. \end{aligned}$$

Thus, S_{k+1} holds and S_n is true for integers $n > 1$.

10. False, when $n = 1$ one gets $1^2 - 1 > 0$, which is inconsistent.

8.7 Exercises

1. If $n = 1$, then $3 - 1 = 2$ and $\dfrac{3 \cdot 1^2 + 1}{2} = \dfrac{4}{2}$.

If $n = 2$, then $(3 - 1) + (6 - 1) = 7$ and $\dfrac{3 \cdot 2^2 + 2}{2} = \dfrac{14}{2} = 7$.

If $n = 3$, then $(3 - 1) + (6 - 1) + (9 - 1) = 15$ and $\dfrac{3 \cdot 3^2 + 3}{2} = \dfrac{30}{2} = 15$.

It is true for $n = 1, 2$, and 3.

3. If $n = 1$, then $\dfrac{1}{1(1+1)} = \dfrac{1}{1+1}$.

If $n = 2$, then $\dfrac{1}{2} + \dfrac{1}{6} = \dfrac{4}{6}$ and $\dfrac{2}{2+1} = \dfrac{2}{3}$.

If $n = 3$, then $\dfrac{1}{2} + \dfrac{1}{6} + \dfrac{1}{12} = \dfrac{9}{12}$ and $\dfrac{3}{3+1} = \dfrac{3}{4}$.

It is true for $n = 1, 2$, and 3.

5. If $n = 1$, one has $1 = 4 - 3$.

   If $n = 2$, then $1 + 4 = 5$ and $4(2) - 3 = 5$.

   If $n = 3$, then $1 + 4 + 9 = 14$ and $4(3) - 3 = 9$.

   It is true for $n = 1, 2$ and false for $n = 3$.

7. If $n = 1$, one has $1 < 1$.

   If $n = 2$, one has $4 < 8$. If $n = 3$, $9 < 27$.

   It is true for $n = 2, 3$ and false for $n = 1$.

9. $S_1 : 2(1) = 1(1 + 1)$

   $$S_k : \sum_{i=1}^{k} 2i = k(k + 1)$$

   $$S_{k+1} : \sum_{i=1}^{k+1} 2i = (k + 1)(k + 2)$$

11. $S_1 : 2 = 2 \cdot 1^2$
    $S_k : 2 + 6 + \ldots + (4k - 2) = 2k^2$

    $S_{k+1} :$

    $$2 + 6 + \ldots + (4(k + 1) - 2) =$$
    $$2 + 6 + \ldots + (4k + 2) = 2(k + 1)^2$$

13. $S_1 : 2 = 2^2 - 2$
    $$S_k : \sum_{i=1}^{k} 2^i = 2^{k+1} - 2$$
    $$S_{k+1} : \sum_{i=1}^{k+1} 2^i = 2^{k+2} - 2$$

15. $S_1 : (ab)^1 = a^1 b^1$
    $S_k : (ab)^k = a^k b^k$
    $S_{k+1} : (ab)^{k+1} = a^{k+1} b^{k+1}$

17. $S_1 :$ If $0 < a < 1$ then $0 < a^1 < 1$
    $S_k :$ If $0 < a < 1$ then $0 < a^k < 1$
    $S_{k+1} :$ If $0 < a < 1$ then $0 < a^{k+1} < 1$

19. Let $T_n : 1 + 2 + \ldots + n = \dfrac{n(n + 1)}{2}$.

    Step 1: If $n = 1$ then $T_1 : 1 = \dfrac{1(2)}{2}$.

    So T_1 is true.

    Step 2: Assume $T_k : 1 + 2 + \ldots + k = \dfrac{k(k + 1)}{2}$
    is true. Add $(k + 1)$ to both sides. Then we

obtain

$$
\begin{aligned}
1 + 2 + \ldots + k + (k + 1) &= \frac{k(k + 1)}{2} + (k + 1) \\
&= \frac{k(k + 1) + 2(k + 1)}{2} \\
&= \frac{(k + 1)(k + 2)}{2}.
\end{aligned}
$$

Then the truth of T_k implies the truth of T_{k+1}. T_n is true for every positive integer n.

21. Let $T_n : 3 + 7 + \ldots + (4n - 1) = n(2n + 1)$.

    Step 1: If $n = 1$ then $T_1 : 3 = 1(2 + 1)$.

    Thus, T_1 is true.

    Step 2: Assume T_k is true i.e.
    $3 + 7 + \ldots + (4k - 1) = k(2k + 1)$.
    Note $4(k + 1) - 1 = 4k + 3$. Adding $4k + 3$ to both sides, we get

    $$3 + \ldots + (4k - 1) + (4k + 3) =$$

    $$
    \begin{aligned}
    &= k(2k + 1) + (4k + 3) \\
    &= 2k^2 + k + 4k + 3 \\
    &= 2k^2 + 5k + 3 \\
    &= (k + 1)(2k + 3) \\
    &= (k + 1)(2(k + 1) + 1).
    \end{aligned}
    $$

The truth of T_k implies the truth of T_{k+1}. T_n is true for every positive integer n.

23. Let $T_n : \sum_{i=1}^{n} 2^i = 2^{n+1} - 2$.

    Step 1: If $n = 1$ then $T_1 : 2 = 2^2 - 2$.

    So T_1 is true.

    Step 2: Assume $T_k : \sum_{i=1}^{k} 2^i = 2^{k+1} - 2$ is true.

    Then add 2^{k+1} to both sides.

    $$
    \begin{aligned}
    \sum_{i=1}^{k} 2^i + 2^{k+1} &= 2^{k+1} - 2 + 2^{k+1} \\
    &= 2 \cdot 2^{k+1} - 2 \\
    &= 2^{k+2} - 2 \\
    &= 2^{(k+1)+1} - 2
    \end{aligned}
    $$

Thus, the truth of T_k implies the truth of T_{k+1}. T_n is true for every positive integer n.

25. Let $T_n : \sum_{i=1}^{n} (3i - 1) = \dfrac{3n^2 + n}{2}$.

Step 1: If $n = 1$ then $T_1 : 3 - 1 = \dfrac{3 + 1}{2}$.
So T_1 is true.

Step 2: Assume $T_k : \sum_{i=1}^{k} (3i - 1) = \dfrac{3k^2 + k}{2}$
is true. Note $3(k + 1) - 1 = 3k + 2$. Then add $3k + 2$ to both sides.

$$\sum_{i=1}^{k} (3i - 1) + 3k + 2 = \dfrac{3k^2 + k}{2} + (3k + 2)$$
$$= \dfrac{3k^2 + k + 2(3k + 2)}{2}$$
$$= \dfrac{3k^2 + 7k + 4}{2}$$
$$= \dfrac{3(k^2 + 2k + 1) + (k + 1)}{2}$$
$$= \dfrac{3(k + 1)^2 + (k + 1)}{2}$$

Thus, the truth of T_k implies the truth of T_{k+1}. T_n is true for every positive integer n.

27. Let $T_n : 1^2 + 2^2 + ... + n^2 = \dfrac{n(n + 1)(2n + 1)}{6}$.

Step 1: If $n = 1$ then $T_1 : 1 = \dfrac{1(2)(3)}{6}$.
So T_1 is true.

Step 2: Assume T_k is true i.e. we
assume $1^2 + 2^2 + ... + k^2 = \dfrac{k(k + 1)(2k + 1)}{6}$
Add $(k + 1)^2$ to both sides.
$1^2 + 2^2 + ... + k^2 + (k + 1)^2 =$

$$= \dfrac{k(k + 1)(2k + 1)}{6} + (k + 1)^2$$
$$= \dfrac{k(k + 1)(2k + 1) + 6(k + 1)^2}{6}$$
$$= \dfrac{(k + 1)[k(2k + 1) + 6(k + 1)]}{6}$$
$$= \dfrac{(k + 1)[2k^2 + 7k + 6]}{6}$$
$$= \dfrac{(k + 1)[k + 2][2k + 3]}{6}$$
$$= \dfrac{(k + 1)[(k + 1) + 1][2(k + 1) + 1]}{6}$$

Thus, the truth of T_k implies the truth of T_{k+1}. T_n is true for every positive integer n.

29. Let $T_n : 1 \cdot 3 + 2 \cdot 4 + ... + n \cdot (n + 2) = \dfrac{n}{6}(n + 1)(2n + 7)$.

Step 1: If $n = 1$ then $T_1 : 1 \cdot 3 = \dfrac{1}{6}(2)(2 + 7)$.
So T_1 is true.

Step 2: Assume T_k is true i.e. we assume
$1 \cdot 3 + 2 \cdot 4 + ... + k \cdot (k + 2) = \dfrac{k}{6}(k + 1)(2k + 7)$.
Add $(k + 1)(k + 3)$ to both sides. Then
$1 \cdot 3 + 2 \cdot 4 + ... + k \cdot (k + 2) + (k + 1)(k + 3) =$

$$= \dfrac{k}{6}(k + 1)(2k + 7) + (k + 1)(k + 3)$$
$$= \dfrac{k(k + 1)(2k + 7) + 6(k + 1)(k + 3)}{6}$$
$$= \dfrac{(k + 1)(2k^2 + 13k + 18)}{6}$$
$$= \dfrac{(k + 1)(2k + 9)(k + 2)}{6}$$
$$= \dfrac{(k + 1)}{6}[k + 2][2k + 9]$$
$$= \dfrac{(k + 1)}{6}[(k + 1) + 1][2(k + 1) + 7]$$

So the truth of T_k implies the truth of T_{k+1}. Then T_n is true for every positive integer n.

31. Let $T_n : \dfrac{1}{1 \cdot 3} + \dfrac{1}{3 \cdot 5} + ... + \dfrac{1}{(2n - 1)(2n + 1)} = \dfrac{n}{2n + 1}$.

Step 1: If $n = 1$ then $T_1 : \dfrac{1}{1 \cdot 3} = \dfrac{1}{2 + 1}$.
So T_1 is true.

Step 2: Assume T_k is true i.e. we assume
$\dfrac{1}{1 \cdot 3} + ... + \dfrac{1}{(2k - 1)(2k + 1)} = \dfrac{k}{2k + 1}$.
Note $\dfrac{1}{[2(k + 1) - 1][2(k + 1) + 1]} = \dfrac{1}{(2k + 1)(2k + 3)}$. Then add $\dfrac{1}{(2k + 1)(2k + 3)}$ to both sides.

So $\dfrac{1}{1 \cdot 3} + \ldots$

$.. + \dfrac{1}{(2k-1)(2k+1)} + \dfrac{1}{(2k+1)(2k+3)} =$

$$= \frac{k}{2k+1} + \frac{1}{(2k+1)(2k+3)}$$

$$= \frac{k(2k+3)+1}{(2k+1)(2k+3)}$$

$$= \frac{2k^2+3k+1}{(2k+1)(2k+3)}$$

$$= \frac{(2k+1)(k+1)}{(2k+1)(2k+3)}$$

$$= \frac{k+1}{2(k+1)+1}.$$

Thus, the truth of T_k implies the truth of T_{k+1}. T_n is true for every positive integer n.

33. Let T_n : If $0 < a < 1$ then $0 < a^n < 1$.

Step 1: Clearly, T_1 is true for $n = 1$.

Step 2: Assume $0 < a < 1$ implies $0 < a^k < 1$, i.e., assume T_k is true. Multiply $0 < a^k < 1$ by a to get $0 < a^{k+1} < a$. Since $a < 1$ and by transitivity, we get $0 < a^{k+1} < 1$.

Then the truth of T_k implies the truth of T_{k+1}. T_n is true for every positive integer n.

35. Let $T_n : n < 2^n$.

Step 1: If $n = 1$, then $1 < 2^1$. So T_1 is true.

Step 2: Assume $k < 2^k$, i.e., assume T_k is true. Add 1 to both sides.

$$\begin{aligned} k+1 &< 2^k+1 \\ &< 2^k+2^k \\ &= 2 \cdot 2^k \\ &= 2^{k+1} \end{aligned}$$

Then $k+1 < 2^{k+1}$.
So, the truth of T_k implies the truth of T_{k+1}. T_n is true for every positive integer n.

37. Let $T_n : 5^n - 1$ is divisible by 4.

Step 1: If $n = 1$, then $5^1 - 1$ is divisible by 4. So T_1 is true.

Step 2: Assume $5^k - 1$ is divisible by 4, i.e., assume T_k is true.
Observe that $5^{k+1} - 1 = 5(5^k - 1) + 4$. Since sums of multiples of 4 are again multiples of 4, we get that $5^{k+1} - 1$ is a multiple of 4.

Then the truth of T_k implies the truth of T_{k+1}. T_n is true for every positive integer n.

39. Let $T_n : (ab)^n = a^n b^n$.

Step 1: If $n = 1$ then $(ab)^1 = a^1 b^1$. So T_1 is true.

Step 2: Assume $(ab)^k = a^k b^k$, i.e., assume T_k is true. Then

$$\begin{aligned} (ab)^{k+1} &= (ab)^k (ab) \\ &= a^k b^k (ab) \text{ since } T_k \text{ is true} \\ &= a^{k+1} b^{k+1}. \end{aligned}$$

Thus, the truth of T_k implies the truth of T_{k+1}. T_n is true for every positive integer n.

41. Let $T_n : \sum\limits_{i=0}^{n} x^i = \dfrac{x^{n+1}-1}{x-1}$, $x \neq 1$.

Step 1: If $n = 1$, then $T_1 : 1 + x = \dfrac{x^2-1}{x-1}$.

T_1 is true since $(x^2 - 1) = (x-1)(x+1)$.

Step 2: Assume $T_k : \sum\limits_{i=0}^{k} x^i = \dfrac{x^{k+1}-1}{x-1}$ is true. Add x^{k+1} to both sides.

$$\begin{aligned} \sum_{i=0}^{k} x^i + x^{k+1} &= \frac{x^{k+1}-1}{x-1} + x^{k+1} \\ &= \frac{x^{k+1}-1+(x-1)x^{k+1}}{x-1} \\ &= \frac{x^{k+1}-1+x^{k+2}-x^{k+1}}{x-1} \\ &= \frac{x^{k+2}-1}{x-1} \end{aligned}$$

Then the truth of T_k implies the truth of T_{k+1}. T_n is true for every positive integer n.

Review Exercises

1. One gets $a_1 = 2^0 = 1$, $a_2 = 2^1 = 2$,
$a_3 = 2^2 = 4$, $a_4 = 2^3 = 8$, $a_5 = 2^4 = 16$

First five terms are $1, 2, 4, 8, 16$

3. $a_1 = \dfrac{(-1)^1}{1!} = -1$, $a_2 = \dfrac{(-1)^2}{2!} = \dfrac{1}{2}$,
$a_3 = \dfrac{(-1)^3}{3!} = \dfrac{-1}{6}$, $a_4 = \dfrac{(-1)^4}{4!} = \dfrac{1}{24}$.

First four terms are $-1, \dfrac{1}{2}, -\dfrac{1}{6}, \dfrac{1}{24}$.

5. $a_1 = 3(0.5)^0 = 3$, $a_2 = 3(0.5)^1 = 1.5$,
and $a_3 = 3(0.5)^2 = 0.75$.

First three terms are $3, 1.5, 0.75$.

7. $a_1 = -3 + 6 = 3$, $a_2 = -6 + 6 = 0$,
and $a_3 = -9 + 6 = -3$.

First three terms are $3, 0, -3$.

9. Since $S = \dfrac{a_1(1 - r^n)}{1 - r}$ is the sum of a geometric

series, $S = \dfrac{0.5(1 - 0.5^4)}{1 - 0.5} = 0.9375$.

11. Since $S = \dfrac{n}{2}(a_1 + a_n)$ is the sum of an

arithmetic series, $S = \dfrac{50}{2}(11 + 207) = 5450$.

13. Since $S = \dfrac{a_1}{1 - r}$ is the sum of a geometric

series, $S = \dfrac{0.3}{1 - 0.1} = \dfrac{0.3}{0.9} = \dfrac{1}{3}$.

15. Since $S = \dfrac{a_1(1 - r^n)}{1 - r}$ is the sum of a geometric

series, $S = \dfrac{1000(1 - 1.05^{20})}{1 - 1.05} \approx 33{,}065.9541$

17. $a_n = \dfrac{(-1)^n}{n + 2}$

19. $a_n = 6\left(\dfrac{1}{6}\right)^{n-1}$

21. $\displaystyle\sum_{i=1}^{\infty} \dfrac{(-1)^{i+1}}{i + 1}$

23. $\displaystyle\sum_{i=1}^{14} 2i$

25. Note, $a_n = a_1 r^{n-1}$. Since $256 = 4r^6$, we get
$64 = r^6$ and the common ratio is $r = \pm 2$.

27. $100\left(1 + \dfrac{0.09}{4}\right)^{40} = \$243.52$

29. $\displaystyle\sum_{i=1}^{10} 1000(1.06)^i = \dfrac{1000(1.06)(1 - 1.06^{10})}{1 - 1.06}$
$= \$13{,}971.64$

31. If the pattern continues, the total number
of Tummy Masters that could be sold is
$\displaystyle\sum_{i=0}^{\infty} 100{,}000(0.9)^i = \dfrac{100{,}000}{1 - 0.9} = 1$ million.

33. $a^4 + \dbinom{4}{1}a^3(2b) +$
$\dbinom{4}{2}a^2(2b)^2 + \dbinom{4}{3}a(2b)^3 + (2b)^4 =$
$a^4 + 8a^3 b + 24a^2 b^2 + 32ab^3 + 16b^4$

35. $(2a)^5 + \dbinom{5}{1}(2a)^4(-b) + \dbinom{5}{2}(2a)^3(-b)^2 +$
$\dbinom{5}{3}(2a)^2(-b)^3 + \dbinom{5}{4}(2a)(-b)^4 + (-b)^5 =$
$32a^5 - 80a^4 b + 80a^3 b^2 - 40a^2 b^3 + 10ab^4 - b^5$

37. $a^{10} + \dbinom{10}{1}a^9 b + \dbinom{10}{2}a^8 b^2 + \ldots =$
$a^{10} + 10a^9 b + 45a^8 b^2 + \ldots$

39. $(2x)^8 + \dbinom{8}{1}(2x)^7\left(\dfrac{y}{2}\right) + \dbinom{8}{2}(2x)^6\left(\dfrac{y}{2}\right)^2 +$
$\ldots$
$= 256x^8 + 512x^7 y + 448x^6 y^2 + \ldots$

41. $\dbinom{13}{9} = 715$

43. $\dfrac{11!}{2!3!6!}2^3 = 36{,}960$

45. 24 terms

47. $5^9 = 1{,}953{,}125$ ways to mark the answers

49. $P(7, 3) = 210$ possible three-letter 'words'

51. $3 \cdot 5 \cdot 4 = 60$ ways to place advertisements

53. $C(8,5) = 56$ possible vacations where order is not taken into account

55. (a) $C(8,4) = 70$ possible councils,
(b) $C(5,4) = 5$ possible councils from Democrats,
(c) $C(5,2) \cdot C(3,2) = 30$ possible councils with two Democrats and two Republicans

57. For KANSAS, $\dfrac{6!}{2!2!} = 180$ arrangements;

for TEXAS, $5! = 120$ arrangements.

59. $2^7 = 128$ different families

61. Since there are 2^{10} ways to answer

the test, the probabilities are

$$P(all \ 10 \ correct) = \frac{1}{1024} \text{ and}$$

$$P(all \ 10 \ wrong) = \frac{1}{1024}.$$

63. (a) 5/13, (b) 8/13, (c) 0, (d) 1

65. Assuming the probabilities of a boy or girl

being born are equal, $P(3 \ boys) = \dfrac{1}{8}$. Then

the odds in favor of 3 boys is $\dfrac{1/8}{7/8}$, i.e., 1 to 7.

67. Odds in favor of catching a perch is

$\dfrac{0.9}{0.1}$, i.e., 9 to 1.

69. By the Addition Rule, $P(Math \ or \ English) =$
$0.7 + 0.6 - 0.4 = 0.9$, i.e., 90%.

71. $40,320$ **73.**   20

75. 1680   **77.** $\dfrac{8!}{2!6!} = 28$

79. $\dfrac{8!}{4!} = 1680$ **81.** $\dfrac{8!}{7!1!} = 8$

83. Let $T_n : 3 + 6 + 9 + \ldots + 3n = \dfrac{3}{2}(n^2 + n)$.

Step 1: If $n = 1$, then $T_1 : 3 = \dfrac{3}{2}(1 + 1)$.

So T_1 is true.

Step 2: Assume $T_k : 3 + 6 + \ldots + 3k = \dfrac{3}{2}(k^2 + k)$

is true. Add $3(k+1)$ to both sides.

$3 + 6 + \ldots + 3k + 3(k+1) =$

$$= \frac{3}{2}(k^2 + k) + 3(k+1)$$

$$= \frac{3(k^2 + k) + 6(k+1)}{2}$$

$$= \frac{3k^2 + 9k + 6}{2}$$

$$= \frac{3}{2}(k^2 + 3k + 2)$$

$$= \frac{3}{2}((k+1)^2 + (k+1))$$

Then the truth of T_k implies the truth of T_{k+1}. T_n is true for every positive integer n.

Chapter 8 Test

1. $a_1 = 2.3$, $a_2 = 2.3 + 0.5$,
$a_3 = 2.3 + 1$, and $a_4 = 2.3 + 1.5$.
First four terms are $2.3, 2.8, 3.3, 3.8$.

2. One finds $c_2 = \dfrac{1}{2} \cdot 20 = 10$,

$c_3 = \dfrac{1}{2} \cdot 10 = 5$, and $c_4 = \dfrac{1}{2} \cdot 5 = 2.5$.

First four terms are $20, 10, 5, 2.5$.

3. $a_n = (-1)^{n-1}(n-1)^2$

4. $a_n = 7 + 3(n-1) = 3n + 4$

5. $a_n = \dfrac{1}{3}\left(-\dfrac{1}{2}\right)^{n-1}$

6. Since $S = \dfrac{n}{2}(a_1 + a_n)$, the

sum is $\dfrac{54}{2}(-5 + 154) = 4023$.

7. Since $S = \dfrac{a_1(1 - r^n)}{1 - r}$, the sum

is $\dfrac{300(1.05)(1 - 1.05^{23})}{1 - 1.05} \approx 13,050.5997$.

8. Since $S = \dfrac{a_1}{1 - r}$, the sum is $\dfrac{0.98}{1 - 0.98} = 49$.

9. Since $a_n = a_1 + d(n-1)$, $9 = -3 + 8d$.
Solving for d, one gets $d = 1.5$.
So $a_n = -3 + 1.5(n-1)$ or $a_n = 1.5n - 4.5$.

10. $\dfrac{9!}{4!2!2!} = 3780$ possible nine-letter 'words'

11. Since $S = \dfrac{n}{2}(a_1 + a_n)$ is the sum of an arithmetic series, the mean daily

sales is $\dfrac{1}{30} \sum\limits_{i=0}^{29} 300 + 10i =$

$\dfrac{1}{30} \dfrac{30}{2}(300 + 590) = \$445$

12. There are $10^4 = 10,000$ possible secret numbers. The probability of guessing the correct number in 3 tries is the probability of getting it in the 1st try plus the probability of getting it in the 2nd try plus the probability of getting it in the 3rd try. Assuming the person remembers what he tries and tries a different number after a failure, this probability is

$\dfrac{1}{10,000} + \dfrac{9,999}{10,000} \cdot \dfrac{1}{9,999} + \dfrac{9,999}{10,000} \cdot \dfrac{9,998}{9,999} \cdot$

$\dfrac{1}{9,998} = \dfrac{3}{10,000}.$

13. At the end of the 300th month the value

of the annuity is $\sum\limits_{i=1}^{300} 700\left(1 + \dfrac{0.06}{12}\right)^i =$

$\sum\limits_{i=1}^{300} 700(1.005)^i = 700(1.005)\left(\dfrac{1 - 1.005^{300}}{1 - 1.005}\right)$

$= \$487,521.25 .$

The cost of the house at the end of 25 years is $120,000(1.06)^{25} = \$515,024.49 .$

No, they will not have enough money to buy the house.

14. $a^5 + \begin{pmatrix} 5 \\ 1 \end{pmatrix}a^4(-2x) + \begin{pmatrix} 5 \\ 2 \end{pmatrix}a^3(-2x)^2 +$

$\begin{pmatrix} 5 \\ 3 \end{pmatrix}a^2(-2x)^3 + \begin{pmatrix} 5 \\ 4 \end{pmatrix}a(-2x)^4 + (-2x)^5 =$

$a^5 - 10a^4x + 40a^3x^2 - 80a^2x^3 + 80ax^4 - 32x^5$

15. $x^{24} + \begin{pmatrix} 24 \\ 1 \end{pmatrix}x^{23}(y^2) + \begin{pmatrix} 24 \\ 2 \end{pmatrix}x^{22}(y^2)^2 + \ldots =$

$x^{24} + 24x^{23}y^2 + 276x^{22}y^4 + \ldots$

16. $\sum\limits_{i=0}^{30} \begin{pmatrix} 30 \\ i \end{pmatrix} m^{30-i} y^i$

17. Note, $P(sum\ is\ 7) = \dfrac{6}{36} = \dfrac{1}{6}.$ The odds in

favor of 7 is $\dfrac{6/36}{30/36} = \dfrac{1}{5},$ i.e., 1 to 5.

18. $C(12,3) = 220$ selections

19. $C(12,2) \cdot C(10,2) = 66 \cdot 45 = 2970$ outcomes

20. $1/P(8,3) = \dfrac{1}{336}$ is the probability of getting the horses and the order of finish correctly.

21. Let $T_n : \sum\limits_{i=1}^{n}\left(\dfrac{1}{2}\right)^i = 1 - 2^{-n}.$

Step 1: If $n = 1$, then $T_1 : \dfrac{1}{2} = 1 - 2^{-1}.$ So T_1 is true.

Step 2: Assume $T_k : \sum\limits_{i=1}^{k}\left(\dfrac{1}{2}\right)^i = 1 - 2^{-k}$ is true.

Add $\left(\dfrac{1}{2}\right)^{k+1}$ to both sides.

$\sum\limits_{i=1}^{k}\left(\dfrac{1}{2}\right)^i + \left(\dfrac{1}{2}\right)^{k+1} = 1 - 2^{-k} + \left(\dfrac{1}{2}\right)^{k+1}$

$= 1 - 2 \cdot 2^{-(k+1)} + 2^{-(k+1)}$

$= 1 - 2^{-(k+1)}$

Then the truth of T_k implies the truth of T_{k+1}. Thus, T_n is true for every positive integer n. Since $1 - 2^{-n} < 1$ for all positive integers n,

then by transitivity we obtain $\sum\limits_{i=1}^{n}\left(\dfrac{1}{2}\right)^i < 1$

for all positive integers n.